SCHAUM'S OUTLINE OF

THEORY AND PROBLEMS

of

PHYSICAL SCIENCE

•

by

ARTHUR BEISER, Ph.D.

SCHAUM'S OUTLINE SERIES

McGRAW-HILL BOOK COMPANY

New York   St. Louis   San Francisco   Auckland   Düsseldorf   Johannesburg
Kuala Lumpur   London   Mexico   Montreal   New Delhi   Panama
Paris   São Paulo   Singapore   Sydney   Tokyo   Toronto

07-004376-0

1 2 3 4 5 6 7 8 9 10 11 12 13 14 15 16 17 18 19 20   SH SH   7 9 8 7 6 5 4

Library of Congress Cataloging in Publication Data

Beiser, Arthur.
   Schaum's outline of theory and problems of physical
science.

   (Schaum's outline series)
    1. Science.   2. Science — Problems, exercises, etc.
I. Title.   II. Title: Theory and problems of physical
science.
[Q161.2.B44]   500.2   74-23356
ISBN 0-07-004376-0

Preface

This book is intended to provide students of physical science with help in mastering elementary physics, chemistry, earth science, and astronomy. A wide spectrum of topics is covered, so that the reader may select those which correspond to his particular needs. Both SI (metric) and British units are used.

Each chapter begins with an outline of its subject. The solved problems that follow are of two kinds: those that show how numerical answers are obtained to typical questions in physics and chemistry, and those that review important facts and ideas in all the physical sciences. The supplementary problems give the reader both a chance for practice and a means to gauge his progress.

ARTHUR BEISER

November 1974

CONTENTS

CONTENTS

CONTENTS

CONTENTS

CONTENTS

CONTENTS

Page

CONTENTS

CONTENTS

CONTENTS

Chapter 1

Physical Quantities

POWERS OF TEN

Very small and very large numbers are common in physical science and are best expressed with the help of powers of 10. Any number in decimal form can be written as a number between 1 and 10 multiplied by a power of 10:

$$834 = 8.34 \times 10^2 \qquad 0.00072 = 7.2 \times 10^{-4}$$

The powers of 10 from 10^{-6} to 10^6 are as follows:

10^0	$= 1$	$= 1$ with decimal point moved 0 places
10^{-1}	$= 0.1$	$= 1$ with decimal point moved 1 place to the left
10^{-2}	$= 0.01$	$= 1$ with decimal point moved 2 places to the left
10^{-3}	$= 0.001$	$= 1$ with decimal point moved 3 places to the left
10^{-4}	$= 0.0001$	$= 1$ with decimal point moved 4 places to the left
10^{-5}	$= 0.00001$	$= 1$ with decimal point moved 5 places to the left
10^{-6}	$= 0.000001$	$= 1$ with decimal point moved 6 places to the left

10^0	$= 1$	$= 1$ with decimal point moved 0 places
10^1	$= 10$	$= 1$ with decimal point moved 1 place to the right
10^2	$= 100$	$= 1$ with decimal point moved 2 places to the right
10^3	$= 1000$	$= 1$ with decimal point moved 3 places to the right
10^4	$= 10,000$	$= 1$ with decimal point moved 4 places to the right
10^5	$= 100,000$	$= 1$ with decimal point moved 5 places to the right
10^6	$= 1,000,000$	$= 1$ with decimal point moved 6 places to the right

CALCULATIONS USING POWERS OF TEN

When numbers written in powers-of-10 notation are to be added or subtracted, they must all be expressed in terms of the *same* power of 10:

$$3 \times 10^2 + 4 \times 10^3 = 0.3 \times 10^3 + 4 \times 10^3 = 4.3 \times 10^3$$

To multiply two powers of 10, add their exponents; to divide one power of 10 by another, subtract the exponent of the latter from that of the former:

$$10^n \times 10^m = 10^{n+m} \qquad \frac{10^n}{10^m} = 10^{n-m}$$

Reciprocals follow the pattern

$$\frac{1}{10^n} = 10^{-n}$$

The rules for finding powers and roots of powers of 10 are

$$(10^n)^m = 10^{n \times m} \qquad \sqrt[m]{10^n} = (10^n)^{1/m} = 10^{n/m}$$

In taking the mth root, the power of 10 should be chosen to be a multiple of m. Thus

$$\sqrt{10^{15}} = \sqrt{10} \times \sqrt{10^{14}} = \sqrt{10} \times 10^7 = 3.16 \times 10^7$$

UNITS

Some common British and SI (metric) units of length and time are

<table>
<tr><td colspan="2" align="center">LENGTH</td><td align="center">TIME</td></tr>
<tr><td>1 foot (ft) = 12 in. = 0.305 m</td><td></td><td>1 minute (min) = 60 seconds (s)</td></tr>
<tr><td>1 inch (in.) = 0.083 ft = 2.54 cm</td><td></td><td>1 hour (hr) = 60 min = 3600 s</td></tr>
<tr><td>1 statute mile (mi) = 5280 ft = 1.61 km</td><td></td><td>1 day = 24 hr = 86,400 s</td></tr>
<tr><td>1 meter = 100 cm = 39.4 in. = 3.28 ft</td><td></td><td></td></tr>
<tr><td>1 centimeter (cm) = 0.01 m = 0.394 in.</td><td></td><td></td></tr>
<tr><td>1 kilometer (km) = 1000 m = 0.621 mi</td><td></td><td></td></tr>
</table>

Subdivisions and multiples of metric units are designated by prefixes according to the corresponding power of 10.

Prefix	Power	Abbreviation	Example
pico-	10^{-12}	p	1 pf = 1 picofarad = 10^{-12} farad
nano-	10^{-9}	n	1 ns = 1 nanosecond = 10^{-9} second
micro-	10^{-6}	μ	1 μA = 1 microampere = 10^{-6} ampere
milli-	10^{-3}	m	1 mm = 1 millimeter = 10^{-3} meter
centi-	10^{-2}	c	1 cl = 1 centiliter = 10^{-2} liter
kilo-	10^{3}	k	1 kg = 1 kilogram = 10^{3} grams
mega-	10^{6}	M	1 MW = 1 megawatt = 10^{6} watts
giga-	10^{9}	G	1 GeV = 1 gigaelectron-volt = 10^{9} electron volts

CONVERTING UNITS

Units are algebraic quantities and may be multiplied and divided by one another. To convert a quantity expressed in a certain unit to its equivalent in a different unit of the same kind, we use the fact that multiplying or dividing anything by 1 does not affect its value. For instance, 12 in. = 1 ft, so 12 in./ft = 1, and we can convert a length s expressed in ft to its value in inches by multiplying s by 12 in./ft:

$$4 \, \text{ft} = 4 \, \cancel{\text{ft}} \times 12 \, \frac{\text{in.}}{\cancel{\text{ft}}} = 48 \, \text{in.}$$

SCALAR AND VECTOR QUANTITIES

A *scalar quantity* has only magnitude and is completely specified by a number and a unit. Examples are mass (a stone has a mass of 2 kg), volume (a bottle has a volume of 12 oz), and frequency (house current has a frequency of 60 cycles/s). Symbols of scalar quantities are printed in italic type (m = mass, V = volume). Scalar quantities of the same kind are added using ordinary arithmetic.

A *vector quantity* has both magnitude and direction. Examples are displacement (an airplane has flown 200 mi to the southwest), velocity (a car is moving at 60 mi/hr to the north), and force (a man applies an upward force of 15 lb to a package). Symbols of vector quantities are printed in boldface type (**v** = velocity, **F** = force) and expressed in handwriting by arrows over the letters ($\vec{v}$, $\vec{F}$). The magnitude of a vector quantity is printed in italic type (F is the magnitude of the force **F**). When vector quantities are added, their directions must be taken into account.

VECTOR ADDITION

A *vector* is an arrowed line whose length is proportional to a certain vector quantity and whose direction indicates the direction of the quantity.

To add the vector **B** to the vector **A**, draw **B** so that its tail is at the head of **A**. The vector sum **A + B** is the vector **R** that joins the tail of **A** and the head of **B** (Fig. 1-1). **R** is usually called the *resultant* of **A** and **B**.

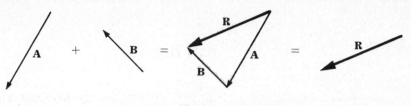

Fig. 1-1

The order in which **A** and **B** are added is not significant, so that **A + B = B + A** (Figs. 1-1 and 1-2).

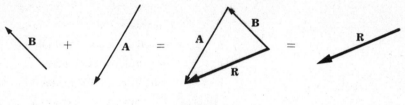

Fig. 1-2

Exactly the same procedure is followed when more than two vectors of the same kind are to be added. The vectors are strung together head to tail (being careful to preserve their correct lengths and directions), and the resultant **R** is the vector drawn from the tail of the first vector to the head of the last. The order in which the vectors are added does not matter (Fig. 1-3).

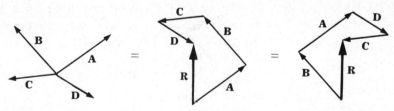

Fig. 1-3

Solved Problems

1.1.   Examples of powers-of-10 notation.

$$20 = 2 \times 10 = 2 \times 10^1$$

$$3043 = 3.043 \times 1000 = 3.043 \times 10^3$$

$$8,700,000 = 8.7 \times 1,000,000 = 8.7 \times 10^6$$

$$0.22 = 2.2 \times 0.1 = 2.2 \times 10^{-1}$$

$$0.000035 = 3.5 \times 0.00001 = 3.5 \times 10^{-5}$$

1.2.   Examples of addition and subtraction.

$$6 \times 10^2 + 5 \times 10^4 \; = \; 0.06 \times 10^4 + 5 \times 10^4 \; = \; 5.06 \times 10^4$$

$$2 \times 10^{-2} + 3 \times 10^{-3} \; = \; 2 \times 10^{-2} + 0.3 \times 10^{-2} \; = \; 2.3 \times 10^{-2}$$

$$7 + 2 \times 10^{-2} \; = \; 7 + 0.02 \; = \; 7.02$$

$$6 \times 10^4 - 4 \times 10^2 \; = \; 6 \times 10^4 - 0.04 \times 10^4 \; = \; 5.96 \times 10^4$$

$$3 \times 10^{-2} - 5 \times 10^{-3} \; = \; 3 \times 10^{-2} - 0.5 \times 10^{-2} \; = \; 2.5 \times 10^{-2}$$

$$7 \times 10^{-5} - 2 \times 10^{-4} \; = \; 0.7 \times 10^{-4} - 2 \times 10^{-4} \; = \; -1.3 \times 10^{-4}$$

$$6.23 \times 10^{-3} - 6.28 \times 10^{-3} \; = \; -0.05 \times 10^{-3} \; = \; -5 \times 10^{-5}$$

1.3.   Examples of multiplication and division.

$$10^5 \times 10^{-2} = 10^{5-2} = 10^3 \qquad \frac{10^4}{10^{-3}} = 10^{4-(-3)} = 10^{4+3} = 10^7$$

$$\frac{10^3}{10^6} = 10^{3-6} = 10^{-3} \qquad \frac{10^5 \times 10^{-7}}{10^2} = 10^{5-7-2} = 10^{-4}$$

1.4.   A sample calculation.

$$\frac{460 \times 0.00003 \times 100{,}000}{9000 \times 0.0062} \; = \; \frac{(4.6 \times 10^2) \times (3 \times 10^{-5}) \times (10^5)}{(9 \times 10^3) \times (6.2 \times 10^{-3})}$$

$$= \; \frac{4.6 \times 3}{9 \times 6.2} \times \frac{10^2 \times 10^{-5} \times 10^5}{10^3 \times 10^{-3}}$$

$$= \; 0.25 \times \frac{10^{2-5+5}}{10^{3-3}} \; = \; 0.25 \times \frac{10^2}{10^0} \; = \; 25$$

1.5.   Examples of powers of numbers.

$$(10^2)^4 \; = \; 10^{2 \times 4} \; = \; 10^8$$

$$(10^{-3})^5 \; = \; 10^{-3 \times 5} \; = \; 10^{-15}$$

$$(10^{-4})^{-3} \; = \; 10^{-4 \times -3} \; = \; 10^{12}$$

$$(3 \times 10^3)^2 \; = \; 3^2 \times (10^3)^2 \; = \; 9 \times 10^6$$

$$(4 \times 10^{-5})^3 \; = \; 4^3 \times (10^{-5})^3 \; = \; 64 \times 10^{-15} \; = \; 6.4 \times 10^{-14}$$

$$(2 \times 10^{-2})^{-4} \; = \; \frac{1}{2^4} \times (10^{-2})^{-4} \; = \; \frac{1}{16} \times 10^8 \; = \; 0.0625 \times 10^8 \; = \; 6.25 \times 10^6$$

1.6.   Examples of square roots.

Even powers of 10:

$$\sqrt{10^6} \; = \; 10^{6/2} \; = \; 10^3$$

$$\sqrt{5 \times 10^4} \; = \; \sqrt{5} \times \sqrt{10^4} \; = \; 2.24 \times 10^2$$

Odd powers of 10:

$$\sqrt{3 \times 10^5} \; = \; \sqrt{30 \times 10^4} \; = \; \sqrt{30} \times \sqrt{10^4} \; = \; 5.48 \times 10^2$$

$$\sqrt{0.000025} \; = \; \sqrt{2.5 \times 10^{-5}} \; = \; \sqrt{25 \times 10^{-6}} \; = \; \sqrt{25} \times \sqrt{10^{-6}} \; = \; 5 \times 10^{-3}$$

1.7.   Examples of cube roots.

$$\sqrt[3]{10^9} \; = \; 10^{9/3} \; = \; 10^3$$

$$\sqrt[3]{10^8} \; = \; \sqrt[3]{10^2 \times 10^6} \; = \; \sqrt[3]{100} \times \sqrt[3]{10^6} \; = \; 4.64 \times 10^2$$

$$\sqrt[3]{3.8 \times 10^{19}} \; = \; \sqrt[3]{38 \times 10^{18}} \; = \; \sqrt[3]{38} \times \sqrt[3]{10^{18}} \; = \; 3.36 \times 10^6$$

$$\sqrt[3]{2.7 \times 10^{-5}} \; = \; \sqrt[3]{27 \times 10^{-6}} \; = \; \sqrt[3]{27} \times \sqrt[3]{10^{-6}} \; = \; 3 \times 10^{-2}$$

1.8.   Rome is 1440 km by road from Paris.   How far is this in miles?

Since 1 km = 0.621 mi,

$$1440 \text{ km} = 1440 \cancel{\text{ km}} \times 0.621 \frac{\text{mi}}{\cancel{\text{km}}} = 894 \text{ mi}$$

1.9.   A man is 6 ft 2 in. tall.   How many cm is this?

Since 1 ft = 12 in. and 1 in. = 2.54 cm,

$$6 \text{ ft } 2 \text{ in.} = (6 \times 12 + 2) \text{ in.} = 74 \cancel{\text{ in.}} \times 2.54 \frac{\text{cm}}{\cancel{\text{in.}}} = 188 \text{ cm}$$

1.10.   How many ft² are there in one m²?

Since 1 m = 3.28 ft, $1 \text{ m}^2 = 1 \text{ m}^2 \times \left(3.28 \frac{\text{ft}}{\text{m}}\right)^2 = 10.76 \text{ ft}^2$

1.11.   Express a velocity of 60 mi/hr in ft/s.

There are 5280 ft in a mile and 3600 s in an hour, and so

$$60 \frac{\text{mi}}{\text{hr}} = 60 \frac{\cancel{\text{mi}}}{\cancel{\text{hr}}} \times 5280 \frac{\text{ft}}{\cancel{\text{mi}}} \times \frac{1}{3600 \text{ s/}\cancel{\text{hr}}} = 88 \frac{\text{ft}}{\text{s}}$$

1.12.   A man walks eastward for 5 miles and then northward for 10 miles.   How far is he from his starting point?   If he had walked directly to his destination, in what direction would he have headed?

From Fig. 1-4, the length of the resultant vector **R** corresponds to a distance of 11.2 miles and a protractor shows that its direction is 27° east of north.

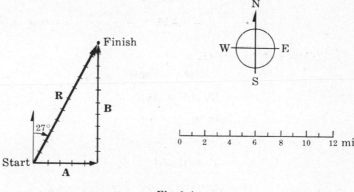

Fig. 1-4

1.13.   Two tugboats are towing a ship.   Each exerts a force of 6 tons, and the angle between the towropes is 60°.   What is the resultant force on the ship?

To add the force vectors **A** and **B**, **B** is shifted parallel to itself so that its tail is at the head of **A**.   The length of the resultant **R** corresponds to a force of 10.4 tons.   (See Fig. 1-5.)

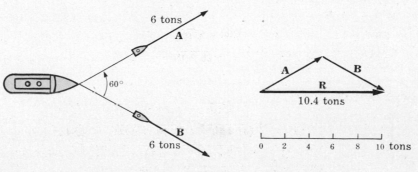

Fig. 1-5

1.14. A boat moving at 5 mi/hr is to cross a river in which the current is flowing at 3 mi/hr. In what direction should the boat head in order to reach a point on the other bank of the river directly opposite its starting point?

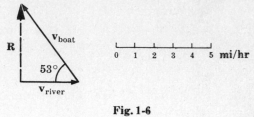

Fig. 1-6

The procedure here is to first draw the vector that represents v_{river}, the velocity of the current. Then the vector v_{boat} is drawn from the head of v_{river} so that its head is directly opposite the tail of v_{river} (Fig. 1-6). A protractor shows the angle between v_{river} and v_{boat} to be 53°.

1.15. In going from one city to another, a car whose driver tends to get lost goes 30 miles north, 50 miles west, and 20 miles southeast. Approximately how far apart are the cities?

The vectors representing the displacements are strung together head-to-tail, and their resultant is found to be 39 miles (Fig. 1-7).

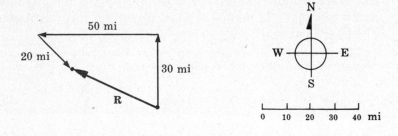

Fig. 1-7

Supplementary Problems

1.16. Express the following numbers in powers-of-10 notation:

(a) 720	(d) 0.000062	(g) 49,527	(j) 49,000,000,000
(b) 890,000	(e) 3.6	(h) 0.002943	(k) 0.000000011
(c) 0.02	(f) 0.4	(i) 0.0014	(l) 1.4763

1.17. Express the following numbers in decimal notation:

(a) 3×10^{-4}	(c) 8.126×10^{-5}	(e) 5×10^2	(g) 4.32145×10^3	(i) 5.7×10^0
(b) 7.5×10^3	(d) 1.01×10^8	(f) 3.2×10^{-2}	(h) 6×10^6	(j) 6.9×10^{-5}

1.18. Perform the following additions and subtractions:

(a) $3 \times 10^2 + 4 \times 10^3$	(f) $6.32 \times 10^2 + 5$	(k) $4.6 \times 10^5 - 3.2 \times 10^7$
(b) $2 \times 10^4 + 5 \times 10^6$	(g) $4 \times 10^3 - 3 \times 10^2$	(l) $3 \times 10^5 - 2.98 \times 10^5$
(c) $7 \times 10^{-2} + 2 \times 10^{-3}$	(h) $5 \times 10^7 - 9 \times 10^4$	(m) $4.76 \times 10^{-3} - 4.81 \times 10^{-3}$
(d) $4 \times 10^{-5} + 5 \times 10^{-3}$	(i) $3.2 \times 10^{-4} - 5 \times 10^{-5}$	(n) $7 \times 10^3 + 5 \times 10^2 - 9 \times 10^2$
(e) $2 \times 10^1 + 2 \times 10^{-1}$	(j) $7 \times 10^4 - 2 \times 10^5$	(o) $3 \times 10^{-4} + 6 \times 10^{-5} - 7 \times 10^{-3}$

1.19. Perform the following multiplications and divisions using powers-of-10 notation:

(a) 5000×0.005

(b) $\dfrac{5000}{0.005}$

(c) $\dfrac{500,000 \times 18,000}{9,000,000}$

(d) $\dfrac{30 \times 80,000,000,000}{0.0004}$

(e) $\dfrac{30,000 \times 0.0000006}{1000 \times 0.02}$

(f) $\dfrac{0.0001}{60,000 \times 200}$

(g) $\dfrac{200 \times 0.00004}{400,000}$

(h) $\dfrac{0.002 \times 0.00000005}{0.000004}$

(i) $\dfrac{400 \times 0.00006}{0.2 \times 20,000}$

(j) $\dfrac{0.06 \times 0.0001}{0.00003 \times 40,000}$

1.20. Evaluate the following and express the results in powers-of-10 notation:

(a) $(4 \times 10^9)^3$

(b) $(2 \times 10^7)^2$

(c) $(2 \times 10^7)^{-2}$

(d) $(2 \times 10^{-2})^5$

(e) $(3 \times 10^{-8})^2$

(f) $(5 \times 10^{11})^{-2}$

(g) $(3 \times 10^{-4})^{-3}$

1.21. Evaluate the following and express the results in powers-of-10 notation. Note that $\sqrt{4} = 2$, $\sqrt{40} = 6.3$, $\sqrt[3]{4} = 1.6$, $\sqrt[3]{40} = 3.4$ and $\sqrt[3]{400} = 7.4$.

(a) $\sqrt{4 \times 10^6}$

(b) $\sqrt{4 \times 10^7}$

(c) $\sqrt{4 \times 10^8}$

(d) $\sqrt{4 \times 10^{-4}}$

(e) $\sqrt{4 \times 10^{-5}}$

(f) $\sqrt[3]{4 \times 10^{12}}$

(g) $\sqrt[3]{4 \times 10^{13}}$

(h) $\sqrt[3]{4 \times 10^{14}}$

(i) $\sqrt[3]{4 \times 10^{15}}$

(j) $\sqrt[3]{4 \times 10^{-6}}$

(k) $\sqrt[3]{4 \times 10^{-7}}$

(l) $\sqrt[3]{4 \times 10^{-8}}$

(m) $\sqrt[3]{4 \times 10^{-9}}$

1.22. The earth is an average of 9.3×10^7 miles from the sun. How far is this in km? In meters?

1.23. How many cubic feet are there in a cubic meter?

1.24. The speed limit in many European towns in 60 km/hr. How many mi/hr is this?

1.25. The speed of light is 3.00×10^8 m/s. What is this speed in ft/s? In mi/s? In mi/hr?

1.26. A nautical mile is 6076 feet long, and a knot is a unit of speed equal to 1 nautical mile/hr. How fast is an 8-knot boat going in mi/hr? In ft/s?

1.27. A man drives 200 miles south and then 60 miles east. What is his displacement from his starting point?

1.28. What is the resultant force produced by a vertically upward force of 50 lb and a horizontal force of 70 lb?

1.29. A boat heads northwest at 10 mi/hr in a river that flows east at 3 mi/hr. What is the magnitude and direction of the boat's velocity relative to the earth's surface?

1.30. An airplane flies 400 miles west from city A to city B, then 300 miles northeast to city C, and finally 100 miles north to city D. How far is it from city A to city D? In what direction must the airplane head to return directly to city A from city D?

Answers to Supplementary Problems

1.16.   (a) 7.2×10^2 (g) 4.9527×10^4

       (b) 8.9×10^5 (h) 2.943×10^{-3}

       (c) 2×10^{-2} (i) 1.4×10^{-3}

       (d) 6.2×10^{-5} (j) 4.9×10^{10}

       (e) 3.6×10^0 (k) 1.1×10^{-8}

       (f) 4×10^{-1} (l) 1.4763×10^0

1.17.   (a) 0.0003   (f) 0.032

       (b) 7500   (g) 4321.45

       (c) 0.00008126   (h) 6,000,000

       (d) 101,000,000   (i) 5.7

       (e) 500   (j) 0.000069

1.18.   (a) 4.3×10^3 (i) 2.7×10^{-4}

       (b) 5.02×10^6 (j) -1.3×10^5

       (c) 7.2×10^{-2} (k) -3.154×10^7

       (d) 5.04×10^{-3} (l) 2×10^3

       (e) 2.02×10^1 (m) -5×10^{-5}

       (f) 6.37×10^2 (n) 6.6×10^3

       (g) 3.7×10^3 (o) -6.64×10^{-3}

       (h) 4.991×10^7

1.19.   (a) 2.5×10^1 (f) 8.3×10^{-12}

       (b) 10^6 (g) 2×10^{-8}

       (c) 10^3 (h) 2.5×10^{-5}

       (d) 6×10^{15} (i) 6×10^{-6}

       (e) 9×10^{-4} (j) 5×10^{-6}

1.20.   (a) 6.4×10^{28} (e) 9×10^{-16}

       (b) 4×10^{14} (f) 4×10^{-24}

       (c) 2.5×10^{-15} (g) 3.7×10^{10}

       (d) 3.2×10^{-9}

1.21.   (a) 2×10^3 (h) 7.4×10^4

       (b) 6.3×10^3 (i) 1.6×10^5

       (c) 2×10^4 (j) 1.6×10^{-2}

       (d) 2×10^{-2} (k) 7.4×10^{-3}

       (e) 6.3×10^{-3} (l) 3.4×10^{-3}

       (f) 1.6×10^4 (m) 1.6×10^{-3}

       (g) 3.4×10^4

1.22. 1.5×10^8 km; 1.5×10^{11} m

1.23.   35.3 ft^3

1.24.   37 mi/hr

1.25. 9.84×10^8 ft/sec; 1.86×10^5 mi/s; 6.72×10^8 mi/hr

1.26.   9.2 mi/hr; 13.5 ft/s

1.27.   209 miles

1.28.   86 lb

1.29.   8.2 mi/hr in a direction 30° west of north

1.30.   364 miles; 31° east of south

Chapter 2

Motion in a Straight Line

VELOCITY

The *velocity* of a body is a vector quantity that describes both how fast it is moving and the direction in which it is headed.

In the case of a body traveling in a straight line, its velocity is simply the rate at which it covers distance. If the distance s covered in the time t is proportional to t, the body has the *constant velocity* of

$$v = \frac{s}{t}$$

$$\text{Velocity} = \frac{\text{distance}}{\text{time}}$$

Given the velocity of a body, we can find how far it travels in a time interval t from the formula

$$s = vt$$

provided the velocity does not change in the time interval.

ACCELERATION

A body whose velocity is changing is *accelerated*. A body is accelerated when its velocity is increasing, decreasing, or changing in direction. Accelerations that involve a change in direction are discussed in Chapter 4.

The *acceleration* of a body is the rate at which its velocity is changing. If a body's velocity is v_0 at the start of a certain time interval t and is v at the end, its acceleration is

$$a = \frac{v - v_0}{t}$$

$$\text{Acceleration} = \frac{\text{velocity change}}{\text{time}}$$

A positive acceleration means an increase in velocity; a negative acceleration means a decrease in velocity. Only constant accelerations are considered here.

Velocity has the dimensions of distance/time. Acceleration has the dimensions of velocity/time or distance/time². Typical acceleration units are the ft/s² and the m/s². Sometimes two different time units are used; for instance, the acceleration of a car that can go from rest to 50 mi/hr in 10 s might be expressed as $a = 5$ (mi/hr)/s.

DISTANCE, VELOCITY, AND ACCELERATION

Let us consider a body whose velocity is v_0 when it starts to be accelerated at a constant rate. After the time t the final velocity of the body will be

$$v = v_0 + at$$

$$\text{Final velocity} = \text{initial velocity} + \text{velocity change}$$

9

How far does the body go during the time interval t? The average velocity $\bar{v}$ of the body is

$$\bar{v} = \frac{v_0 + v}{2}$$

and so, since $v = v_0 + at$,

$$\bar{v} = \frac{v_0 + v_0 + at}{2} = v_0 + \frac{1}{2}at$$

The distance traveled during t is therefore

$$s = \bar{v}t = v_0 t + \frac{1}{2}at^2$$

If the body is accelerated starting from rest, $v_0 = 0$ and

$$s = \frac{1}{2}at^2$$

ACCELERATION OF GRAVITY

All bodies in free fall near the earth's surface have the same downward acceleration of

$$g = 32 \text{ ft/s}^2 = 9.8 \text{ m/s}^2$$

A body falling from rest in a vacuum thus has a velocity of 32 ft/s at the end of the first second, 64 ft/s at the end of the next second, and so forth. The farther the body falls, the faster it moves.

A body in free fall has the same downward acceleration whether it starts from rest or has an initial velocity in some direction.

The presence of air affects the motion of falling bodies partly through buoyancy and partly through air resistance. Thus two different objects falling in air from the same height will not, in general, reach the ground at exactly the same time. Because air resistance increases with velocity, eventually a falling body reaches a *terminal velocity* that depends upon its mass, size, and shape, and it cannot fall any faster than that.

FALLING BODIES

When buoyancy and air resistance can be neglected, a falling body has the constant acceleration g and the formulas for uniformly accelerated motion apply to it. Thus a body dropped from rest has the velocity

$$v = gt$$

after the time t and has fallen through a vertical distance of

$$h = \frac{1}{2}gt^2$$

From the latter formula we see that

$$t = \sqrt{\frac{2h}{g}}$$

and so the velocity of the body is related to the distance it has fallen by $v = gt$ or

$$v = \sqrt{2gh}$$

To reach a certain height h, a body thrown upward must have the same initial velocity as the final velocity of a body falling from that height, namely $v = \sqrt{2gh}$.

Solved Problems

[Air resistance is assumed negligible in these problems.]

2.1. A ship travels 9 miles in 45 min. What is its speed in mi/hr?

Since   45 min = 3/4 hr,

$$v = \frac{s}{t} = \frac{9 \text{ mi}}{3/4 \text{ hr}} = 12 \frac{\text{mi}}{\text{hr}}$$

2.2. The velocity of sound in air at sea level is about 1100 ft/s. If a man hears a clap of thunder 3 s after seeing a lightning flash, how far away was the lightning?

The velocity of light is so great compared with the velocity of sound that the time needed for the light of the flash to reach the man can be neglected.   Hence

$$s = vt = 1100 \frac{\text{ft}}{\text{s}} \times 3 \text{ s} = 3300 \text{ ft}$$

2.3. The velocity of light is 3×10^8 m/s. How long does it take light to reach the earth from the sun, which is 1.5×10^{11} m away?

$$t = \frac{s}{v} = \frac{1.5 \times 10^{11} \text{ m}}{3 \times 10^8 \text{ m/s}} = 500 \text{ s} = 8\tfrac{1}{3} \text{ min}$$

2.4. A car travels 270 miles in 4.5 hours. (*a*) What is its average speed? (*b*) How far will it go in 7 hours at this average speed? (*c*) How long will it take to travel 300 mi at this average speed?

(*a*)
$$v = \frac{s}{t} = \frac{270 \text{ mi}}{4.5 \text{ hr}} = 60 \frac{\text{mi}}{\text{hr}}$$

(*b*)
$$s = vt = 60 \frac{\text{mi}}{\text{hr}} \times 7 \text{ hr} = 420 \text{ mi}$$

(*c*)
$$t = \frac{s}{v} = \frac{300 \text{ mi}}{60 \text{ mi/hr}} = 5 \text{ hr}$$

2.5. An airplane whose air speed is 400 mi/hr has a tail wind of 120 mi/hr. How long will it take the airplane to cover 1000 miles relative to the ground?

The ground speed of the airplane is

$$v = 400 \frac{\text{mi}}{\text{hr}} + 120 \frac{\text{mi}}{\text{hr}} = 520 \frac{\text{mi}}{\text{hr}}$$

Hence the time needed to cover 1000 miles over the ground is

$$t = \frac{s}{v} = \frac{1000 \text{ mi}}{520 \text{ mi/hr}} = 1.9 \text{ hr}$$

2.6. A car moves at 40 mi/hr for 2 hr and then at 30 mi/hr for $1\tfrac{1}{2}$ hr. (*a*) How far did it go? (*b*) What was its average speed for the whole trip?

(*a*)
$$s = v_1 t_1 + v_2 t_2 = 40 \frac{\text{mi}}{\text{hr}} \times 2 \text{ hr} + 30 \frac{\text{mi}}{\text{hr}} \times 1.5 \text{ hr} = 125 \text{ mi}$$

(*b*) A total distance of 125 miles was covered in $3\tfrac{1}{2}$ hours, and so

$$\bar{v} = \frac{s}{t} = \frac{125 \text{ mi}}{3.5 \text{ hr}} = 36 \frac{\text{mi}}{\text{hr}}$$

2.7. A car starts from rest and reaches a velocity of 40 ft/s in 10 s. (*a*) What is its acceleration? (*b*) If its acceleration remains the same, what will its velocity be after 15 s?

Here $v_0 = 0$. Hence

(*a*)
$$a = \frac{v}{t} = \frac{40 \text{ ft/s}}{10 \text{ s}} = 4 \text{ ft/s}^2$$

(*b*)
$$v = at = 4\frac{\text{ft}}{\text{s}^2} \times 15 \text{ s} = 60\frac{\text{ft}}{\text{s}}$$

2.8. (*a*) What is the acceleration of a car that goes from 20 mi/hr to 30 mi/hr in 1.5 s? (*b*) At the same acceleration, how long will it take the car to go from 30 mi/hr to 36 mi/hr?

(*a*)
$$a = \frac{v - v_0}{t} = \frac{30 \text{ mi/hr} - 20 \text{ mi/hr}}{1.5 \text{ s}} = 6.7 \text{ (mi/hr)/s}$$

(*b*)
$$t = \frac{v - v_0}{a} = \frac{36 \text{ mi/hr} - 30 \text{ mi/hr}}{6.7 \text{ (mi/hr)/s}} = 0.9 \text{ s}$$

2.9. A car has an acceleration of 8 m/s². (*a*) How much time is needed for it to reach a velocity of 24 m/s starting from rest? (*b*) How far does it go during this period of time?

(*a*)
$$t = \frac{v}{a} = \frac{24 \text{ m/s}}{8 \text{ m/s}^2} = 3 \text{ s}$$

(*b*) Since the car starts from rest, $v_0 = 0$ and
$$s = \frac{1}{2}at^2 = \frac{1}{2} \times 8\frac{\text{m}}{\text{s}^2} \times (3 \text{ s})^2 = 36 \text{ m}$$

2.10. The brakes of a certain car can produce an acceleration of 6 m/s². (*a*) How long does it take the car to come to a stop from a velocity of 30 m/s? (*b*) How far does the car travel during the time the brakes are applied?

(*a*)
$$t = \frac{v}{a} = \frac{30 \text{ m/s}}{6 \text{ m/s}^2} = 5 \text{ s}$$

(*b*) Here the signs of v_0 and a are important. The initial velocity of the car is $v_0 = +30$ m/s and its acceleration is -6 m/s², so that
$$s = v_0 t + \frac{1}{2}at^2 = 30\frac{\text{m}}{\text{s}} \times 5 \text{ s} - \frac{1}{2} \times 6\frac{\text{m}}{\text{s}^2} \times (5 \text{ s})^2 = 75 \text{ m}$$

2.11. A stone dropped from a bridge strikes the water 2.5 s later. (*a*) What is its final velocity in m/s? (*b*) How high is the bridge?

(*a*)
$$v = gt = 9.8\frac{\text{m}}{\text{s}^2} \times 2.5 \text{ s} = 24.5\frac{\text{m}}{\text{s}}$$

(*b*)
$$h = \frac{1}{2}gt^2 = \frac{1}{2} \times 9.8\frac{\text{m}}{\text{s}^2} \times (2.5 \text{ s})^2 = 30.6 \text{ m}$$

2.12. A ball is dropped from a window 64 ft above the ground. (*a*) How long does it take to reach the ground? (*b*) What is its final velocity?

(*a*) Since $h = \frac{1}{2}gt^2$,
$$t = \sqrt{\frac{2h}{g}} = \sqrt{\frac{2 \times 64 \text{ ft}}{32 \text{ ft/s}^2}} = \sqrt{4 \text{ s}^2} = 2 \text{ s}$$

(b) $$v = gt = 32\frac{\text{ft}}{\text{s}^2} \times 2\text{ s} = 64\frac{\text{ft}}{\text{s}}$$

2.13. What velocity must a ball have when thrown upward if it is to reach a height of 100 ft?

The upward velocity the ball must have is the same as the downward velocity a ball would have if dropped from that height.   Hence

$$v = \sqrt{2gh} = \sqrt{2 \times 32\frac{\text{ft}}{\text{s}^2} \times 100\text{ ft}} = \sqrt{6400\frac{\text{ft}^2}{\text{s}^2}} = 80\frac{\text{ft}}{\text{s}}$$

2.14. A ball is rolled off the edge of a table 0.78 m high with a horizontal velocity of 0.5 m/s. (a) How long does it take to reach the floor?   (b) How far from the foot of the table does the ball strike the floor?

(a) The horizontal velocity of the ball has no effect on its vertical motion.   Therefore the ball reaches the floor at the same time as a ball dropped from a height of 0.78 m, namely

$$t = \sqrt{\frac{2h}{g}} = \sqrt{\frac{2 \times 0.78\text{ m}}{9.8\text{ m/s}^2}} = \sqrt{0.16\text{ s}^2} = 0.4\text{ s}$$

(b) In 0.4 s the ball will travel a horizontal distance of

$$s = vt = 0.5\frac{\text{m}}{\text{s}} \times 0.4\text{ s} = 0.2\text{ m}$$

2.15. A ball is thrown downward with an initial velocity of 20 ft/s.   (a) How fast is it moving after 2 s?   (b) How far does it fall in 2 s?

(a) $$v = v_0 + gt = 20\frac{\text{ft}}{\text{s}} + 32\frac{\text{ft}}{\text{s}^2} \times 2\text{ s} = 20\frac{\text{ft}}{\text{s}} + 64\frac{\text{ft}}{\text{s}} = 84\frac{\text{ft}}{\text{s}}$$

(b) $$s = v_0t + \frac{1}{2}gt^2 = 20\frac{\text{ft}}{\text{s}} \times 2\text{ s} + \frac{1}{2} \times 32\frac{\text{ft}}{\text{s}^2} \times (2\text{ s})^2$$

$$= 40\text{ ft} + 64\text{ ft} = 104\text{ ft}$$

2.16. A ball is thrown upward with an initial velocity of 20 ft/s.   (a) How fast is it moving after $\frac{1}{2}$ s and in what direction?   (b) How fast is it moving after 2 s and in what direction?

We consider upward as + and downward as −.   Then $v_0 = +20$ ft/s and $g = -32$ ft/s², so that

(a) $$v = v_0 + gt = 20\frac{\text{ft}}{\text{s}} - 32\frac{\text{ft}}{\text{s}^2} \times \frac{1}{2}\text{ s} = 20\frac{\text{ft}}{\text{s}} - 16\frac{\text{ft}}{\text{s}} = 4\frac{\text{ft}}{\text{s}}$$

which is positive and hence upward.   After 2 s

(b) $$v = v_0 + gt = 20\frac{\text{ft}}{\text{s}} - 32\frac{\text{ft}}{\text{s}^2} \times 2\text{ s} = 20\frac{\text{ft}}{\text{s}} - 64\frac{\text{ft}}{\text{s}} = -44\frac{\text{ft}}{\text{s}}$$

which is negative and hence downward.

2.17. A man in a closed elevator cab with no floor indicator does not know whether the cab is stationary, moving upward at constant velocity, or moving downward at constant velocity.   To try to find out, he drops a coin from a height of 6 ft and times its fall with a stopwatch.   What would he find in each case?

Since the coin has exactly the same velocity as the elevator cab when it is dropped, this experiment would give the same time of fall in each case, namely

$$t = \sqrt{\frac{2h}{g}} = \sqrt{\frac{2 \times 6 \text{ ft}}{32 \text{ ft/s}^2}} = \sqrt{0.375 \text{ s}^2} = 0.61 \text{ s}$$

However, if the elevator cab were *accelerated* upward or downward, the time of fall would be respectively less or more than this.

Supplementary Problems

2.18.   A ship steams at a constant velocity of 15 mi/hr.   (a) How far does it travel in a day?   (b) How long does it take to travel 500 miles?

2.19.   A car moves at 50 mi/hr for 1/2 hr and then at 60 mi/hr for 2 hr.   (a) How far did it go?   (b) What was its average velocity for the entire trip?

2.20.   An airplane whose air speed is 500 mi/hr covers a distance of 1000 mi in $2\frac{1}{2}$ hr.   How strong was the headwind against it?

2.21.   How long does it take an echo to return to a man standing 300 ft from a cliff?   The velocity of sound in air is about 1100 ft/s.

2.22.   A pitcher takes 0.1 s to throw a baseball, which leaves his hand at a velocity of 90 ft/s.   What is the acceleration of the baseball while it is being thrown?

2.23.   A car comes to a stop in 6 s from a velocity of 30 m/s.   (a) What is its acceleration?   (b) At the same acceleration, how long would it take the car to come to a stop from a velocity of 40 m/s?

2.24.   The brakes of a car can slow it down from 60 mi/hr to 40 mi/hr in 2 s.   How long will it take to bring the car to a stop from an initial velocity of 25 mi/hr at the same acceleration?

2.25.   An object starts from rest with an acceleration of 10 m/s².   (a) How far does it go in 0.5 s?   (b) What is its velocity after 0.5 s?

2.26.   A car has an initial velocity of 20 ft/s when it begins to be accelerated at 5 ft/s².   (a) How long does it take to reach a velocity of 50 ft/s?   (b) How far does it go during this period of time?

2.27.   A stone is dropped from the edge of a cliff.   (a) What is its velocity 3 s later?   (b) How far does it fall in this time?   (c) How far will it fall in the next second?

2.28.   A boy throws a ball 60 ft vertically into the air.   (a) How long does he have to wait to catch it on the way down?   (b) What was its initial velocity?   (c) What will be its final velocity?

2.29.   The Empire State Building is 1472 ft high.   (a) How long would it take an object dropped from the top of the building to reach the ground?   (b) What would the object's final velocity be?

2.30.   A body in free fall reaches the ground in 5 s.   (a) From what height in m was it dropped?   (b) What is its final velocity?   (c) How far did it fall in the last second of its descent?

2.31.   A rifle with a muzzle velocity of 200 m/s is fired with its barrel horizontal at a height of 1.5 m above the ground.   (a) How long a time is the bullet in the air?   (b) How far away from the rifle does the bullet strike the ground?   (c) If the muzzle velocity were 150 m/s, would there be any difference in these answers?

2.32.   A ball is thrown vertically upward with a velocity of 12 m/s. (*a*) At what height is the ball 1 s later? (*b*) 2 s later? (*c*) What is the maximum height the ball reaches?

2.33.   A ball is thrown horizontally with a velocity of 12 m/s. (*a*) How far has the ball fallen 1 s later? (*b*) 2 s later?

2.34.   A ball is thrown vertically downward with a velocity of 12 m/s. (*a*) How far has the ball fallen 1 s later? (*b*) 2 s later?

Answers to Supplementary Problems

2.18.   360 miles; 33 hr

2.19.   145 miles; 58 mi/hr

2.20.   100 mi/hr

2.21.   0.55 s

2.22.   900 ft/s^2

2.23.   5 m/s^2; 8 s

2.24.   2.5 s

2.25.   1.25 m; 5 m/s

2.26.   6 s; 210 ft

2.27.   96 ft/s; 144 ft; 112 ft

2.28.   3.9 s; 62 ft/s; 62 ft/s

2.29.   9.6 s; 307 ft/s

2.30.   123 m; 49 m/s; 44 m

2.31.   0.56 s; 112 m; the time would be unchanged since it depends only on the height of the rifle, but the distance would be reduced to 84 m.

2.32.   7.1 m; 4.4 m; 7.3 m

2.33.   4.9 m; 19.6 m

2.34.   16.9 m; 43.6 m

Chapter 3

The Laws of Motion

FIRST LAW OF MOTION

According to Newton's first law of motion, a body at rest will remain at rest and a body in motion will remain in motion at constant velocity in a straight line if no net force acts on it.

This law thus provides a definition of force: a force is any influence that can change the velocity of a body. In order to accelerate something, a net force must be applied to it. Conversely, every acceleration is due to the action of a net force. (Since it is possible for two or more forces acting on the same body to cancel out with a vector sum of zero, a "net force" or "unbalanced force" must be specified.)

MASS

The property a body has of resisting any change in its state of rest or of uniform motion in a straight line is called *inertia*. The inertia of a body is related to what can be loosely thought of as the "amount of matter" it contains. A quantitative measure of inertia is *mass*: the more mass a body has, the less its acceleration when a net force acts on it.

SECOND LAW OF MOTION

Newton's second law of motion states that the net force acting on a body is proportional to the mass of the body and to its acceleration; the direction of the force is the same as that of the body's acceleration.

In a properly chosen set of units, the proportionality between force and the product of mass and acceleration is an equality, so that

$$F = ma$$

Force = mass × acceleration

The second law of motion is the key to understanding the behavior of moving bodies since it links cause (force) and effect (acceleration) in a definite way.

UNITS OF MASS AND FORCE

In the SI system, the unit of mass is the *kilogram* (kg) and the unit of force is the *newton* (N); a net force of 1 N acting on a mass of 1 kg produces an acceleration of 1 m/s².

In the British system, the unit of mass is the *slug* and the unit of force is the *pound* (lb); a net force of 1 lb acting on a mass of 1 slug produces an acceleration of 1 ft/s².

The second law of motion in SI and British units is as follows:

SI units: F(newtons) = m(kilograms) × a(m/s²)

British units: F(pounds) = m(slugs) × a(ft/s²)

16

WEIGHT AND MASS

The *weight* of a body is the gravitational force with which the earth attracts it. If a person weighs 150 lb, this means that the earth pulls him down with a force of 150 lb. Weight is different from mass, which is a measure of the response of the body to an applied force. The weight of a body varies with its location near the earth (or other astronomical body) whereas its mass is the same everywhere in the universe.

The weight of a body is the force that causes it to be accelerated downward with the acceleration of gravity g. Hence, from the second law of motion with $F = w$ and $a = g$,

$$w = mg$$

Weight = mass × acceleration of gravity

Because g is constant near the earth's surface, the weight of a body there is proportional to its mass — a large mass is heavier than a small one.

System of units	To find mass m given weight w	To find weight w given mass m
SI	$m(\text{kg}) = \dfrac{w(\text{N})}{9.8 \text{ m/s}^2}$	$w(\text{N}) = m(\text{kg}) \times 9.8 \text{ m/s}^2$
British	$m(\text{slugs}) = \dfrac{w(\text{lb})}{32 \text{ ft/s}^2}$	$w(\text{lb}) = m(\text{slugs}) \times 32 \text{ ft/s}^2$

Conversion of units

1 kg = 10^3 g = 0.0685 slug [1 kg corresponds to 2.21 lb in the sense that the *weight* of 1 kg is 2.21 lb]

1 slug = 14.6 kg [1 slug corresponds to 32 lb in the sense that the *weight* of 1 slug is 32 lb]

1 newton = 0.225 lb

1 lb = 4.45 newtons

THIRD LAW OF MOTION

Newton's third law of motion states that when one body exerts a force on another body, the second exerts an equal force in the opposite direction on the first.

Thus for every action force, there is an equal and opposite reaction force; no force can occur all by itself. Action and reaction forces never balance out because they act on *different* bodies.

Solved Problems

3.1. Is there any connection between the first and second laws of motion? Between the second and third?

The first law of motion is included in the second, since a body is not accelerated when there is no net force acting on it and it must therefore remain at rest or in motion at constant velocity. The second and third laws of motion are independent of each other.

3.2. A book rests on a table. (*a*) Show the forces acting on the table and the corresponding reaction forces. (*b*) Why don't the forces acting on the table cause it to move?

(*a*) See Fig. 3-1.

(*b*) The forces that act on the table have a vector sum of zero, so there is no net force acting on it.

3.3. In the process of walking, what force makes a person move forward?

The person's foot exerts a backward force on the ground; the forward reaction force of the ground on the foot produces the forward motion.

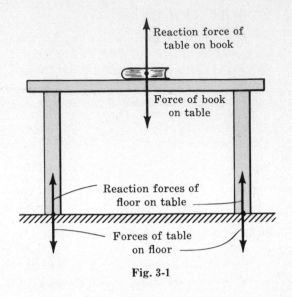

Fig. 3-1

PROBLEMS IN SI UNITS

3.4. (*a*) What is the weight of an object whose mass is 5 kg? (*b*) What is its acceleration when a net force of 100 N acts on it?

(*a*) $$w = mg = 5 \text{ kg} \times 9.8 \, \frac{\text{m}}{\text{s}^2} = 49 \text{ N}$$

(*b*) From the second law of motion, $F = ma$, we have

$$a = \frac{F}{m} = \frac{100 \text{ N}}{5 \text{ kg}} = 20 \, \frac{\text{m}}{\text{s}^2}$$

3.5. A force of 1 N acts on (*a*) a body whose mass is 1 kg, and (*b*) a body whose weight is 1 N. Find their respective accelerations.

(*a*) $$a = \frac{F}{m} = \frac{1 \text{ N}}{1 \text{ kg}} = 1 \, \frac{\text{m}}{\text{s}^2}$$

(*b*) The mass of a body whose weight is w is $m = w/g$. Hence, in general,

$$a = \frac{F}{m} = \frac{F}{w/g} = \frac{F}{w} g$$

Here $F = w = 1$ N, so the acceleration is $a = g = 9.8$ m/s^2.

3.6. A 10-kg body is observed to have an acceleration of 5 m/s^2. What is the net force acting on it?

$$F = ma = 10 \text{ kg} \times 5 \, \frac{\text{m}}{\text{s}^2} = 50 \text{ N}$$

3.7. A force of 80 N gives an object of unknown mass an acceleration of 20 m/s^2. What is its mass?

$$m = \frac{F}{a} = \frac{80 \text{ N}}{20 \text{ m/s}^2} = 4 \text{ kg}$$

3.8. A force of 3000 N is applied to a 1500-kg car at rest. (*a*) What is its acceleration? (*b*) What will its velocity be 5 s later?

$$(a) \qquad a = \frac{F}{m} = \frac{3000 \text{ N}}{1500 \text{ kg}} = 2\frac{\text{m}}{\text{s}^2} \qquad\qquad (b) \qquad v = at = 2\frac{\text{m}}{\text{s}^2} \times 5 \text{ s} = 10\frac{\text{m}}{\text{s}}$$

3.9. An empty truck whose mass is 2000 kg has a maximum acceleration of 1 m/s². What will its maximum acceleration be when it carries a load of 1000 kg?

The maximum force available is

$$F = ma = 2000 \text{ kg} \times 1\frac{\text{m}}{\text{s}^2} = 2000 \text{ N}$$

When this force is applied to the total mass of the loaded truck, which is 3000 kg, the resulting acceleration will be

$$a = \frac{F}{m} = \frac{2000 \text{ N}}{3000 \text{ kg}} = \frac{2}{3}\frac{\text{m}}{\text{s}^2}$$

3.10. A 1000-kg car goes from 10 m/s to 20 m/s in 5 s. What force is acting on it?

$$a = \frac{v - v_0}{t} = \frac{20 \text{ m/s} - 10 \text{ m/s}}{5 \text{ s}} = 2\frac{\text{m}}{\text{s}^2}$$

$$F = ma = 1000 \text{ kg} \times 2\frac{\text{m}}{\text{s}^2} = 2000 \text{ N}$$

3.11. The brakes of a 1000-kg car exert 3000 N. (a) How long will it take the car to come to a stop from a velocity of 30 m/s? (b) How far will the car travel during this time?

(a) The acceleration the brakes can produce is

$$a = \frac{F}{m} = \frac{3000 \text{ N}}{1000 \text{ kg}} = 3\frac{\text{m}}{\text{s}^2}$$

Here the initial velocity is $v_0 = 30$ m/s, the final velocity is $v = 0$, and the acceleration is -3 m/s², which is negative since the car is slowing down. Hence

$$v = v_0 + at$$

$$0 = 30\frac{\text{m}}{\text{s}} - 3\frac{\text{m}}{\text{s}^2} \times t$$

$$t = \frac{30 \text{ m/s}}{3 \text{ m/s}^2} = 10 \text{ s}$$

$$(b) \qquad s = v_0 t + \frac{1}{2}at^2 = 30\frac{\text{m}}{\text{s}} \times 10 \text{ s} - \frac{1}{2} \times 3\frac{\text{m}}{\text{s}^2} \times (10 \text{ s})^2$$

$$= 300 \text{ m} - 150 \text{ m} = 150 \text{ m}$$

PROBLEMS IN BRITISH UNITS

3.12. (a) What is the weight of an object whose mass is 50 slugs? (b) What is the mass of an object whose weight is 50 lb?

$$(a) \qquad w = mg = 50 \text{ slugs} \times 32\frac{\text{ft}}{\text{s}^2} = 1600 \text{ lb}$$

$$(b) \qquad m = \frac{w}{g} = \frac{50 \text{ lb}}{32 \text{ ft/s}^2} = 1.56 \text{ slug}$$

3.13. (a) What is the mass of a 160-lb man? (b) With what force is he attracted to the earth? (c) What is the reaction force to this force? (d) If he jumps from a diving board, what will his downward acceleration be?

(a)
$$m = \frac{w}{g} = \frac{160 \text{ lb}}{32 \text{ ft/s}^2} = 5 \text{ slugs}$$

(b) The attractive force exerted on the man by the earth is his weight of 160 lb.

(c) The reaction force is an upward pull he exerts on the earth of 160 lb. Because the earth is so much more massive than the man, the presence of this reaction force is not obvious, but it nevertheless exists.

(d) His downward acceleration is the acceleration of gravity, $g = 32 \text{ ft/s}^2$.

3.14. A net force of 75 lb acts on a body of mass 25 slugs which is initially at rest. (a) Find its acceleration. (b) How fast will the body be moving 12 s later?

(a)
$$a = \frac{F}{m} = \frac{75 \text{ lb}}{25 \text{ slugs}} = 3\frac{\text{ft}}{\text{s}^2}$$

(b)
$$v = at = 3\frac{\text{ft}}{\text{s}^2} \times 12 \text{ s} = 36\frac{\text{ft}}{\text{s}}$$

3.15. A net force of 150 lb acts on a body whose weight is 96 lb. What is its acceleration?

The mass of the body is
$$m = \frac{w}{g} = \frac{96 \text{ lb}}{32 \text{ ft/s}^2} = 3 \text{ slugs}$$

Its acceleration is therefore
$$a = \frac{F}{m} = \frac{150 \text{ lb}}{3 \text{ slugs}} = 50\frac{\text{ft}}{\text{s}^2}$$

3.16. How much force is needed to bring a 3200-lb car from rest to a velocity of 44 ft/s (30 mi/hr) in 8 s?

The car's mass is
$$m = \frac{w}{g} = \frac{3200 \text{ lb}}{32 \text{ ft/s}^2} = 100 \text{ slugs}$$

and its acceleration is
$$a = \frac{v}{t} = \frac{44 \text{ ft/s}}{8 \text{ s}} = 5.5\frac{\text{ft}}{\text{s}^2}$$

Hence
$$F = ma = 100 \text{ slugs} \times 5.5\frac{\text{ft}}{\text{s}^2} = 550 \text{ lb}$$

3.17. The brakes of a certain 2400-lb car can exert a maximum force of 750 lb. (a) What is the minimum time needed to slow the car down from 60 ft/s to 20 ft/s? (b) How far does the car travel in this time?

(a) The mass of the car is
$$m = \frac{w}{g} = \frac{2400 \text{ lb}}{32 \text{ ft/s}^2} = 75 \text{ slugs}$$

Its maximum acceleration is therefore
$$a = \frac{F}{m} = \frac{750 \text{ lb}}{75 \text{ slugs}} = 10\frac{\text{ft}}{\text{s}^2}$$

Here $v_0 = 60$ ft/s, $v = 20$ ft/s, and $a = -10$ ft/s². Since $a = (v - v_0)/t$,
$$t = \frac{v - v_0}{a} = \frac{20 \text{ ft/s} - 60 \text{ ft/s}}{-10 \text{ ft/s}^2} = \frac{-40 \text{ ft/s}}{-10 \text{ ft/s}^2} = 4 \text{ s}$$

(b)
$$s = v_0 t + \frac{1}{2}at^2 = 60\frac{\text{ft}}{\text{s}} \times 4 \text{ s} - \frac{1}{2} \times 10\frac{\text{ft}}{\text{s}^2} \times (4 \text{ s})^2$$

$$= 240 \text{ ft} - 80 \text{ ft} = 160 \text{ ft}$$

3.18. A 3200-lb elevator cab is supported by a cable in which the maximum safe tension is 4000 lb. (*a*) What is the greatest upward acceleration the elevator cab can have? (*b*) The greatest downward acceleration?

(*a*) The mass of the cab is

$$m = \frac{w}{g} = \frac{3200 \text{ lb}}{32 \text{ ft/s}^2} = 100 \text{ slugs}$$

The maximum net upward force the cable can exert is the difference between the maximum tension of 4000 lb and the 3200-lb weight of the cab, or 800 lb. Hence the greatest upward acceleration is

$$a = \frac{F}{m} = \frac{800 \text{ lb}}{100 \text{ slugs}} = 8 \frac{\text{ft}}{\text{s}^2}$$

(*b*) The cable is flexible and therefore cannot push down on the cab. The maximum downward acceleration therefore corresponds to free fall with an acceleration of $g = 32 \text{ ft/s}^2$.

3.19. A weight of 50 lb and another of 30 lb are suspended by a rope on either side of a frictionless pulley (Fig. 3-2). What is the acceleration of each weight?

The net force acting on the system of two objects is the *difference* between their weights:

$$F = 50 \text{ lb} - 30 \text{ lb} = 20 \text{ lb}$$

The mass to be accelerated is the total mass of the system, namely

$$m = \frac{w_1 + w_2}{g} = \frac{50 \text{ lb} + 30 \text{ lb}}{32 \text{ ft/s}^2} = 2.5 \text{ slugs}$$

Hence the upward acceleration of the system is

$$a = \frac{F}{m} = \frac{20 \text{ lb}}{2.5 \text{ slugs}} = 8 \frac{\text{ft}}{\text{s}^2}$$

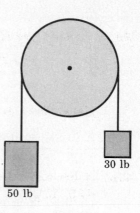

30 lb

50 lb

The 50-lb weight moves downward with this acceleration, and the 30-lb weight moves upward with the same acceleration.

Fig. 3-2

Supplementary Problems

3.20. Since action and reaction forces are always equal in magnitude and opposite in direction, how can anything ever be accelerated?

3.21. A horse is pulling a cart. (*a*) What is the force that causes the horse to move forward? (*b*) What is the force that causes the cart to move forward?

3.22. Is it possible for something to have a downward acceleration greater than g? If so, how can this be accomplished?

3.23. (*a*) When a horizontal force equal to its weight is applied to an object on a frictionless surface, what is its acceleration? (*b*) What is its acceleration when the force is applied vertically upward?

3.24. (*a*) What is the weight of 6 kg of potatoes? (*b*) What is the mass of 6 N of potatoes?

3.25. A force of 10 N is applied to (*a*) a body of mass 5 kg, and (*b*) a body of weight 5 N. Find their accelerations.

3.26. (*a*) What is the weight of 2 slugs of salami? (*b*) What is the mass of 2 lb of salami?

3.27. (*a*) How much upward force is needed to support a 20-kg object at rest? (*b*) To give it an upward acceleration of 2 m/s²? (*c*) To give it a downward acceleration of 2 m/s²?

3.28. (*a*) What is the acceleration of a 5-kg object suspended by a string when an upward force of 39 N is applied to the string? (*b*) When an upward force of 49 N is applied to the string? (*c*) When an upward force of 59 N is applied to the string?

3.29.   (a) How much applied force is needed to give an 8-lb object an upward acceleration of 10 ft/s²? (b) A downward acceleration of 10 ft/s²? (c) In what direction must the latter force act?

3.30.   A net force of 12 N gives an object an acceleration of 4 m/s². (a) What net force is needed to give it an acceleration of 1 m/s²? (b) An acceleration of 10 m/s²?

3.31.   A certain net force gives a 2-kg object an acceleration of 0.5 m/s². What acceleration would the same force give a 10-kg object?

3.32.   When an 8-lb rifle is fired, the 0.02-lb bullet receives an acceleration of 10^5 ft/s² while it is in the barrel. (a) How much force acts on the bullet? (b) Does any force act on the rifle? If so, how much and in what direction? (c) The bullet is accelerated for 0.007 s. How fast does it leave the barrel of the rifle?

3.33.   How much force is needed to accelerate a train whose mass is 1000 metric tons (1 metric ton = 1000 kg) from rest to a velocity of 6 m/s in 2 min?

3.34.   (a) How much force is needed to increase the velocity of a 6400-lb truck from 20 ft/s to 30 ft/s in 5 s? (b) How far does the truck travel in this time?

3.35.   (a) How much force is needed to decrease the velocity of a 6400-lb truck from 30 ft/s to 20 ft/s in 5 s? (b) How far does the truck travel in this time?

3.36.   A car strikes a stone wall at a velocity of 40 ft/s (27 mi/hr). (a) The car is rigidly built and the 160-lb driver comes to a stop in a time of 0.05 s. How much force acts on him? (b) The car is built so that its front end collapses gradually and the driver comes to a stop in 0.1 s. How much force acts on him in this case?

3.37.   A 0.05-kg snail goes from rest to a velocity of 0.01 m/s in 5 s. (a) How much force does it exert? (b) How far does it go during this time?

3.38.   A 160-lb man stands on a scale in an elevator cab. What does the scale read when the cab is (a) ascending at a constant velocity of 10 ft/s? (b) Ascending at a constant acceleration of 2 ft/s²? (c) Descending at a constant velocity of 10 ft/s? (d) Descending at a constant acceleration of 2 ft/s²? (e) In free fall because the cable has broken?

3.39.   An 80-kg man stands on a scale in an elevator cab. When it starts to move, the scale reads 700 N. (a) Is the cab moving upward or downward? (b) Is its velocity constant? If so, what is it. If not, what is its acceleration?

Answers to Supplementary Problems

3.20.   The action and reaction forces always act on different bodies.

3.21.   (a) The reaction force the ground exerts on its feet.
        (b) The force the horse exerts on it.

3.22.   Yes, by an applied downward force in addition to the downward force of gravity.

3.23. g; 0

3.24.   59 N; 0.61 kg

3.25.   2 m/s²; 19.6 m/s²

3.26.   64 lb; 0.0625 slug

3.27.   196 N; 236 N; 156 N

3.28.   2 m/s² downward; 0; 2 m/s² upward

3.29.   10.5 lb; 5.5 lb; upward

3.30.   3 N; 30 N

3.31.   0.1 m/s²

3.32.   62.5 lb; 62.5 lb backward; 700 ft/s

3.33. 5×10^4 N

3.34.   400 lb; 125 ft

3.35.   400 lb; 125 ft

3.36.   4000 lb; 2000 lb

3.37. 10^{-4} N; 0.025 m

3.38.   (a) 160 lb; (b) 170 lb; (c) 160 lb;
        (d) 150 lb; (e) 0

3.39.   Downward; no; 1.05 m/s²

Chapter 4

Circular Motion and Gravitation

UNIFORM CIRCULAR MOTION

A body that moves in a circular path at a velocity whose magnitude is constant is said to undergo *uniform circular motion.*

CENTRIPETAL ACCELERATION

Although the velocity of a body in uniform circular motion is constant in magnitude, its direction changes continually. The body is therefore accelerated. The direction of this *centripetal acceleration* is toward the center of the circle in which the body moves, and its magnitude is

$$a_c = \frac{v^2}{r}$$

$$\text{Centripetal acceleration} = \frac{(\text{velocity of body})^2}{\text{radius of circular path}}$$

Because the acceleration is perpendicular to the path followed by the body, its velocity changes only in direction and not in magnitude.

CENTRIPETAL FORCE

The inward force that must be applied to keep a body moving in a circle is called *centripetal force.* Without centripetal force, circular motion cannot occur. Since $F = ma$, the magnitude of the centripetal force on a body in uniform circular motion is

$$\text{Centripetal force} = F_c = \frac{mv^2}{r}$$

GRAVITATION

According to Newton's *law of universal gravitation,* every body in the universe attracts every other body with a force that is directly proportional to each of their masses and inversely proportional to the square of the distance between them. In equation form,

$$\text{Gravitational force} = F = G\frac{m_1 m_2}{r^2}$$

where m_1 and m_2 are the masses of any two bodies, r is the distance between them, and G is a constant whose values in SI and British units are respectively

$$\text{SI units:} \quad G = 6.67 \times 10^{-11} \frac{\text{N-m}^2}{\text{kg}^2}$$

$$\text{British units:} \quad G = 3.44 \times 10^{-8} \frac{\text{lb-ft}^2}{\text{slug}^2}$$

Gravitation provides the centripetal forces that keep the planets in their orbits around the sun and the moon in its orbit around the earth.   A spherical body behaves gravitationally as though its entire mass were concentrated at its center.

FUNDAMENTAL FORCES

Gravitational force is a *fundamental force* in the sense that it cannot be explained in terms of any other force.   There are only four fundamental forces: gravitational, electromagnetic, weak nuclear, and strong nuclear.   These forces are responsible for all physical processes in the universe.   The two types of nuclear force have very short ranges and act within atomic nuclei; it is possible that the weak force is related to the electromagnetic force. Electromagnetic forces, which are unlimited in range, act between electrically charged particles and determine the structures of atoms, molecules, solids, and liquids; when one object is in contact with another object, electromagnetic forces are ultimately responsible for the forces they exert on each other.   Gravitational forces act between all masses and are important in determining the structures of planets, stars, and galaxies of stars.

Solved Problems

4.1. How much centripetal force is needed to keep an 0.5-kg stone moving in a horizontal circle of radius 1 m at a velocity of 4 m/s?

$$F_c = \frac{mv^2}{r} = \frac{0.5 \text{ kg} \times (4 \text{ m/s})^2}{1 \text{ m}} = 8 \text{ N}$$

4.2. A centripetal force of 1 N is used to keep a 0.1-kg yoyo moving in a horizontal circle of radius 0.7 m.   What is the yoyo's velocity?

Since $F_c = mv^2/r$,

$$v = \sqrt{\frac{F_c r}{m}} = \sqrt{\frac{0.7 \text{ m} \times 1 \text{ N}}{0.1 \text{ kg}}} = \sqrt{7} \text{ m/s} = 2.6 \text{ m/s}$$

4.3. How much centripetal force is needed to keep a 160-lb skater moving in a circle 20 ft in radius at a velocity of 10 ft/s?

The skater's mass is $m = \dfrac{w}{g} = \dfrac{160 \text{ lb}}{32 \text{ ft/s}^2} = 5$ slugs.   Hence

$$F_c = \frac{mv^2}{r} = \frac{5 \text{ slugs} \times (10 \text{ ft/s})^2}{20 \text{ ft}} = 25 \text{ lb}$$

4.4. A 1000-kg car rounds a turn of radius 30 m at a velocity of 9 m/s.   (*a*) How much centripetal force is required?   (*b*) Where does this force come from?

(*a*) $$F_c = \frac{mv^2}{r} = \frac{1000 \text{ kg} \times (9 \text{ m/s})^2}{30 \text{ m}} = 2700 \text{ N}$$

(*b*) The centripetal force on a car making a turn on a level road is provided by the road acting via friction on the car's tires.

4.5. The maximum force a road can exert on the tires of a certain 3200-lb car is 2000 lb. What is the maximum velocity at which the car can round a turn of radius 320 ft?

The car's mass is $m = \dfrac{w}{g} = \dfrac{3200 \text{ lb}}{32 \text{ ft/s}^2} = 100$ slugs. Solving the formula $F_c = mv^2/r$ for v gives

$$v = \sqrt{\frac{F_c\, r}{m}} = \sqrt{\frac{2000 \text{ lb} \times 320 \text{ ft}}{100 \text{ slugs}}} = \sqrt{6400} \text{ ft/s} = 80 \text{ ft/s}$$

which is

$$80 \text{ ft/s} \times 0.682 \, \frac{\text{mi/hr}}{\text{ft/s}} = 55 \text{ mi/hr}$$

4.6. A string 0.5 m long is used to whirl a 1-kg stone in a vertical circle at a uniform velocity of 5 m/s. (*a*) What is the tension in the string when the stone is at the top of the circle? (*b*) When the stone is at the bottom of the circle?

(*a*) The centripetal force needed to keep the stone moving at 5 m/s is

$$F_c = \frac{mv^2}{r} = \frac{1 \text{ kg} \times (5 \text{ m/s})^2}{0.5 \text{ m}} = 50 \text{ N}$$

At the top of the circle, the weight of the stone is a downward force on it acting toward the center of the circle:

$$w = mg = 1 \text{ kg} \times 9.8 \text{ m/s}^2 = 9.8 \text{ N}$$

From Fig. 4-1(*a*) it is seen that the tension in the string is

$$T = F_c - w = 50 \text{ N} - 9.8 \text{ N} = 40.2 \text{ N}$$

(*b*) At the bottom of the circle, the weight of the stone acts away from the center of the circle, so the string must provide a force equal to w plus F_c, as indicated in Fig. 4-1(*b*). Hence

$$T = F_c + w = 50 \text{ N} + 9.8 \text{ N} = 59.8 \text{ N}$$

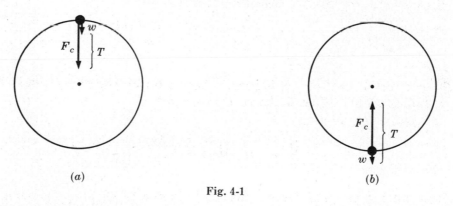

(*a*)   (*b*)

Fig. 4-1

4.7. What is the gravitational force a 1-ton lead sphere exerts on another identical sphere 10 ft away?

The mass of each sphere is

$$m = \frac{w}{g} = \frac{2000 \text{ lb}}{32 \text{ ft/s}^2} = 62.5 \text{ slugs}$$

Hence

$$F = G\frac{m_1 m_2}{r^2} = 3.44 \times 10^{-8} \frac{\text{lb-ft}^2}{\text{slug}^2} \times \frac{62.5 \text{ slugs} \times 62.5 \text{ slugs}}{(10 \text{ ft})^2} = 1.3 \times 10^{-6} \text{ lb}$$

which is less than the force that would result from blowing gently on one of the spheres.

4.8. (*a*) Find the gravitational force the earth exerts on the moon. The earth's mass is 5.98×10^{24} kg, the moon's mass is 7.36×10^{22} kg, and the moon's orbit is an average of 3.84×10^8 m in radius. (*b*) Find the orbital velocity of the moon on the assumption that its orbit is circular. (*c*) How long does it take the moon to complete an orbit around the earth?

(a) $$F = G\frac{m_1 m_2}{r^2} = 6.67 \times 10^{-11}\frac{\text{N-m}^2}{\text{kg}^2} \times \frac{5.98 \times 10^{24}\text{ kg} \times 7.36 \times 10^{22}\text{ kg}}{(3.84 \times 10^8\text{ m})^2}$$

$$= 1.99 \times 10^{20}\text{ N}$$

(b) The gravitational force the earth exerts on the moon provides the centripetal force that keeps it in its orbit.   Hence

$$F_c = F_{\text{grav}}$$

$$\frac{m_m v^2}{r} = G\frac{m_m m_e}{r^2}$$

and so, solving for v,

$$v = \sqrt{\frac{Gm_e}{r}} = \sqrt{\frac{6.67 \times 10^{-11}\text{ N-m}^2/\text{kg}^2 \times 5.98 \times 10^{24}\text{ kg}}{3.84 \times 10^8\text{ m}}} = 1.02 \times 10^3\text{ m/s}$$

(c) The circumference of the moon's orbit is $s = 2\pi r = 2\pi \times 3.84 \times 10^8\text{ m} = 2.41 \times 10^9\text{ m}$.   Hence the time needed for the moon to circle the earth is

$$t = \frac{s}{v} = \frac{2.41 \times 10^9\text{ m}}{1.02 \times 10^3\text{ m/s}} = 2.36 \times 10^6\text{ s} = 27.3\text{ days}$$

This agrees with the observed period of the moon's orbit.

4.9. Find the velocity an artificial satellite must have in order to pursue a circular orbit around the earth just above the surface.

In a stable orbit, the gravitational force mg on the satellite must be equal to the centripetal force mv^2/r required.   Hence

$$\frac{mv^2}{r} = mg$$

$$v^2 = rg$$

$$v = \sqrt{rg}$$

The mass of the satellite is irrelevant.   To find v we use the radius of the earth r_e for r and the acceleration of gravity at the earth's surface for g.   This gives

$$v = \sqrt{r_e g} = \sqrt{6.4 \times 10^6\text{ m} \times 9.8\text{ m/s}^2} = 7.9 \times 10^3\text{ m/s}$$

which is about 18,000 mi/hr.   With a smaller velocity than this, a space vehicle projected horizontally above the earth will fall to the surface; with a larger velocity, it will have an elliptical rather than a circular orbit.

4.10. A girl weighs 128 lb on the earth's surface.   (a) What would she weigh at a height above the earth's surface of one earth radius?   (b) What would her mass be there?

(a) Since the gravitational force the earth exerts on an object a distance r from its center varies as $1/r^2$, the gravitational force on the girl relative to its value mg at the earth's surface (where $r = r_e$) is

$$F = \left(\frac{r_e}{r}\right)^2 mg$$

When the girl is a distance r_e above the earth, $r = 2r_e$, and

$$F = \left(\frac{r_e}{r}\right)^2 mg = \left(\frac{r_e}{2r_e}\right)^2 w = \frac{1}{4} \times 128\text{ lb} = 32\text{ lb}$$

(b) Her mass of $m = \dfrac{w}{g} = \dfrac{128\text{ lb}}{32\text{ ft/s}^2} = 4\text{ slugs}$ is the same everywhere.

4.11. The radius of the earth is 6.4×10^6 m, $g = 9.8$ m/s^2 at the earth's surface, and $G = 6.67 \times 10^{-11}$ N-m^2/kg^2.   From these figures find the mass of the earth.

The force on an object of mass m at the earth's surface is, from Newton's law of gravitation, $F = Gmm_e/r_e^2$. This force is equal to the object's weight of mg. Hence

$$mg = G\frac{mm_e}{r_e^2}$$

The mass m cancels out, and we find that

$$m_e = \frac{gr_e^2}{G} = \frac{9.8 \text{ m/s}^2 \times (6.4 \times 10^6 \text{ m})^2}{6.67 \times 10^{-11} \text{ N-m}^2/\text{kg}^2} = 6.0 \times 10^{24} \text{ kg}$$

4.12. The moon's mass is 7.36×10^{22} kg and its radius is 1.74×10^6 m. (a) Find the acceleration of gravity g_m at its surface. (b) How much would a 75-kg man weigh there? What percentage of his weight on the earth's surface is this?

(a) From Problem 4.11 we have

$$g_m = \frac{Gm_m}{r_m^2} = \frac{6.67 \times 10^{-11} \text{ N-m}^2/\text{kg}^2 \times 7.36 \times 10^{22} \text{ kg}}{(1.74 \times 10^6 \text{ m})^2} = 1.6 \text{ m/s}^2$$

(b) The man's weight on the moon would be

$$w = mg_m = 75 \text{ kg} \times 1.6 \text{ m/s}^2 = 120 \text{ N}$$

which is 16% of his weight of $mg = 735$ N on the earth's surface.

4.13. (a) Find the velocity an artificial satellite must have to pursue a stable orbit just above the moon's surface. (b) Is this more or less than the velocity of a similar earth satellite? By how much?

(a) $$v = \sqrt{rg_m} = \sqrt{1.74 \times 10^6 \text{ m} \times 1.6 \text{ m/s}^2} = 1.7 \times 10^3 \text{ m/s}$$

(b) The velocity of an earth satellite is, from Problem 4.9, 7.9×10^3 m/s, so the corresponding velocity for a moon satellite is only 22% as great.

Supplementary Problems

4.14. Are there any circumstances under which a body can move in a curved path without being accelerated?

4.15. In what way, if any, do g and G change with increasing height above the earth's surface?

4.16. The moon's mass is approximately 1% of the earth's mass. How does the gravitational pull of the earth on the moon compare with the gravitational pull of the moon on the earth?

4.17. A hole is drilled to the center of the earth and a stone whose mass is 1 kg at the earth's surface is dropped into it. What is the mass of the stone when it is at the earth's center? What is its weight?

4.18. A 0.02-kg ball is whirled in a horizontal circle at the end of a string 0.5 m long whose breaking strength is 1 N. Neglecting gravity, what is the maximum velocity the ball can have?

4.19. How much centripetal force is needed to keep a 4-lb iron ball moving in a horizontal circle of radius 5 ft at a velocity of 15 ft/s?

4.20. A 5000-kg airplane makes a horizontal turn 1 km (1000 m) in radius at a velocity of 50 m/s. How much centripetal force is required?

4.21. A motorcycle begins to skid when it makes a turn of 50-ft radius at a velocity of 40 ft/s. What is the highest velocity at which it can make a turn of 100-ft radius?

4.22.   A string 1 m long is used to whirl a ball of unknown mass in a vertical circle. What is the minimum velocity of the ball if the string is to be just taut when the ball is at the top of the circle?

4.23.   A man swings a pail of water in a vertical circle 3.1 ft in radius. (a) If the water is not to spill, what is the minimum velocity the pail can have? (b) How much time per revolution is this equivalent to?

4.24.   What is the gravitational attraction between an 80-kg man and a 50-kg woman who are 2 m apart?

4.25.   What is the gravitational attraction between two 3200-lb elephants when they are 20 ft apart?

4.26.   The acceleration of gravity on the surface of Mars is $0.4\,g$. How much would a man weigh on the surface of Mars whose weight on the earth's surface is 180 lb?

4.27.   A man weighs 180 lb on the earth's surface. (a) What would he weigh at a distance from the center of the earth of three times the radius of the earth? (b) What would his mass there be?

4.28.   A 10-kg monkey is 10^6 m above the earth's surface. (a) What is his weight there? (b) What is his mass there?

4.29.   Find the velocity in m/s an artificial earth satellite must have in order to pursue a circular orbit at an altitude of half an earth radius.

4.30.   (a) If a satellite is placed in orbit about the earth halfway between the earth and the moon, what will its velocity be? (b) How many days will it take to circle the earth?

4.31.   The sun's mass is 2×10^{30} kg and the average radius of Jupiter's orbit is 7.8×10^{11} m. Find the orbital velocity of Jupiter.

4.32.   Jupiter has a mass of 1.9×10^{27} kg and a radius of 7.0×10^7 m. (a) What is the acceleration of gravity on Jupiter's surface? (b) What would a 60-kg girl weigh there? (c) How does this compare with her weight on the earth's surface?

4.33.   (a) Find the velocity an artificial satellite must have to pursue a stable orbit just above the surface of Jupiter. (b) Is this more or less than the velocity of a similar earth satellite? By how much?

Answers to Supplementary Problems

4.14.   No.

4.15.   The value of g decreases; the value of G is the same everywhere in the universe.

4.16.   They are equal.

4.17.   1 kg; 0

4.18.   5 m/s

4.19.   5.6 lb

4.20. 1.25×10^4 N

4.21.   56 ft/s

4.22.   3.1 m/s

4.23.   10 ft/s; 1.9 s

4.24. 6.67×10^{-8} N

4.25. 8.6×10^{-7} lb

4.26.   72 lb

4.27.   20 lb; 5.6 slugs

4.28.   73 N; 10 kg

4.29. 6.5×10^3 m/s

4.30. 1.4×10^3 m/s; 10 days

4.31. 1.3×10^4 m/s

4.32.   2.6 m/s^2; 156 N; she weighs 588 N on the earth, so her weight on Jupiter is only 27% as great

4.33. 1.3×10^4 m/s; 1.6 times more

Chapter 5

Energy

WORK

Work is a measure of the amount of change (in a general sense) a force produces when it acts upon a body. The change may be in the velocity of the body, in its position, in its size or shape, and so forth.

By definition, the work done by a force acting on a body is equal to the product of the force and the distance through which the force acts, provided that **F** and **s** are in the same direction. Thus

$$W = F \times s$$

Work = force × distance

If **F** is perpendicular to **s**, no work is done. Work is a scalar quantity.

UNITS OF WORK

The unit of work is the product of a force unit and a length unit. In the SI system the unit of work is the *joule* (J).

SI units: 1 joule (J) = 1 newton-meter = 0.738 ft-lb

British units: 1 foot-pound (ft-lb) = 1.36 J

POWER

Power is the rate at which work is done by a force. Thus

$$P = \frac{W}{t}$$

Power = $\dfrac{\text{work done}}{\text{time}}$

The more power something has, the more work it can perform in a given time.

UNITS OF POWER

Two special units of power are in wide use, the *watt* and the *horsepower*, where

1 watt (W) = 1 joule/second = 1.34×10^{-3} hp

1 horsepower (hp) = 550 ft-lb/s = 746 W

A *kilowatt* (kW) is equal to 10^3 W or 1.34 hp. A *kilowatt-hour* is the work done in 1 hr by an agency whose power output is 1 kW; hence 1 kw-hr = 3.6×10^6 J.

29

ENERGY

Energy is that property something has which enables it to do work. The more energy something has, the more work it can perform. Every kind of energy falls into one of three general categories: kinetic energy, potential energy, and rest energy.

The units of energy are the same as those of work, namely the joule and the foot-pound.

KINETIC ENERGY

The energy a body has by virtue of its motion is called *kinetic energy*. If the body's mass is m and its velocity is v, its kinetic energy is

$$\text{Kinetic energy} = \text{KE} = \frac{1}{2}mv^2$$

POTENTIAL ENERGY

The energy a body has by virtue of its position is called *potential energy*. A book held above the floor has gravitational potential energy because the book can do work on something else as it falls; a nail held near a magnet has magnetic potential energy because the nail can do work as it moves toward the magnet; the wound spring in a watch has elastic potential energy because the spring can do work as it unwinds.

The gravitational potential energy of a body of mass m that is at a height h above some reference level is

$$\text{Gravitational potential energy} = \text{PE} = mgh$$

where g is the acceleration of gravity. In terms of the weight w of the body,

$$\text{PE} = wh$$

REST ENERGY

Matter can be converted into energy, and energy into matter. The *rest energy* of a body is the energy it has by virtue of its mass alone. Thus mass can be regarded as a form of energy. The rest energy of a body is in addition to any KE or PE it might also have.

If the mass of a body is m_0 when it is at rest, its rest energy is

$$\text{Rest energy} = E_0 = m_0 c^2$$

In this formula c is the velocity of light, whose value is

$$c = 3.00 \times 10^8 \text{ m/s} = 9.83 \times 10^8 \text{ ft/s} = 186{,}000 \text{ mi/s}$$

The rest mass m_0 is specified here because the mass of a moving body increases with its velocity; the increase is only significant at extremely high velocities, however.

CONSERVATION OF ENERGY

According to the law of *conservation of energy*, energy cannot be created or destroyed, although it can be transformed from one kind into another. The total amount of energy in the universe is constant. A falling stone provides a simple example: more and more of its initial potential energy turns into kinetic energy as its velocity increases, until finally all of its PE has become KE when it strikes the ground. The KE of the stone is then transferred to the ground by the impact.

Solved Problems

5.1. A force of 60 lb is used to push a 150-lb crate a distance of 10 ft across a level warehouse floor. (a) How much work is done? (b) What is the change in the crate's potential energy?

(a) The weight of the crate does not matter here since its height does not change. The work done is

$$W = Fs = 60 \text{ lb} \times 10 \text{ ft} = 600 \text{ ft-lb}$$

(b) The crate's height does not change, so its potential energy remains the same.

5.2. (a) How much work is done in raising a 2000-lb elevator cab to a height of 80 ft? (b) How much potential energy does the cab have in its new position?

(a) The force needed is equal to the weight of the cab. Hence

$$W = Fs = wh = 2000 \text{ lb} \times 80 \text{ ft} = 1.6 \times 10^5 \text{ ft-lb}$$

(b)
$$PE = wh = 1.6 \times 10^5 \text{ ft-lb}$$

5.3. (a) How much work is done in raising a 2-kg book from the ground to a height of 1.5 m? (b) How much potential energy does the book have in its new position?

(a)
$$W = Fs = mgh = 2 \text{ kg} \times 9.8 \text{ m/s}^2 \times 1.5 \text{ m} = 29.4 \text{ J}$$

(b)
$$PE = mgh = 29.4 \text{ J}$$

5.4. (a) How much force must be applied to hold a 2-kg book 1.5 m above the ground? (b) How much work is done when it is held there for 10 min?

(a) The force needed is the book's weight of

$$w = mg = 2 \text{ kg} \times 9.8 \text{ m/s}^2 = 19.6 \text{ N}$$

(b) No work is done since the book does not move.

5.5. Ten thousand joules of work is performed in raising a 200-kg bronze statue. How high is it raised?

Since $W = Fs = mgh$,

$$h = \frac{W}{mg} = \frac{10^4 \text{ J}}{200 \text{ kg} \times 9.8 \text{ m/s}^2} = 5.1 \text{ m}$$

5.6. A 150-lb man runs up a staircase 10 ft high in 5 s. Find his minimum power output in hp.

The minimum downward force the man's legs must exert is equal to his weight of 150 lb. Hence

$$P = \frac{W}{t} = \frac{Fs}{t} = \frac{150 \text{ lb} \times 10 \text{ ft}}{5 \text{ s}} = 300 \text{ ft lb/s}$$

Since 1 hp = 550 ft-lb/s,

$$P = \frac{300 \text{ ft-lb/s}}{550 \text{ (ft-lb/s)/hp}} = 0.55 \text{ hp}$$

5.7. A motorboat requires 80 hp to move at the constant velocity of 10 mi/hr. How much resistive force does the water exert on the boat at this velocity?

Since $v = s/t$, $P = \dfrac{W}{t} = \dfrac{Fs}{t} = Fv$. Here

$$v = 10 \text{ mi/hr} \times 1.47 \frac{\text{ft/s}}{\text{mi/hr}} = 14.7 \text{ ft/s}$$

and
$$P = 80 \text{ hp} \times 550 \frac{\text{ft-lb/s}}{\text{hp}} = 44{,}000 \text{ ft-lb/s}$$

Hence
$$F = \frac{P}{v} = \frac{44{,}000 \text{ ft-lb/s}}{14.7 \text{ ft/s}} = 3000 \text{ lb}$$

5.8. A horse has a power output of 1 hp when it pulls a wagon with a force of 300 N. What is the wagon's velocity?

Since $P = 1 \text{ hp} = 746 \text{ W} = Fv$, $\quad v = \dfrac{P}{F} = \dfrac{746 \text{ W}}{300 \text{ N}} = 2.5 \text{ m/s}$.

5.9. Find the kinetic energy of a 1000-kg car whose velocity is 20 m/s.

$$\text{KE} = \tfrac{1}{2}mv^2 = \tfrac{1}{2} \times 1000 \text{ kg} \times (20 \text{ m/s})^2 = 2 \times 10^5 \text{ J}$$

5.10. What velocity does a 1-kg object have when its kinetic energy is 1 J?

Since $\text{KE} = \tfrac{1}{2}mv^2$,
$$v = \sqrt{\frac{2\text{KE}}{m}} = \sqrt{\frac{2 \times 1 \text{ J}}{1 \text{ kg}}} = \sqrt{2} \text{ m/s} = 1.4 \text{ m/s}$$

5.11. A 128-lb girl skates at a velocity of 15 ft/s. What is her kinetic energy?

The girl's mass is $\quad m = \dfrac{w}{g} = \dfrac{128 \text{ lb}}{32 \text{ ft/s}^2} = 4 \text{ slugs}$, and so her kinetic energy is

$$\text{KE} = \tfrac{1}{2}mv^2 = \tfrac{1}{2} \times 4 \text{ slugs} \times (15 \text{ ft/s})^2 = 450 \text{ ft-lb}$$

5.12. What is the kinetic energy of a 2500-lb car whose velocity is 40 mi/hr?

The mass of the car is
$$m = \frac{w}{g} = \frac{2500 \text{ lb}}{32 \text{ ft/s}^2} = 78 \text{ slugs}$$
and its velocity is
$$v = 40 \text{ mi/hr} \times 1.47 \frac{\text{ft/s}}{\text{mi/hr}} = 59 \text{ ft/s}$$
Hence the car's kinetic energy is
$$\text{KE} = \tfrac{1}{2}mv^2 = \tfrac{1}{2} \times 78 \text{ slugs} \times (59 \text{ ft/s})^2 = 1.36 \times 10^5 \text{ ft-lb}$$

5.13. At her highest point, a girl on a swing is 7 ft above the ground, and at her lowest point she is 3 ft above the ground. What is her maximum velocity?

The girl's maximum velocity v occurs at the lowest point. Her kinetic energy there equals her loss of potential energy in descending through a height of $\quad h = 7 \text{ ft} - 3 \text{ ft} = 4 \text{ ft}$. Hence

$$\text{KE} = \text{PE}$$
$$\tfrac{1}{2}mv^2 = mgh$$
$$v = \sqrt{2gh} = \sqrt{2 \times 32 \text{ ft/s}^2 \times 4 \text{ ft}} = \sqrt{256} \text{ ft/s} = 16 \text{ ft/s}$$

This result is independent of the girl's mass.

5.14. A man skis down a slope 200 m high. If his velocity at the bottom of the slope is 20 m/s, what percentage of his initial potential energy was lost due to friction and air resistance?

$$\frac{\text{Final KE}}{\text{Initial PE}} = \frac{\tfrac{1}{2}mv^2}{mgh} = \frac{v^2}{2gh} = \frac{(20 \text{ m/s})^2}{2 \times 9.8 \text{ m/s}^2 \times 200 \text{ m}} = 0.102 = 10.2\%$$

which means 89.8% of the initial PE was lost.

5.15. Approximately 4×10^9 kg of matter is converted into energy in the sun each second. What is the power output of the sun?

The energy produced by the sun per second is

$$E_0 = m_0 c^2 = 4 \times 10^9 \text{ kg} \times (3 \times 10^8 \text{ m/s})^2 = 3.6 \times 10^{26} \text{ J}$$

Hence the power output is

$$P = \frac{E_0}{t} = \frac{3.6 \times 10^{26} \text{ J}}{1 \text{ s}} = 3.6 \times 10^{26} \text{ W}$$

5.16. How much mass is converted into energy per day in a nuclear power plant operated at a level of 100 megawatts (100×10^6 W)?

There are $60 \times 60 \times 24 = 86{,}400$ s/day, so the energy liberated per day is

$$E_0 = Pt = 10^8 \text{ W} \times 8.64 \times 10^4 \text{ s} = 8.64 \times 10^{12} \text{ J}$$

Since $E_0 = m_0 c^2$,

$$m_0 = \frac{E_0}{c^2} = \frac{8.64 \times 10^{12} \text{ J}}{(3 \times 10^8 \text{ m/s})^2} = 9.6 \times 10^{-5} \text{ kg}$$

Supplementary Problems

5.17. The earth exerts a gravitational force of 2×10^{20} N on the moon, and the moon travels 2.4×10^9 m each time it orbits the earth. How much work does the earth do on the moon in each orbit?

5.18. How much work must be done to raise a 1100-kg car 2 m above the ground?

5.19. A 20-lb object is raised to a height of 40 ft above the ground. (a) How much work was done? (b) What is the potential energy of the object? (c) If the object is dropped, what will its kinetic energy be just before it strikes the ground?

5.20. A horse exerts a force of 200 lb while pulling a sled for 3 miles. (a) How much work does the horse do? (b) If the trip takes 30 min, what is the power output of the horse in hp?

5.21. In 1970 the population of the world was about 3.5×10^9 and about 2×10^{20} J of work was performed under man's control. Find the average power consumption per person in watts and in hp. (1 year $= 3.15 \times 10^7$ s)

5.22. A certain 80-kg mountain climber has an average power output of 0.1 hp. (a) How much work does he perform in climbing a mountain 2000 m high? (b) How long does he take to climb the mountain? (c) What is his potential energy at the top?

5.23. A man uses a rope and a system of pulleys to raise a 200-lb box to a height of 10 ft. He exerts a force of 60 lb on the rope and pulls a total of 40 ft of rope through the pulleys. (a) How much work does he perform? (b) By how much is the potential energy of the box increased? (c) If these answers are different, what do you think the reason is?

5.24. The four engines of a DC-8 airplane develop a total of 30,000 hp when its velocity is 240 m/s. How much force do the engines exert?

5.25. Neglecting friction and air resistance, is more work needed to accelerate a car from 10 mi/hr to 20 mi/hr or from 20 mi/hr to 30 mi/hr?

5.26. Find the kinetic energy of a 2-g (0.002-kg) insect when it is flying at 0.4 m/s.

5.27. The electrons in a television picture tube whose impacts on the screen produce the flashes of light that make up the image have masses of 9.1×10^{-31} kg and typical velocities of 3×10^7 m/s. What is the kinetic energy of such an electron?

5.28.   (*a*) What is the kinetic energy of a 3200-lb car traveling at 100 ft/s (68 mi/hr)?   (*b*) If the car can reach this velocity in 12 s starting from rest, what is the power output of its engine?

5.29.   (*a*) What velocity does a 1-slug object have when its kinetic energy is 1 ft-lb?   (*b*) What velocity does a 1-lb object have when its kinetic energy is 1 ft-lb?

5.30.   A stone is dropped from a height of 100 ft. At what height is half of its energy potential and half kinetic?

5.31.   From what height would a car have to fall to the ground in order to do as much work (that is, damage — largely to itself) as a car striking a wall at 88 ft/s (60 mi/hr)?

5.32.   This book weighs about 1.5 lb. What is its rest energy in ft-lb? In joules?

5.33.   Approximately 4 million ft-lb of energy is liberated when 1 lb of dynamite explodes. How much matter is converted into energy in this process?

5.34.   A sedentary person uses energy at an average rate of about 70 W. (*a*) How many joules of energy does he use per day? (*b*) All of this energy originates in the sun. How much matter is converted into energy per day to supply such a person?

Answers to Supplementary Problems

5.17.   No work is done because the force on the moon is perpendicular to its direction of motion.

5.18. 2.16×10^4 J

5.19.   800 ft-lb; 800 ft-lb; 800 ft-lb

5.20. 3.17×10^6 ft-lb; 3.2 hp

5.21.   1800 W; 2.4 hp

5.22. 1.57×10^6 J; 5 hr 50 min; 1.57×10^6 J

5.23.   2400 ft-lb; 2000 ft-lb; 400 ft-lb was expended in doing work against frictional forces in the pulleys

5.24. 9.3×10^4 N

5.25.   From 20 mi/hr to 30 mi/hr.

5.26. 1.6×10^{-4} J

5.27. 4.1×10^{-16} J

5.28. 5×10^5 ft-lb; 4.17×10^4 ft-lb/s = 76 hp

5.29.   1.4 ft/s; 8 ft/s

5.30.   50 ft

5.31.   121 ft

5.32. 4.5×10^{16} ft-lb; 6.1×10^{16} J

5.33. 1.32×10^{-10} lb

5.34. 6.05×10^6 J; 6.72×10^{-11} kg

Chapter 6

Momentum

LINEAR MOMENTUM

Work and energy are scalar quantities that have no directions associated with them. When two or more bodies interact with one another, or a single body breaks up into two or more others, the various directions of motion cannot be related by energy considerations alone. The vector quantity called *linear momentum* is important in analyzing such events.

The linear momentum (usually called simply *momentum*) of a body of mass m and velocity $\mathbf{v}$ is the product of m and $\mathbf{v}$:

$$\text{Momentum} = m\mathbf{v}$$

The units of momentum are the kg-m/s and the slug-ft/s. The direction of the momentum of a body is the same as the direction in which it is moving.

The greater the momentum of a body, the greater its tendency to continue in motion. Thus a baseball struck by a bat (v large) is harder to stop than a baseball thrown by hand (v small), and an iron shot (m large) is harder to stop than a baseball (m small) of the same velocity.

CONSERVATION OF LINEAR MOMENTUM

According to the law of conservation of linear momentum, the total linear momentum of a system of bodies isolated from the rest of the universe remains constant regardless of what happens within the system. The total linear momentum of the system is the vector sum of the momenta of the various bodies included in the system; by "isolated" is meant that no net force of external origin acts on the system.

In particular, momentum is conserved in collisions. If a billiard ball strikes a stationary one, the two move off in such a way that the vector sum of their momenta is the same as the initial momentum of the first ball (Fig. 6-1). This is true even if the balls move in different directions.

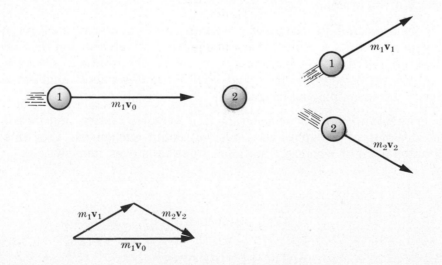

Fig. 6-1

35

ROCKET PROPULSION

Momentum conservation underlies the operation of a rocket. The momentum of a rocket on the ground is zero. When the fuel is ignited, exhaust gases shoot downward and the rocket body rises to balance their momentum so that the total remains zero. A rocket does not "push" against the ground or the atmosphere, and indeed is most efficient in space where there is no air to cause friction.

The final velocity of a rocket depends upon the velocity of its exhaust gases and upon the amount of fuel it can carry. When extremely high velocities are needed, as in space exploration, it is more efficient to use two or more rocket stages. The first stage has as its payload another, smaller rocket. When the fuel of the first stage has been burnt up, the second stage is released from the shell of the first and is fired. Since it is already moving rapidly and does not have the burden of the motor and fuel tanks of the first stage, the second stage can reach a much higher final velocity. This process can be repeated a number of times, depending upon the required velocity.

ANGULAR MOMENTUM

The equivalent of linear momentum in rotational motion is *angular momentum*. The greater the angular momentum of a spinning object such as a top, the greater its tendency to continue to spin.

The angular momentum of a rotating body is proportional to its mass and to how rapidly it is turning. In addition, the angular momentum depends upon how the mass is distributed relative to the axis of rotation: the farther away the mass is located from the axis of rotation, the greater the angular momentum. A flat disk such as a record turntable has more angular momentum than a tall cylinder of the same mass and velocity of rotation.

CONSERVATION OF ANGULAR MOMENTUM

According to the law of conservation of angular momentum, the total angular momentum of a system of bodies isolated from the rest of the universe remains constant regardless of what happens within the system.

A skater performing a spin makes use of conservation of angular momentum in an interesting way. He begins to turn with his arms and one leg outstretched, and then brings them in close to his body to distribute his mass nearer to the axis of rotation. Because no external force is involved in moving his arms and leg, the skater's total angular momentum cannot be changed by this action, and he spins faster as a result.

Like linear momentum, angular momentum is a vector quantity, and its conservation implies that the direction of the spin axis tends to remain unchanged. For this reason a spinning top remains upright whereas a stationary one falls over immediately.

Solved Problems

6.1. Find the momentum of a 50-kg girl running at 6 m/s.

$$mv = 50 \text{ kg} \times 6 \text{ m/s} = 300 \text{ kg-m/s}$$

6.2. A 160-lb man runs a mile in 4 min. What is his average momentum?

The man's mass m and average velocity $\bar{v}$ are respectively

$$m = \frac{w}{g} = \frac{160 \text{ lb}}{32 \text{ ft/s}^2} = 5 \text{ slugs} \qquad \bar{v} = \frac{1 \text{ mile} \times 5280 \text{ ft/mi}}{4 \text{ min} \times 60 \text{ s/min}} = 22 \text{ ft/s}$$

Hence his average momentum is $m\bar{v} = 5 \text{ slugs} \times 22 \text{ ft/s} = 110 \text{ slug-ft/s}$.

6.3. An airplane's velocity is doubled. (*a*) What happens to its momentum? Is the law of conservation of momentum obeyed? (*b*) What happens to its kinetic energy? Is the law of conservation of energy obeyed?

(*a*) The airplane's momentum of mv also doubles. Momentum is conserved because the increase in the airplane's velocity is accompanied by the backward motion of air through the action of its engines, so the total momentum of airplane + air remains the same when their opposite directions are taken into account.

(*b*) The airplane's kinetic energy of $\frac{1}{2}mv^2$ increases fourfold. Energy is conserved because the additional KE comes from chemical potential energy released in the airplane's engines.

6.4. A 2000-lb car moving at 50 mi/hr collides head-on with a 3000-lb car moving at 20 mi/hr, and the two cars stick together. Which way does the wreckage move?

The 2000-lb car had the greater initial momentum, so the wreckage moves in the same direction it had.

6.5. A 5-kg rifle fires a 15-g (0.015 kg) bullet at a muzzle velocity of 600 m/s. Find the recoil velocity of the rifle.

From conservation of momentum, $m_r v_r = m_b v_b$, and so

$$v_r = \frac{m_b}{m_r} \times v_b = \frac{0.015 \text{ kg}}{5 \text{ kg}} \times 600 \text{ m/s} = 1.8 \text{ m/s}$$

6.6. An astronaut is in space at rest relative to an orbiting spacecraft. His total weight is 300 lb and he throws away a 1-lb wrench at a velocity of 15 ft/s relative to the spacecraft. How fast does he move off in the opposite direction?

From conservation of momentum,

$$m_a v_a = m_w v_w$$

$$\frac{w_a}{g} \times v_a = \frac{w_w}{g} \times v_w$$

$$v_a = \frac{w_w}{w_a} \times v_w = \frac{1 \text{ lb}}{300 \text{ lb}} \times 15 \text{ ft/s} = 0.05 \text{ ft/s}$$

We notice that the g's have canceled out, so that it was not necessary to find the mass values first; the ratio of two masses is always the same as the ratio of the corresponding weights.

6.7. A 0.5-kg snowball moving at 20 m/s strikes and sticks to a 70-kg man standing on the frictionless surface of a frozen pond. What is the man's final velocity?

Let $v_1 =$ snowball's velocity   and $v_2 =$ final velocity of man + snowball.   Then

Initial momentum of snowball   =   final momentum of man + snowball

$$m_s v_1 = (m_m + m_s)v_2$$

$$v_2 = \left(\frac{m_s}{m_m + m_s}\right)v_1 = \frac{0.5 \text{ kg}}{70.5 \text{ kg}} \times 20 \text{ m/s} = 0.14 \text{ m/s}$$

6.8. A 40-kg skater traveling at 4 m/s overtakes a 60-kg skater traveling at 2 m/s in the same direction and collides with him. (*a*) If the two skaters remain in contact, what is their final velocity? (*b*) How much kinetic energy is lost?

(*a*) Let v_1 = initial velocity of 40-kg skater, v_2 = initial velocity of 60-kg skater, and v_3 = final velocity of the two skaters. Then

$$\text{Initial total momentum} = \text{final total momentum}$$

$$m_1v_1 + m_2v_2 = (m_1 + m_2)v_3$$

$$v_3 = \frac{m_1v_1 + m_2v_2}{m_1 + m_2} = \frac{(40 \text{ kg} \times 4 \text{ m/s}) + (60 \text{ kg} \times 2 \text{ m/s})}{40 \text{ kg} + 60 \text{ kg}} = 2.8 \text{ m/s}$$

(*b*)

$$\text{Initial KE} = \tfrac{1}{2}m_1v_1^2 + \tfrac{1}{2}m_2v_2^2$$

$$= \tfrac{1}{2} \times 40 \text{ kg} \times (4 \text{ m/s})^2 + \tfrac{1}{2} \times 60 \text{ kg} \times (2 \text{ m/s})^2 = 440 \text{ J}$$

$$\text{Final KE} = \tfrac{1}{2}(m_1 + m_2)v_3^2 = \tfrac{1}{2} \times 100 \text{ kg} \times (2.8 \text{ m/s})^2 = 392 \text{ J}$$

Therefore 48 J of energy is lost, 11% of the original amount.

6.9. The two skaters of Problem 6.8 are moving in opposite directions and collide head-on. (*a*) If they remain in contact, what is their final velocity? (*b*) How much kinetic energy is lost?

(*a*) We take into account the opposite directions of motion by letting $v_1 = +4$ m/s and $v_2 = -2$ m/s. Then

$$v_3 = \frac{m_1v_1 + m_2v_2}{m_1 + m_2} = \frac{(40 \text{ kg} \times 4 \text{ m/s}) - (60 \text{ kg} \times 2 \text{ m/s})}{40 \text{ kg} + 60 \text{ kg}} = +0.4 \text{ m/s}$$

Since v_3 is +0.4 m/s, the two skaters move off in the same direction the 40-kg skater had originally, which is to be expected since he had the greater initial momentum.

(*b*)

$$\text{Initial KE} = 440 \text{ J} \quad (\text{as in Problem 6.8})$$

$$\text{Final KE} = \tfrac{1}{2}(m_1 + m_2)v_3^2 = \tfrac{1}{2} \times 100 \text{ kg} \times (0.4 \text{ m/s})^2 = 8 \text{ J}$$

Therefore 432 J of energy is lost, 98% of the original amount. This is the reason why head-on collisions of automobiles produce much more damage than overtaking collisions.

6.10. Why do all helicopters have two propellers?

If a single propeller were used, conservation of angular momentum would require the helicopter itself to rotate in the opposite direction.

6.11. If the icecaps at the North and South Poles were to melt and the resulting water to be added to the oceans, why would you expect the length of the day to change? Would the length of the day increase or decrease?

If the water from the icecaps leaves the vicinity of the earth's axis and becomes distributed farther away from it, the earth's rotational velocity must decrease in order that angular momentum be conserved. Hence the length of the day would increase.

Supplementary Problems

6.12. Find the momentum of a 100-kg ostrich running at 15 m/s.

6.13. Find the momentum of a 3200-lb car moving at 60 mi/hr (88 ft/s).

6.14. An object at rest breaks up into two parts which fly off. Must they move in opposite directions?

6.15. A moving object strikes a stationary one. After the collision, must they move in the same direction? Must they move in opposite directions?

6.16. An 8-lb rifle and a 10-lb rifle fire identical bullets with the same muzzle velocities. Compare the recoil momenta and recoil velocities of the two rifles.

6.17. An empty dump truck is coasting with its engine off along a level road when rain starts to fall. Neglecting friction, what if anything happens to the velocity of the truck?

6.18. A 160-lb man dives horizontally from a 640-lb boat with a velocity of 6 ft/s. What is the recoil velocity of the boat?

6.19. Four 50-kg girls simultaneously dive horizontally at 2.5 m/s from the same side of a boat, whose recoil velocity is 0.1 m/s. What is the mass of the boat?

6.20. An unoccupied 2400-lb car has coasted down a hill and is moving along a level road at 40 ft/s. In order to stop the car, a 12,000-lb truck moving in the opposite direction collides head-on with it. What should the truck's velocity be in order that both vehicles come to a stop after the collision?

6.21. A 50-kg boy at rest on roller skates catches a 0.6-kg ball moving toward him at 30 m/s. How fast does he move backward as a result?

6.22. A 1200-kg car traveling at 10 m/s overtakes a 1000-kg car traveling at 8 m/s and collides with it. (*a*) If the two cars stick together, what is their final velocity? (*b*) How much kinetic energy is lost? What percentage of the original KE is this?

6.23. The cars of Problem 6.22 are moving in opposite directions and collide head-on. (*a*) If they stick together, what is their final velocity? (*b*) How much kinetic energy is lost? What percentage of the original KE is this?

Answers to Supplementary Problems

6.12. 1500 kg-m/s   **6.13.** 8800 slug-ft/s   **6.14.** Yes

6.15. No; no; they can move in any directions provided that the vector sum of their momenta equals the initial momentum of the first object.

6.16. The recoil momenta are the same, but the lighter rifle has a higher recoil velocity.

6.17. The truck's velocity decreases as rainwater accumulates in it, since the total momentum must remain constant despite the increase in mass.

6.18. 1.5 ft/s   **6.21.** 0.36 m/s

6.19. 5000 kg   **6.22.** 9.09 m/s; 1100 J; 1.2%

6.20. 8 ft/s   **6.23.** 1.82 m/s; 88,356 J; 96%

Chapter 7

Relativity

SPECIAL THEORY OF RELATIVITY

The special theory of relativity, which was developed by Albert Einstein in 1905, is concerned with frames of reference that move at constant velocities with respect to one another. The special theory is built upon two postulates which have been verified in countless experiments:

1. The laws of nature are the same in all frames of reference in relative motion at constant velocity. This postulate expresses the absence of a universal frame of reference. If the laws of nature were different for different observers in relative motion, they could establish from these differences which objects are "really stationary" and which are "really moving". But the laws of nature are the same, hence motion is a relative quantity, not an absolute one.

2. The velocity of light in free space has the same value to all observers, regardless of their state of motion. If an observer is in relative motion toward or away from a source of light, the velocity of light he measures is always the same as the velocity when he is at rest relative to the source.

LENGTH CONTRACTION

Measurements of many physical quantities are affected by relative motion between an observer and what he observes. Thus the length L of an object in motion is always smaller than its length L_0 when it is at rest, a phenomenon often called the *Lorentz contraction*. This contraction occurs only in the direction of the relative motion; a spacecraft in flight is shorter to an observer on the ground than it is when at rest, but not narrower. Since motion is relative, the effect works both ways: to an observer in a moving spacecraft, objects on the ground appear shorter than they did when he was at rest on the ground.

The relationship between L and L_0 is

$$L = L_0 \sqrt{1 - \frac{v^2}{c^2}}$$

where L_0 = length measured when object is at rest relative to observer ("proper length")

L = length measured when object is in motion relative to observer

v = velocity of relative motion

c = velocity of light

Because the velocity of light is so great (3×10^8 m/s, which is 186,000 mi/s) relativistic length contractions are not evident in everyday life. However, they are very significant in events that involve elementary particles such as the electrons, protons, and neutrons of which atoms are composed.

TIME DILATION

A clock moving with respect to an observer appears to tick less rapidly than it does when at rest with respect to him. If someone in a spacecraft finds that the time interval between two events in the spacecraft is t_0, a person on the ground would find that the same interval has the longer duration t. This effect is called *time dilation* (to dilate = to become larger), and like length contraction it is reciprocal: a person on a spacecraft in flight finds time intervals on the ground to be longer than someone on the ground finds them to be.

The relationship between t and t_0 is

$$t = \frac{t_0}{\sqrt{1 - v^2/c^2}}$$

where t_0 = time interval on clock at rest relative to observer ("proper time")

t = time interval on clock in motion relative to observer

v = velocity of relative motion

c = velocity of light

RELATIVITY OF MASS

The mass of an object moving with respect to an observer appears greater than it does when at rest with respect to him. The relationship between the mass m_0 measured at rest and the mass m measured in relative motion is

$$m = \frac{m_0}{\sqrt{1 - v^2/c^2}}$$

where m_0 = mass measured when object is at rest relative to observer ("rest mass")

m = mass measured when object is in motion relative to observer

v = velocity of relative motion

c = velocity of light

MASS AND ENERGY

Another of the conclusions of special relativity is the mass-energy relationship $E_0 = m_0 c^2$ mentioned in Chapter 5.

GENERAL RELATIVITY

The general theory of relativity concerns frames of reference that are accelerated. One of the starting points of the theory is the *principle of equivalence*: there is no way for an observer in a windowless laboratory to distinguish between the effects produced by a gravitational field and those produced by an acceleration of the laboratory. General relativity provides an explanation for gravitation in terms of a distortion of space.

Three important conclusions of general relativity, all confirmed by experiment, are that light is deflected by a gravitational field; that a clock ticks more slowly in a strong gravitational field than in a weak one; and that the planet Mercury's orbit around the sun exhibits a certain feature that cannot be accounted for from Newton's law of gravitation alone.

Solved Problems

7.1. Can an observer in a windowless laboratory determine by experiments performed entirely in the laboratory whether or not he is moving through space in a curved path?

Yes. Motion along a curved path is accelerated, and accelerations can be detected by an isolated observer. To verify this, close your eyes while riding in a car: it is easy to tell when the car goes around a curve or changes the magnitude of its velocity.

7.2. What happens to the mass of an object as its velocity approaches closer and closer to the velocity of light? Can its velocity ever equal the velocity of light?

As $v \to c$, $\sqrt{1 - v^2/c^2} \to 0$, and the object's mass approaches infinity. Since nothing can have an infinite mass, no material object can have the velocity of light.

7.3. An astronaut is lying down in a spacecraft parallel to the direction in which it is moving. To an observer on earth, the spacecraft's velocity is 2×10^8 m/s and the astronaut is 4 ft tall. What is the astronaut's height as measured when he is at rest?

Here $L = 4$ ft, $v/c = 2/3$, and $v^2/c^2 = 0.444$. Hence, from $L = L_0\sqrt{1 - v^2/c^2}$,

$$L_0 = \frac{L}{\sqrt{1 - v^2/c^2}} = \frac{4 \text{ ft}}{\sqrt{1 - 0.444}} = \frac{4 \text{ ft}}{0.746} = 5.36 \text{ ft} = 5 \text{ ft } 4 \text{ in.}$$

7.4. The muon is an elementary particle that decays into an electron and two neutrinos (the neutrino is another kind of elementary particle) in an average of 2.2×10^{-6} s when it is at rest. What average muon lifetime does an observer find when a beam of them moves past him at a velocity of 2.9×10^8 m/s?

Here $t_0 = $ lifetime at rest $= 2.2 \times 10^{-6}$ s and $v = 2.9 \times 10^8$ m/s, $v/c = 2.9/3 = 0.967$, $v^2/c^2 = 0.934$. Hence the lifetime that is found when the muons move relative to an observer is

$$t = \frac{t_0}{\sqrt{1 - v^2/c^2}} = \frac{2.2 \times 10^{-6} \text{ s}}{\sqrt{1 - 0.934}} = \frac{2.2 \times 10^{-6} \text{ s}}{0.256} = 8.6 \times 10^{-6} \text{ s}$$

7.5. What must be the velocity of a spacecraft if 1 hour on its clock is to correspond to 1 hour + 1 second on a clock on the earth?

Here $t_0 = 3600$ s and $t = 3601$ s. We proceed as follows:

$$t = \frac{t_0}{\sqrt{1 - v^2/c^2}}$$

$$\sqrt{1 - \frac{v^2}{c^2}} = \frac{t_0}{t}$$

$$1 - \frac{v^2}{c^2} = \frac{t_0^2}{t^2}$$

$$\frac{v^2}{c^2} = 1 - \frac{t_0^2}{t^2}$$

$$v = c\sqrt{1 - \frac{t_0^2}{t^2}} = 3 \times 10^8 \text{ m/s} \times \sqrt{1 - \left(\frac{3600 \text{ s}}{3601 \text{ s}}\right)^2}$$

$$= 3 \times 10^8 \text{ m/s} \times \sqrt{0.000555} = 3 \times 10^8 \text{ m/s} \times 0.0236 = 7.1 \times 10^6 \text{ m/s}$$

7.6. An electron is traveling at 99.9% of the velocity of light. What is its mass?

Its rest mass is $m_0 = 9.1 \times 10^{-31}$ kg, $v = 0.999\,c$, and $v^2/c^2 = (0.999)^2 = 0.998$. Hence

$$m = \frac{m_0}{\sqrt{1 - v^2/c^2}} = \frac{9.1 \times 10^{-31}\ \text{kg}}{\sqrt{1 - 0.998}} = \frac{9.1 \times 10^{-31}\ \text{kg}}{0.0447} = 2.03 \times 10^{-29}\ \text{kg}$$

7.7. A spacecraft in flight is observed to have a mass 1% greater than its rest mass. What is its velocity in mi/s?

Here $m_0/m = 100/101 = 0.990$, $(m_0/m)^2 = 0.980$, and $c = 186{,}000$ mi/s. Starting from

$$m = \frac{m_0}{\sqrt{1 - v^2/c^2}}$$

we proceed as in Problem 7.5 to obtain

$$v = c\sqrt{1 - m_0^2/m^2} = c\sqrt{1 - 0.980} = 186{,}000\ \text{mi/s} \times 0.14 = 26{,}040\ \text{mi/s}$$

Supplementary Problems

7.8. The laws of physics are the same in all frames of reference at rest with respect to one another. Are they also the same in frames of reference moving relative to one another at constant velocities? At constant accelerations?

7.9. If the velocity of light were 300 m/s instead of 3×10^8 m/s, would relativistic phenomena be more or less conspicuous in everyday life than they are now?

7.10. Two observers, Fred on the earth and Alice in a spacecraft, both set their watches to the same time when the spacecraft passes the earth. To Fred, Alice's watch seems to run slow. What does Alice find regarding Fred's watch?

7.11. An astronaut whose height on the earth is 6 ft is lying parallel to the axis of a spacecraft moving at $0.9\,c$ relative to the earth. (*a*) What is his height as measured by an observer in the same spacecraft? (*b*) What is his height as measured by an observer in another spacecraft moving parallel to his at the same velocity? (*c*) What is his height as measured by an observer on the earth?

7.12. A spacecraft is 100 m long on the ground. In flight, its length is 99 m as measured by an observer on the ground. How fast is it moving?

7.13. A certain elementary particle has a lifetime of 10^{-7} s when measured at rest. How far does it go before decaying if its velocity is $0.99\,c$ when it is created?

7.14. At what velocity of a spacecraft does one year on board correspond to two years on the earth?

7.15. What is the velocity of an electron whose mass is ten times its rest mass?

7.16. An astronaut in flight measures his mass to be 80 kg and an observer on the earth finds it to be 82 kg. How fast is the spacecraft moving?

Answers to Supplementary Problems

7.8. Yes; no **7.9.** Much more conspicuous.

7.10. She finds that Fred's watch runs slow by exactly the same amount that Fred finds her watch to run slow.

7.11. 6 ft; 6 ft; 2.6 ft **7.13.** 213 m **7.15.** 2.98×10^8 m/s

7.12. 4.2×10^7 m/s **7.14.** 2.6×10^8 m/s **7.16.** 6.59×10^7 m/s

Chapter 8

Fluids

DENSITY

The *density* (*d*) of a substance is its mass per unit volume. The SI unit of density is the kilogram per cubic meter (kg/m³); the density of aluminum, for instance, is 2700 kg/m³. Another common unit of density is the gram per cubic centimeter (g/cm³). Since 1 kg = 1000 g and 1 m³ = (100 cm)³ = 10⁶ cm³,

$$1 \text{ g/cm}^3 = 10^3 \text{ kg/m}^3$$

Hence the density of aluminum can also be given as 2.7 g/cm³.

In British units density is properly expressed in slugs/ft³. The density of aluminum in these units is 5.3 slugs/ft³. Because weight rather than mass is normally specified in this system, the quantity *weight density* is customarily used. The weight density of a substance is its weight per unit volume. Thus the weight density of aluminum is 170 lb/ft³. There is no special symbol for weight density, and either *w/V* or *dg* can be used for it.

SPECIFIC GRAVITY

The *specific gravity* of a substance is its density relative to that of pure water, which is

$$d(\text{water}) = 1000 \text{ kg/m}^3 = 1.00 \text{ g/cm}^3 = 1.94 \text{ slugs/ft}^3$$

The weight density of water is

$$dg(\text{water}) = 62 \text{ lb/ft}^3$$

Since the density of water is 1 g/cm³, the specific gravity of a substance is the same as the numerical value of its density when given in g/cm³. Thus the specific gravity of aluminum is 2.7.

PRESSURE

When a force acts perpendicular to a surface, the *pressure* exerted is the ratio between the magnitude of the force and the area of the surface:

$$p = \frac{F}{A}$$

$$\text{Pressure} = \frac{\text{force}}{\text{area}}$$

Pressures are properly expressed in N/m² or in lb/ft², but other units are often used:

$$1 \text{ lb/in}^2 = 144 \text{ lb/ft}^2$$

1 *atmosphere* (atm) = average pressure exerted by the earth's atmosphere at sea level

$$= 1.013 \times 10^5 \text{ N/m}^2 = 14.7 \text{ lb/in}^2$$

1 *millibar* (mb) = 100 N/m² (widely used in meteorology)

PRESSURE IN A FLUID

Pressure is a useful quantity where fluids (gases and liquids) are concerned because of the following properties of fluids.

1. The forces a fluid exerts on the walls of its container, and those the walls exert on the fluid, always act perpendicular to the walls.

2. The force exerted by the pressure in a fluid is the same in all directions at a given depth.

3. An external pressure exerted on a fluid is transmitted uniformly throughout the fluid. This does not mean that pressures in a fluid are the same everywhere, because the weight of the fluid itself exerts pressures that increase with increasing depth. The pressure at a depth h in a fluid of density d due to the weight of fluid above is

$$p = dgh$$

Hence the total pressure at that depth is

$$p = p_{\text{external}} + dgh$$

When a body of fluid is in an open container, the atmosphere exerts an external pressure on it.

GAUGE PRESSURE

Pressure gauges measure the difference between an unknown pressure and atmospheric pressure. What they measure is known as *gauge pressure*, and the true pressure is known as *absolute pressure*:

$$p = p_{\text{gauge}} + p_{\text{atm}}$$

Absolute pressure = gauge pressure + atmospheric pressure

A tire whose gauge pressure is 30 lb/in² contains air at an absolute pressure of 45 lb/in², since sea-level atmospheric pressure is about 15 lb/in².

ARCHIMEDES' PRINCIPLE

An object immersed in a fluid is acted upon by an upward force that arises because pressures in a fluid increase with depth. Hence the upward force on the bottom of the object is more than the downward force on its top. The difference between the two, called the *buoyant force,* is equal to the weight of a body of the fluid whose volume is the same as that of the object. This is *Archimedes' principle*: The buoyant force on a submerged object is equal to the weight of fluid the object displaces.

If the buoyant force is less than the weight of the object itself, the object sinks; if the buoyant force equals the weight of the object, the object floats in equilibrium at any depth in the fluid; if the buoyant force is more than the weight of the object, the object floats with part of its volume above the surface.

BERNOULLI'S PRINCIPLE

According to *Bernoulli's principle,* the greater the velocity of a fluid, the smaller its pressure. An example is the lift produced by an airplane's wing. Owing to the shape of the wing, air moving past its upper surface must travel faster than air moving past its lower surface. Hence the pressure on the lower surface is greater than that on the upper one, and the net result is an upward force on the wing.

Solved Problems

8.1. An ice cube floats in a glass of water filled to the brim. What happens when the ice melts?

   The water level remains unchanged. The ice cube displaces a mass of water equal to its own mass, and when it melts the volume of water produced equals the volume of water it displaced when frozen.

8.2. A wooden block is on the bottom of a tank when water is poured in. The contact between the block and the tank is so good that no water gets between them. Is there a buoyant force on the block?

   There is no buoyant force since there is no water under the block to exert an upward force on it.

8.3. The specific gravity of gold is 19. (a) What is the mass of a cubic centimeter of gold? (b) What is the weight of a cubic inch of gold?

   (a) Since the density of water is 1 g/cm^3, the density of gold is 19 g/cm^3 and 1 cm^3 has a mass of 19 g.

   (b) Since the weight density of water is 62 lb/ft^3, the weight density of gold is $dg = 19 \times 62$ lb/ft^3 = 1200 lb/ft^3. Because 1 ft^3 = 12 in. $\times$ 12 in. $\times$ 12 in. = 1728 in^3, a cubic inch of gold weighs

$$w = (dg)V = 1200 \frac{\text{lb}}{\text{ft}^3} \times \frac{1 \text{ ft}^3}{1728 \text{ in}^3} = 0.7 \text{ lb}$$

8.4. An oak beam 10 cm by 20 cm by 4 m has a mass of 58 kg. (a) Find the density and specific gravity of oak. (b) Does oak float in water?

   (a) The volume of the beam is $V = 0.1$ m $\times$ 0.2 m $\times$ 4 m = 0.08 m^3 and so its density is

$$d = \frac{m}{V} = \frac{58 \text{ kg}}{0.08 \text{ m}^3} = 725 \text{ kg/m}^3$$

   Since the density of water is 1000 kg/m^3, the specific gravity of oak is

$$\text{sp gr} = \frac{d_{\text{oak}}}{d_{\text{water}}} = \frac{725}{1000} = 0.725$$

   (b) Any material whose specific gravity is less than 1 floats in water, so oak does.

8.5. How much does the air in a room 12 ft square and 10 ft high weigh? The weight density of air is 0.08 lb/ft^3 at sea level.

   The volume of the room is $V = 12$ ft $\times$ 12 ft $\times$ 10 ft = 1440 ft^3. Hence the weight of the air is
$$w = (dg)V = 0.08 \text{ lb/ft}^3 \times 1440 \text{ ft}^3 = 115 \text{ lb}$$

8.6. The density of mammals is roughly the same as that of water. Find the volume of a 500-lb lion.

$$V = \frac{w}{dg} = \frac{500 \text{ lb}}{62 \text{ lb/ft}^3} = 8.06 \text{ ft}^3$$

8.7. A 130-lb woman balances on the heel of her right shoe, which is 1 in. in radius. How much pressure does she exert on the ground? How does this compare with atmospheric pressure?

The area of the heel is $A = \pi r^2 = 3.14$ in^2, so the pressure is

$$p = \frac{F}{A} = \frac{130 \text{ lb}}{3.14 \text{ in}^2} = 41.4 \text{ lb/in}^2$$

Since $p_{\text{atm}} = 14.7$ lb/in^2, this pressure is 2.8 times greater.

8.8. An airplane whose mass is 20,000 kg and whose wing area is 60 m^2 is in level flight. What is the average difference in pressure between the upper and lower surfaces of its wings? Express the answer in N/m^2 and in atm.

The upward force on an airplane in level flight is equal to its weight, which here is

$$F = w = mg = 20{,}000 \text{ kg} \times 9.8 \text{ m/s}^2 = 1.96 \times 10^5 \text{ N}$$

The pressure difference p is therefore

$$p = \frac{F}{A} = \frac{1.96 \times 10^5 \text{ N}}{60 \text{ m}^2} = 3267 \text{ N/m}^2$$

Since 1 atm $= 1.013 \times 10^5$ N/m^2,

$$p = \frac{3.267 \times 10^3 \text{ N/m}^2}{(1.013 \times 10^5 \text{ N/m}^2)/\text{atm}} = 0.0322 \text{ atm}$$

8.9. What is the pressure at the bottom of a swimming pool 6 ft deep that is filled with fresh water? Express the answer in lb/in^2.

$$
\begin{aligned}
p &= p_{\text{atm}} + (dg)h \\
&= 14.7 \frac{\text{lb}}{\text{in}^2} + 62 \frac{\text{lb}}{\text{ft}^3} \times 6 \text{ ft} \times \frac{1 \text{ ft}^2}{144 \text{ in}^2} \\
&= 14.7 + 2.6 \text{ lb/in}^2 = 17.3 \text{ lb/in}^2
\end{aligned}
$$

Note the use of the conversion factor 1 ft^2/144 in^2.

8.10. The interior of a submarine located at a depth of 50 m in seawater is maintained at sea-level atmospheric pressure. Find the force acting on a window 20 cm square. The density of seawater is 1.03×10^3 kg/m^3.

The pressure outside the submarine is $p = p_{\text{atm}} + dgh$ and the pressure inside it is p_{atm}. Hence the net pressure p' acting on the window is

$$p' = dgh = 1.03 \times 10^3 \frac{\text{kg}}{\text{m}^3} \times 9.8 \frac{\text{m}}{\text{s}^2} \times 50 \text{ m} = 5.05 \times 10^5 \frac{\text{N}}{\text{m}^2}$$

Since the area of the window is $A = 0.2 \text{ m} \times 0.2 \text{ m} = 0.04$ m^2, the force acting on it is

$$F = p'A = 5.05 \times 10^5 \text{ N/m}^2 \times 4 \times 10^{-2} \text{ m}^2 = 2.02 \times 10^4 \text{ N}$$

8.11. An iron anchor weighs 200 lb in air. How much force is required to support the anchor when it is immersed in seawater? The weight density of iron is 480 lb/ft^3 and that of seawater is 64 lb/ft^3.

Since $dg = w/V$, the volume of the anchor is

$$V = \frac{w}{dg} = \frac{200 \text{ lb}}{480 \text{ lb/ft}^3} = 0.417 \text{ ft}^3$$

The weight of seawater displaced by the anchor is

$$w = (dg)V = 64 \text{ lb/ft}^3 \times 0.417 \text{ ft}^3 = 27 \text{ lb}$$

Thus the buoyant force on the anchor is 27 lb and the net force needed to support it is

$$200 \text{ lb} - 27 \text{ lb} = 173 \text{ lb}$$

8.12. A 70-kg man jumps off a raft 2 m square moored in a freshwater lake. By how much does the raft rise?

The volume of water that must be displaced by the raft to support the man is

$$V = \frac{m}{d} = \frac{70 \text{ kg}}{10^3 \text{ kg/m}^3} = 0.07 \text{ m}^3$$

The area of the raft is $A = 2\text{ m} \times 2\text{ m} = 4 \text{ m}^2$. Since Volume = height × area, the raft rises by

$$h = \frac{V}{A} = \frac{0.07 \text{ m}^3}{4 \text{ m}^2} = 0.018 \text{ m} = 1.8 \text{ cm}$$

8.13. The density of ice is 920 kg/m³ and that of seawater is 1030 kg/m³. What percentage of the volume of an iceberg is submerged?

When an iceberg of volume V floats, its weight of $d_{ice}gV$ is balanced by the buoyant force on it which is equal to the weight of water displaced. If V_{sub} is the volume of the iceberg that is submerged, the weight of water displaced is $d_{water}gV_{sub}$. Hence

$$\text{Weight of iceberg} = \text{weight of displaced water}$$
$$d_{ice}gV = d_{water}gV_{sub}$$

$$\frac{V_{sub}}{V} = \frac{d_{ice}}{d_{water}} = \frac{920 \text{ kg/m}^3}{1030 \text{ kg/m}^3} = 0.89 = 89\%$$

Eighty-nine percent of the volume of an iceberg is under the water's surface.

8.14. A 100-gallon steel tank weighs 50 lb when empty. Will it float in seawater when filled with gasoline? The weight density of gasoline is 42 lb/ft³, that of seawater is 64 lb/ft³, and 1 gallon = 0.134 ft³.

The volume of the tank is $V = 100 \text{ gal} \times 0.134 \text{ ft}^3/\text{gal} = 13.4 \text{ ft}^3$. The total weight of the tank when filled with gasoline is

$$w = 50 \text{ lb} + (dg)_{gasoline}V = 50 \text{ lb} \times 42 \text{ lb/ft}^3 \times 13.4 \text{ ft}^3$$
$$= 50 \text{ lb} + 563 \text{ lb} = 613 \text{ lb}$$

The maximum buoyant force on the tank is exerted when the tank is completely submerged. Thus

$$F_{max} = (dg)_{water}V = 64 \text{ lb/ft}^3 \times 13.4 \text{ ft}^3 = 858 \text{ lb}$$

Since the weight of the filled tank is less than 858 lb, it will float.

8.15. A certain helium-filled airship provides a lift (structure + payload) of 20 tons. What volume of helium does it contain? The weight density of helium at sea-level pressure is 0.011 lb/ft³ and that of air is 0.08 lb/ft³.

The total weight w that must be supported is the 20 tons = 40,000 lb of the airship itself plus the weight $(dg)_{He}V$ of the helium it contains, where V is the airship's volume:

$$w = 40,000 \text{ lb} + (dg)_{He}V$$

The buoyant force on the airship is the weight of air it displaces:

$$F = (dg)_{air}V$$

Hence

$$\text{Buoyant force} = \text{weight of airship} + \text{helium}$$
$$F = w$$
$$(dg)_{air}V = 40,000 \text{ lb} + (dg)_{He}V$$

$$V = \frac{40,000 \text{ lb}}{(dg)_{air} - (dg)_{He}} = \frac{40,000 \text{ lb}}{(0.08 - 0.011) \text{ lb/ft}^3} = 5.8 \times 10^5 \text{ ft}^3$$

Supplementary Problems

8.16. A sailboat has a lead keel to help keep it upright despite the pressure wind exerts on its sails. What difference, if any, is there between the stability of a sailboat in fresh water and in seawater?

8.17. Dam A and dam B are identical in size and shape, and the water levels at both are the same height above their bases. Dam A holds back a lake that contains 1 cubic mile of water, and dam B holds back a lake that contains 2 cubic miles of water. What is the ratio between the total force exerted on dam A and that exerted on dam B?

8.18. A 50-g gold bracelet is dropped into a full glass of water and 2.6 cm^3 of water overflows. What is the density of gold? What is its specific gravity?

8.19. The weight density of ice is 58 lb/ft^3. What is its specific gravity?

8.20. How much does the water in a swimming pool 20 ft long, 10 ft wide, and 6 ft deep weigh?

8.21. The density of iron is 7.8×10^3 kg/m^3. (*a*) What is the specific gravity of iron? (*b*) How many cubic meters does a metric ton (1000 kg) of iron occupy?

8.22. An airplane whose weight is 50,000 lb and whose wing area is 600 ft^2 is in level flight. What is the average difference in pressure between the upper and lower surfaces of its wings?

8.23. A 70-kg man wears shoes whose area is 200 cm^2 each. How much pressure does he exert on the ground?

8.24. A phonograph needle whose point is 0.1 mm (10^{-4} m) in radius exerts a downward force of 0.02 N. What is the pressure on the record groove? How many atm is this?

8.25. What is the pressure at a depth of 100 m in the ocean? How many atm is this? The density of seawater is 1.03×10^3 kg/m^3.

8.26. What pressure is experienced by a skin diver 20 ft below the surface of a freshwater lake?

8.27. (*a*) How much force is required to raise a 1000-kg block of concrete to the surface of a freshwater lake? (*b*) How much force is needed to lift it out of the water? The density of concrete is 2.3×10^3 kg/m^3.

8.28. An aluminum bar weighs 17 lb in air. How much force is required to support the bar when it is immersed in gasoline? The weight density of aluminum is 170 lb/ft^3 and that of gasoline is 42 lb/ft^3.

8.29. A raft 8 ft wide, 12 ft long, and 2 ft high is made from solid balsa wood ($dg = 8$ lb/ft^3). How much weight can it support in seawater ($dg = 64$ lb/ft^3)?

8.30. People have roughly the same density as fresh water. Find the buoyant force exerted by the atmosphere on a 50-kg woman at sea level where the density of air is 1.3 kg/m^3.

8.31. A balloon weighing 100 kg has a capacity of 1000 m^3. If it is filled with hydrogen, how great a payload in kg can it support? At sea level the density of hydrogen is 0.09 kg/m^3 and that of air is 1.3 kg/m^3.

Answers to Supplementary Problems

8.16. The boat is more stable in fresh water because the buoyancy of the lead keel is less there.

8.17. The forces are the same.

8.18. 19 g/cm^3; 19

8.19. 0.93

8.20. 74,400 lb

8.21. 7.8; 0.128 m^3

8.22. 83.3 lb/ft^2

8.23. 1.72×10^4 N/m^2

8.24. 6.37×10^5 N/m^2; 6.3 atm

8.25. 1.11×10^6 N/m^2; 11.0 atm

8.26. 23.3 lb/in^2

8.27. 5539 N; 9800 N

8.28. 12.8 lb

8.29. 10,752 lb

8.30. 0.64 N

8.31. 1110 kg

Chapter 9

Heat

INTERNAL ENERGY

Every body of matter, whether solid, liquid, or gas, consists of atoms or molecules which are in rapid motion. The kinetic energies of these particles constitute the *internal energy* of the body of matter. The *temperature* of the body is a measure of the average kinetic energy of its particles. *Heat* may be thought of as internal energy in transit. When heat is added to a body, its internal energy increases and its temperature rises; when heat is removed from a body, its internal energy decreases and its temperature falls.

TEMPERATURE

Temperature is familiar as the property of a body of matter responsible for sensations of hot or cold when it is touched. Temperature provides an indicator of the direction of internal energy flow: when two objects are in contact, internal energy goes from the one at the higher temperature to the one at the lower temperature, regardless of the total amounts of internal energy in each one. Thus if hot coffee is poured into a cold cup, the coffee becomes cooler and the cup becomes warmer.

A *thermometer* is a device for measuring temperature. Matter usually expands when heated and contracts when cooled, the relative amount of change being different for different substances. This behavior is the basis of most thermometers, which make use of the different rates of expansion of mercury and glass, or of two metal strips joined together, to indicate temperature.

TEMPERATURE SCALES

The *Celsius* (or *centigrade*) temperature scale assigns 0° to the freezing point of water and 100° to its boiling point. On the *Fahrenheit* scale these points are respectively 32° and 212°. A Fahrenheit degree is therefore 5/9 as large as a Celsius degree. The following formulas give the procedure for converting a temperature expressed in one scale to the corresponding value in the other:

$$T_F = \frac{9}{5}T_C + 32°$$

$$T_C = \frac{5}{9}(T_F - 32°)$$

HEAT

Heat is a form of energy which, when added to a body of matter, increases its internal energy content and thereby causes its temperature to rise. The customary symbol for heat is Q.

Because heat is a form of energy, the proper SI unit of heat is the joule. However, the *kilocalorie* is still widely used with SI units: 1 kilocalorie (kcal) is the amount of heat needed to raise the temperature of 1 kg of water by 1 °C. The calorie itself is the amount of heat needed to raise the temperature of 1 g of water by 1 °C; hence 1 kcal = 1000 calories. (The calorie used by dieticians to measure the energy content of foods is the same as the kilocalorie.)

The British unit of heat is the *British thermal unit* (Btu): 1 Btu is the amount of heat needed to raise the temperature of 1 lb of water by 1 °F. To convert heat figures from one system to the other we note that

$$1 \text{ kcal } = 3.97 \text{ Btu}$$

$$1 \text{ Btu } = 0.252 \text{ kcal}$$

Although weight rather than mass is specified in the British system when dealing with heat, in practice this makes no difference in the various calculations. Whenever m appears in the equations of heat, it is understood to refer to mass in kg when metric units are used and to weight in lb when British units are used.

SPECIFIC HEAT CAPACITY

Different substances respond differently to the addition or removal of heat. For instance, 1 kg of water increases in temperature by 1 °C when 1 kcal of heat is added, but 1 kg of aluminum increases in temperature by 4.5 °C when this is done. The *specific heat capacity* of a substance is the amount of heat needed to change the temperature of a unit quantity of it by 1°. The symbol of specific heat capacity is c; its metric unit is the kcal/kg-°C and its British unit is the Btu/lb-°F. The numerical value of c for a substance is the same in both systems of units.

Among common materials, water has the highest specific heat capacity, namely 1.00 kcal/kg-°C (or Btu/lb-°F). Ice and steam have lower specific heat capacities than water, respectively 0.50 and 0.48 kcal/kg-°C (or Btu/lb-°F). Metals usually have low specific heat capacities; thus lead and iron have $c = 0.03$ and 0.11 kcal/kg-°C (or Btu/lb-°F) respectively.

When an amount of heat Q is transferred to or from a mass m of a substance whose specific heat capacity is c, the resulting temperature change ΔT is related to Q, m, and c by the formula

$$Q = mc \, \Delta T$$

Heat transferred = mass × specific heat capacity × temperature change

CHANGE OF STATE

When heat is continuously added to a solid, it grows hotter and hotter and finally begins to melt. While it is melting, the material remains at the same temperature and the absorbed heat goes into changing its state from solid to liquid. After all the solid is melted, the temperature of the resulting liquid then increases as more heat is supplied until it begins to boil. Now the material again stays at a constant temperature until all of it has become a gas, after which the gas temperature rises.

The amount of heat that must be added to a unit quantity (1 kg or 1 lb) of a substance at its melting point to change it from a solid to a liquid is called its *heat of fusion* (L_f). The same amount of heat must be removed from a unit quantity of the substance when it is a liquid at its melting point to change it to a solid.

The amount of heat that must be added to a unit quantity of a substance at its boiling point to change it from a liquid to a gas is called its *heat of vaporization* (L_v). The same amount of heat must be removed from a unit quantity of the substance when it is a gas at its boiling point to change it to a liquid.

The heat of fusion of water is $L_f = 80$ kcal/kg $= 144$ Btu/lb and its heat of vaporization is $L_v = 540$ kcal/kg $= 972$ Btu/lb.

PRESSURE AND BOILING POINT

The boiling point of a liquid depends upon the pressure applied to it: the higher the pressure, the higher the boiling point. Thus water under a pressure of 2 atm boils at 121 °C instead of at 100 °C as it does at sea-level atmospheric pressure. At high altitudes, where the atmospheric pressure is less than at sea level, water boils at a lower temperature than 100 °C. At an elevation of 2000 m, for instance, atmospheric pressure is about three-quarters of its sea-level value and water boils at 93 °C there.

Solved Problems

9.1. A person is dissatisfied with the rate at which eggs cook in a pan of boiling water. Would they cook faster if he (*a*) turns up the gas flame; (*b*) uses a pressure cooker?

(*a*) No. The maximum temperature that water can have while in the liquid state is its boiling point. Increasing the rate at which heat is supplied to a pan of water increases the rate at which steam is produced, but does not raise the temperature of the water beyond 100 °C (212 °F).

(*b*) Yes. In a pressure cooker, the pressure is greater than normal atmospheric pressure, which elevates the boiling point and so causes the eggs to cook faster.

9.2. What is the Celsius equivalent of 80 °F?

$$T_C = \frac{5}{9}(T_F - 32°) = \frac{5}{9}(80° - 32°) = 26.7 °C$$

9.3. What is the Fahrenheit equivalent of 80 °C?

$$T_F = \frac{9}{5}T_C + 32° = \frac{9}{5} \times 80° + 32° = 176 °F$$

9.4. Oxygen freezes at −362 °F. What is the Celsius equivalent of this temperature?

$$T_C = \frac{5}{9}(T_F - 32°) = \frac{5}{9}(-362° - 32°) = -219 °C$$

9.5. Nitrogen freezes at −210 °C. What is the Fahrenheit equivalent of this temperature?

$$T_F = \frac{9}{5}T_C + 32° = \frac{9}{5}(-210°) + 32° = -346 °F$$

9.6. How much heat must be added to 3 kg of water to raise its temperature from 20 °C to 80 °C?

The temperature change is $\Delta T = 80 °C - 20 °C = 60 °C$. Hence

$$Q = mc\,\Delta T = 3 \text{ kg} \times 1 \text{ kcal/kg-°C} \times 60 °C = 180 \text{ kcal}$$

9.7. Two hundred Btu of heat is removed from a 50-lb block of ice initially at 25 °F. What is its final temperature? ($c_{ice} = 0.5$ Btu/lb-°F)

$$Q = mc\,\Delta T$$

$$\Delta T = \frac{Q}{mc} = \frac{200 \text{ Btu}}{50 \text{ lb} \times 0.5 \text{ Btu/lb-}°\text{F}} = 8 \text{ °F}$$

The final temperature is therefore $25\text{ °F} - 8\text{ °F} = 17\text{ °F}$.

9.8. Ten kcal of heat is added to a 1-kg sample of wood and its temperature is found to rise from 20° C to 44 °C. What is the specific heat capacity of the wood?

$$Q = mc\,\Delta T$$

$$c = \frac{Q}{m\,\Delta T} = \frac{10 \text{ kcal}}{1 \text{ kg} \times 24\text{ °C}} = 0.42 \text{ kcal/kg-}°\text{C}$$

9.9. Three lb of water at 100 °F is added to 5 lb of water at 40 °F. What is the final temperature of the mixture?

If T is the final temperature, then the 5 lb of water initially at 40 °F undergoes a temperature change of $\Delta T_1 = T - 40\text{ °F}$ and the 3 lb of water initially at 100 °F undergoes a temperature change of $\Delta T_2 = 100\text{ °F} - T$. We proceed as follows:

$$\text{Heat gained} = \text{heat lost}$$

$$m_1 c_1\,\Delta T_1 = m_2 c_2\,\Delta T_2$$

$$5 \text{ lb} \times 1 \text{ Btu/lb-}°\text{F} \times (T - 40\text{ °F}) = 3 \text{ lb} \times 1 \text{ Btu/lb-}°\text{F} \times (100\text{ °F} - T)$$

$$(5T - 200) \text{ Btu} = (300 - 3T) \text{ Btu}$$

$$8T = 500$$

$$T = 62.5\text{ °F}$$

9.10. In preparing tea, 600 g of water at 90 °C is poured into a 200-g china pot ($c_{pot} = 0.2$ kcal/kg-°C) at 20 °C. What is the final temperature of the water?

$$\text{Heat gained by pot} = \text{heat lost by water}$$

$$m_{pot}c_{pot}\,\Delta T_{pot} = m_{water}c_{water}\,\Delta T_{water}$$

$$0.2 \text{ kg} \times 0.2 \text{ kcal/kg-}°\text{C} \times (T - 20\text{ °C}) = 0.6 \text{ kg} \times 1 \text{ kcal/kg-}°\text{C} \times (90\text{ °C} - T)$$

$$(0.04T - 0.8) \text{ kcal} = (54 - 0.6T) \text{ kcal}$$

$$0.64T = 54.8$$

$$T = 86\text{ °C}$$

9.11. In order to raise the temperature of 5 kg of water from 20 °C to 30 °C a 2-kg iron bar is heated and then dropped into the water. What should the temperature of the bar be? ($c_{iron} = 0.11$ kcal/kg-°C)

Let the temperature of the iron bar be T. Then the change in the water's temperature is $\Delta T_w = 30\text{ °C} - 20\text{ °C} = 10\text{ °C}$ and the change in the bar's temperature is $\Delta T_{iron} = T - 30\text{ °C}$. We proceed in the usual way:

$$\text{Heat gained by water} = \text{heat lost by bar}$$

$$m_w c_w\,\Delta T_w = m_{iron}c_{iron}\,\Delta T_{iron}$$

$$5 \text{ kg} \times 1 \text{ kcal/kg-}°\text{C} \times 10\text{ °C} = 2 \text{ kg} \times 0.11 \text{ kcal/kg-}°\text{C} \times (T - 30\text{ °C})$$

$$50 \text{ kcal} = (0.22\,T - 6.6) \text{ kcal}$$

$$0.22\,T = 56.6$$

$$T = 257\text{ °C}$$

9.12. How much heat must be added to 200 lb of lead at 70 °F to cause it to melt? The specific heat capacity of lead is 0.03 Btu/lb-°F, it melts at 626 °F, and its heat of fusion is 10.6 Btu/lb.

Here $\Delta T = 626\ °\text{F} - 70\ °\text{F} = 556\ °\text{F}$. Hence

$$
\begin{aligned}
Q &= mc\,\Delta T + mL_f \\
&= 200\ \text{lb} \times 0.03\ \text{Btu/lb-}°\text{F} \times 556\ °\text{F} + 200\ \text{lb} \times 10.6\ \text{Btu/lb} \\
&= 3336\ \text{Btu} + 2120\ \text{Btu} = 5456\ \text{Btu}
\end{aligned}
$$

9.13. Five kg of water at 40 °C is poured on a large block of ice at 0 °C. How much ice melts?

$$
\begin{aligned}
\text{Heat lost by water} &= \text{heat gained by ice} \\
m_w c\,\Delta T &= m_{\text{ice}} L_f \\
5\ \text{kg} \times 1\ \text{kcal/kg-}°\text{C} \times 40\ °\text{C} &= m_{\text{ice}} \times 80\ \text{kcal/kg} \\
m_{\text{ice}} &= \left(\frac{200}{80}\right)\text{kg} = 2.5\ \text{kg}
\end{aligned}
$$

9.14. Five hundred kcal of heat is added to 2 kg of water at 80 °C. How much steam is produced?

The heat needed to raise the temperature of the water from 80 °C to the boiling point of 100 °C is

$$
Q_1 = m_1 c\,\Delta T = 2\ \text{kg} \times 1\ \text{kcal/kg-}°\text{C} \times 20\ °\text{C} = 40\ \text{kcal}
$$

Hence $Q_2 = 500\ \text{kcal} - 40\ \text{kcal} = 460\ \text{kcal}$ of heat is available to convert water at 100 °C to steam at the same temperature. Since $Q_2 = m_{\text{steam}} L_v$, the steam produced is

$$
m_{\text{steam}} = \frac{Q_2}{L_v} = \frac{460\ \text{kcal}}{540\ \text{kcal/kg}} = 0.85\ \text{kg}
$$

9.15. A 30-g ice cube at 0 °C is dropped into 200 g of water at 30 °C. What is the final temperature?

If T is the final temperature, then $\Delta T_{\text{ice}} = T - 0\ °\text{C}$ and $\Delta T_{\text{water}} = 30\ °\text{C} - T$. Therefore

$$
\begin{aligned}
\text{Heat gained by ice} &= \text{heat lost by water} \\
m_{\text{ice}} L_f + m_{\text{ice}} c_{\text{water}}\,\Delta T_{\text{ice}} &= m_{\text{water}} c_{\text{water}}\,\Delta T_{\text{water}} \\
0.03\ \text{kg} \times 80\ \frac{\text{kcal}}{\text{kg}} + 0.03\ \text{kg} \times 1\ \frac{\text{kcal}}{\text{kg-}°\text{C}} \times (T - 0\ °\text{C}) &= 0.2\ \text{kg} \times 1\ \frac{\text{kcal}}{\text{kg-}°\text{C}} \times (30\ °\text{C} - T) \\
(2.4 + 0.03\,T)\ \text{kcal} &= (6 - 0.2\,T)\ \text{kcal} \\
0.23\,T &= 3.6 \\
T &= 15.7\ °\text{C}
\end{aligned}
$$

9.16. How much steam at 292 °F is required to melt 1 lb of ice at 32 °F?

Here $\Delta T_1 = 292\ °\text{F} - 212\ °\text{F} = 80\ °\text{F}$ and $\Delta T_2 = 212\ °\text{F} - 32\ °\text{F} = 180\ °\text{F}$. Therefore, if m_s is the mass of steam,

$$
\begin{aligned}
\text{Heat gained by ice} &= \text{heat lost by steam} \\
m_{\text{ice}} L_f &= m_s c_s\,\Delta T_1 + m_s L_v + m_s c_{\text{water}}\,\Delta T_2 \\
1\ \text{lb} \times 144\ \text{Btu/lb} &= m_s \times 0.48\ \text{Btu/lb-}°\text{F} \times 80\ °\text{F} + m_s \times 972\ \text{Btu/lb} \\
&\quad + m_s \times 1\ \text{Btu/lb-}°\text{F} \times 180\ °\text{F} \\
144\ \text{Btu} &= (38.4 + 972 + 180)m_s\ \text{Btu} = 1190 m_s\ \text{Btu} \\
m_s &= \left(\frac{144}{1190}\right)\text{lb} = 0.12\ \text{lb}
\end{aligned}
$$

Supplementary Problems

9.17. Why is an ice cube at 0 °C more effective in cooling a drink than the same mass of water at 0 °C?

9.18. Ethyl alcohol melts at −114 °C and boils at 78 °C. What are the Fahrenheit equivalents of these temperatures?

9.19. Bromine melts at 19 °F and boils at 140 °F. What are the Celsius equivalents of these temperatures?

9.20. How much heat must be removed from 4 lb of water at 200 °F to reduce its temperature to 50 °F?

9.21. How much heat must be added to a 20-kg block of ice to raise its temperature from −20 °C to −5° C? ($c_{ice} = 0.50$ kcal/kg-°C)

9.22. How much heat is lost by a 50-g silver spoon when it is cooled from 20 °C to 0 °C? ($c_{silver} = 0.056$ kcal/kg-°C)

9.23. Seven hundred kcal of heat is added to a 100-kg marble statue of Isaac Newton initially at 18 °C. What is its final temperature? ($c_{marble} = 0.21$ kcal/kg-°C)

9.24. Two Btu of heat is added to a $\frac{1}{2}$-lb glass vessel at 70 °F and its temperature is found to rise to 90 °F. What is the specific heat capacity of the glass?

9.25. Ten kg of water at 5 °C is added to 100 kg of water at 80 °C. What is the final temperature of the mixture?

9.26. A 600-g copper dish contains 1500 g of water at 20 °C. A 100-g iron bar at 120 °C is dropped into the water. What is the final temperature of the water? ($c_{copper} = 0.093$ kcal/kg-°C; $c_{iron} = 0.11$ kcal/kg-°C)

9.27. Four pounds of soup at 140 °F is poured into a 4-lb china serving dish at 70 °F. What is the final temperature of the soup? ($c_{soup} = 0.9$ Btu/lb-°F; $c_{dish} = 0.2$ Btu/lb-°F)

9.28. How much water at 35 °F must be added to 10 lb of punch at 70 °C that is in a 3-lb silver punch bowl to lower its temperature to 60 °C? ($c_{punch} = 0.7$ Btu/lb-°F; $c_{silver} = 0.056$ Btu/lb-°F)

9.29. How much heat must be removed from 200 g of water at 30 °C to convert it to ice at 0 °C?

9.30. Three lb of water at 100 °F is poured on a large block of ice at 32 °F. How much ice melts?

9.31. Five hundred kcal of heat is added to 10 kg of zinc at 70 °F. How much zinc melts? The specific heat capacity of zinc is 0.092 kcal/kg-°C, it melts at 420 °C, and its heat of fusion is 24 kcal/kg.

9.32. How much ice at 0 °C must be added to 200 g of water at 30 °C to lower its temperature to 20 °C?

9.33. Four thousand Btu of heat is added to 3 lb of water at 100 °F. How much steam is produced? What is the temperature of the steam?

Answers to Supplementary Problems

9.17. The ice absorbs heat from the drink in order to melt to water at 0 °C.

9.18. −173 °F; 172 °F

9.19. −7 °C; 60 °C

9.20. 600 Btu

9.21. 150 kcal

9.22. 0.056 kcal

9.23. 51 °C

9.24. 0.2 Btu/lb-°F

9.25. 73 °C

9.26. 20.7 °C

9.27. 127 °F

9.28. 2.9 lb

9.29. 22 kcal

9.30. 1.4 lb

9.31. 7.4 kg

9.32. 20 g

9.33. 3 lb; 731 °F

Chapter 10

Kinetic Theory of Matter

BOYLE'S LAW

At constant temperature, the volume of a sample of gas is inversely proportional to the absolute pressure applied to the gas. The greater the pressure, the smaller the volume. This relationship is known as *Boyle's law*. If p_1 is the gas pressure when its volume is V_1 and p_2 is its pressure when its volume is V_2, then Boyle's law states that

$$p_1 V_1 = p_2 V_2 \quad (T = \text{constant})$$

ABSOLUTE TEMPERATURE SCALE

When the temperature of a sample of gas is changed while the pressure on it is held constant, its volume changes by 1/273 of its volume at 0 °C for each temperature change of 1 °C. If it were possible to cool a gas sample to −273 °C, its volume would diminish to zero. Since all gases condense into liquids at temperatures above −273 °C, this experiment cannot be carried out; nevertheless, −273 °C is a significant temperature.

On the *absolute temperature scale,* the zero point is set at −273 °C. Temperatures in this scale are expressed in *kelvins* (K); these units are equal to Celsius degrees. Thus

$$T_K = T_C + 273$$

The freezing point of water on the absolute scale is 273 K and its boiling point is 373 K.

CHARLES'S LAW

Because of the way the absolute temperature scale is defined, the relationship between the temperature and volume of a gas sample at constant pressure can be expressed as

$$\frac{V_1}{T_1} = \frac{V_2}{T_2} \quad (p = \text{constant})$$

In this formula, which is called *Charles's law,* V_1 is the volume of the sample at the absolute temperature T_1 and V_2 is its volume at the absolute temperature T_2; the formula only holds when the temperatures are expressed in the absolute scale.

IDEAL GAS LAW

Boyle's law and Charles's law can be combined to form the *ideal gas law*:

$$\frac{p_1 V_1}{T_1} = \frac{p_2 V_2}{T_2}$$

This law is obeyed fairly well by all gases through a wide range of pressures and temperatures. An *ideal gas* is one for which $pV/T = $ constant under all circumstances; though no such gas actually exists, the fact that a real gas behaves approximately like an ideal one provides a specific target for theories of the gaseous state.

The ideal gas law is further discussed in Chapter 25.

KINETIC THEORY OF GASES

The *kinetic theory of gases* holds that a gas is composed of very small particles, called molecules, that are in constant random motion. The molecules are far apart relative to their dimensions and do not interact with one another except in collisions.

The pressure a gas exerts is due to the impacts of its molecules; there are so many molecules in even a small gas sample that the individual blows appear as a continuous force. Boyle's law is readily understood in terms of the kinetic theory of gases. Expanding a gas sample means that its molecules must travel farther between successive impacts on the container walls and that the impacts are spread over a larger area. Hence an increase in volume means a decrease in pressure, and vice versa.

MOLECULAR ENERGY

According to the kinetic theory of gases, the average kinetic energy of the molecules of a gas is proportional to the absolute temperature of the gas. This relationship is usually expressed in the form

$$KE_{av} = \frac{3}{2}kT$$

where k = Boltzmann's constant = 1.38×10^{-23} J/K. Actual molecular energies vary considerably on either side of KE_{av}.

At absolute zero, 0 K, gas molecules would be at rest, which is why this is such a significant temperature. At any temperature, all gases have the same average molecular energy. Therefore, in a gas whose molecules are heavy, the molecules move more slowly on the average than do those in a gas at the same temperature whose molecules are light.

Charles's law follows directly from the above interpretation of temperature. Compressing a gas causes its temperature to rise because molecules rebound from the inward-moving walls of the container with increased energy, just as a tennis ball rebounds with greater energy when struck by a racket. Similarly, expanding a gas causes its temperature to fall because molecules rebound from the outward-moving walls with decreased energy.

SOLIDS AND LIQUIDS

The molecules of a solid are close enough together to exert forces on one another that hold the entire assembly to a definite size and shape. As in the case of a gas, the molecules are in constant motion, but they vibrate about fixed locations instead of moving randomly. The molecules of a liquid continually move around past one another more or less freely, which enables the liquid to flow, but their spacing does not change and so the volume of a given liquid sample does not vary.

When a solid melts, the original ordered arrangement of its molecules changes to the random arrangement of molecules in a liquid. To accomplish the change, the molecules must be pulled apart against the forces holding them in place, which requires energy. The heat of fusion of a solid represents this energy. When a liquid boils, the heat of vaporization represents the energy needed to pull its molecules entirely free of one another so that a gas is formed.

ATOMS AND MOLECULES

Elements are the fundamental substances of which all matter in bulk is composed. There are 105 known elements, of which a number are not found in nature but have been prepared in the laboratory. Elements cannot be transformed into one another by ordinary chemical or physical means, but two or more elements can combine to form a *compound*, which is a substance whose properties are different from those of its constituent elements.

The ultimate particles of an element are called *atoms,* and those of a compound which exists in the gaseous state are called *molecules.* The molecules of a compound consist of the atoms of the elements that compose it joined together in a specific arrangement; each molecule of water, for instance, contains two hydrogen atoms and one oxygen atom, as its symbol H_2O indicates. Many compounds in the solid and liquid states do not consist of individual molecules, as discussed later. Elemental gases may consist of atoms (helium, He; argon, Ar) or of molecules (hydrogen, H_2; oxygen, O_2).

The masses of atoms and molecules are expressed in *atomic mass units* (u), where

$$1 \text{ atomic mass unit } = 1\text{ u} = 1.660 \times 10^{-27} \text{ kg}$$

The mass of a molecule is the sum of the masses of the atoms of which it is composed; thus $m_{H_2O} = 2m_H + m_O$.

Solved Problems

10.1. Gas molecules have velocities comparable with those of rifle bullets, yet we all know that a gas with a strong odor, such as ammonia, takes several seconds to diffuse through a room. Why?

Gas molecules collide frequently with one another, which means that a particular molecule follows a long, very complicated path in going from one place to another.

10.2. Explain the evaporation of a liquid at a temperature below its boiling point on the basis of the kinetic theory of matter.

At any moment in a liquid, some molecules are moving faster and others are moving slower than the average. The fastest ones are able to escape from the liquid surface despite the attractive forces exerted by the other molecules; this constitutes evaporation. The warmer the liquid, the greater the number of very fast molecules, and the more rapidly evaporation takes place. Since the molecules that remain behind are the slower ones, the liquid has a lower temperature than before (unless heat has been added to it from an outside source during the process).

10.3. A 1-liter sample of nitrogen at 0 °C and 1 atm pressure is compressed to a volume of 0.5 liter. If the temperature of the sample is unchanged, what happens to its pressure? To the average velocity of its molecules?

Since $p_2 = p_1V_1/V_2$ and $V_1/V_2 = 2$, the pressure doubles to 2 atm. The average molecular velocity is unchanged since it depends only on the temperature.

10.4. A steel cylinder contains 3 ft³ of oxygen at an absolute pressure of 75 lb/in². What volume would this amount of oxygen occupy at sea-level atmospheric pressure of 15 lb/in²?

If p_1 and V_1 are the pressure and volume of the oxygen in the cylinder and V_2 is the volume at atmospheric pressure p_2,

$$V_2 = \frac{p_1V_1}{p_2} = \frac{75 \text{ lb/in}^2 \times 3 \text{ ft}^3}{15 \text{ lb/in}^2} = 15 \text{ ft}^3$$

10.5. Nitrogen boils at −196 °C. What is this temperature on the absolute scale?

$$T_K = T_C + 273 = -196 + 273 = 77 \text{ K}$$

10.6. The surface temperature of the sun is 6000 K. What is the Celsius equivalent of this temperature?

$$T_C = T_K - 273 = 5727 \text{ °C}$$

10.7. To what temperature must a gas sample initially at 0 °C and atmospheric pressure be heated if its volume is to double while its pressure remains the same?

Since $T_1 = 0 \text{ °C} = 273 \text{ K}$ and $V_2 = 2V_1$, from Charles's law

$$T_2 = \frac{T_1 V_2}{V_1} = \frac{273 \text{ K} \times 2V_1}{V_1} = 546 \text{ K} = 273 \text{ °C}$$

10.8. The tire of a car contains air at an absolute pressure of 35 lb/in² when its temperature is 50 °F. If the tire's volume does not change, what is the pressure when the temperature is 120 °F?

The first step is to convert the temperatures to their absolute equivalents:

$$T_1 = 50 \text{ °F} = 10 \text{ °C} = 283 \text{ K} \quad \text{and} \quad T_2 = 120 \text{ °F} = 49 \text{ °C} = 322 \text{ K}$$

Since $p_1 = 35$ lb/in² and $V_1 = V_2$, from the ideal gas law $p_1 V_1/T_1 = p_2 V_2/T_2$ we have

$$p_2 = \frac{T_2 p_1}{T_1} = \frac{322 \text{ K} \times 35 \text{ lb/in}^2}{283 \text{ K}} = 40 \text{ lb/in}^2$$

10.9. A tank whose capacity is 0.1 m³ contains helium at a pressure of 10 atm and a temperature of 20 °C. A rubber weather balloon is inflated with this helium. (a) The gas cools as it expands, and when the pressure of the helium in the balloon is 1 atm, its temperature is −40 °C. Find the volume of the balloon. (b) Eventually the helium in the balloon absorbs heat from the air around it and returns to 20 °C. Find the volume of the balloon at this time.

(a) $T_1 = 20 \text{ °C} = 293 \text{ K}$, $T_2 = -40 \text{ °C} = 233 \text{ K}$, $V_1 = 0.1$ m³, $p_1 = 10$ atm, $p_2 = 1$ atm. From the ideal gas law,

$$V_2 = \frac{T_2 p_1 V_1}{T_1 p_2} = \frac{233 \text{ K} \times 10 \text{ atm} \times 0.1 \text{ m}^3}{293 \text{ K} \times 1 \text{ atm}} = 0.8 \text{ m}^3$$

The volume of the balloon is therefore 0.7 m³ at this time, since the tank retains 0.1 m³ of the helium after the expansion.

(b) Here $p_1 = 10$ atm, $p_3 = 1$ atm, $V_1 = 0.1$ m³, and, since $T_1 = T_3$, Boyle's law can be used. We have

$$V_3 = \frac{p_1 V_1}{p_3} = \frac{10 \text{ atm} \times 0.1 \text{ m}^3}{1 \text{ atm}} = 1 \text{ m}^3$$

The volume of the balloon is therefore 0.9 m³, assuming it is still attached to the tank.

10.10. Find the mass of (a) the water molecule, H_2O, and (b) the ethyl alcohol molecule, C_2H_6O. The atomic masses of H, C, and O are respectively 1.008 u, 12.01 u, and 16.00 u.

(a)
$$2 \text{ H} = 2 \times 1.008 \text{ u} = 2.02 \text{ u}$$
$$O = 1 \times 16.00 \text{ u} = \underline{16.00 \text{ u}}$$
$$18.02 \text{ u}$$

$$m(H_2O) = 18.02 \text{ u} \times 1.66 \times 10^{-27} \text{ kg/u} = 2.99 \times 10^{-26} \text{ kg}$$

(b)
$$2 \text{ C} = 2 \times 12.01 \text{ u} = 24.02 \text{ u}$$
$$6 \text{ H} = 6 \times 1.008 \text{ u} = 6.05 \text{ u}$$
$$O = 1 \times 16.00 \text{ u} = \underline{16.00 \text{ u}}$$
$$46.07 \text{ u}$$

$$m(C_2H_6O) = 46.07 \text{ u} \times 1.66 \times 10^{-27} \text{ kg/u} = 7.65 \times 10^{-26} \text{ kg}$$

10.11. What is the average kinetic energy of the molecules of any gas at 100 °C?

The absolute temperature corresponding to 100 °C is

$$T_K = T_C + 273 = 373 \text{ K}$$

The average kinetic energy at this temperature is

$$\text{KE}_{av} = \frac{3}{2}kT = \frac{3}{2} \times 1.38 \times 10^{-23} \text{ J/K} \times 373 \text{ K} = 7.72 \times 10^{-21} \text{ J}$$

10.12. What is the average velocity of the molecules in a sample of oxygen at 100 °C? The mass of an oxygen molecule is 5.3×10^{-26} kg.

Since $\text{KE}_{av} = \frac{1}{2}mv_{av}^2 = \frac{3}{2}kT$, $v_{av} = \sqrt{\frac{3kT}{m}}$. Here $T = 100$ °C = 373 K, and so

$$v_{av} = \sqrt{\frac{3kT}{m}} = \sqrt{\frac{3 \times 1.38 \times 10^{-23} \text{ J/K} \times 373 \text{ K}}{5.3 \times 10^{-26} \text{ kg}}} = \sqrt{29.1 \times 10^4} \text{ m/s}$$
$$= 5.4 \times 10^2 \text{ m/s} = 540 \text{ m/s}$$

10.13. A certain tank holds 1 g of hydrogen at 0 °C and another identical tank holds 1 g of oxygen at 0 °C. The mass of an oxygen molecule is 16 times greater than that of a hydrogen molecule. (*a*) Which tank contains more molecules? How many more? (*b*) Which gas exerts the greater pressure? How much greater? (*c*) In which gas do the molecules have greater average energies? How much greater? (*d*) In which gas do the molecules have greater average velocities? How much greater?

(*a*) There are 16 times more hydrogen molecules.

(*b*) The hydrogen pressure is 16 times greater because there are 16 times more molecules to exert force on the container walls.

(*c*) The average molecular energies are the same in both gases because their temperatures are the same.

(*d*) The average velocity of the hydrogen molecules is $\sqrt{16} = 4$ times more than that of the oxygen molecules because the hydrogen molecules are 16 times lighter and $v_{av} = \sqrt{3kT/m}$.

Supplementary Problems

10.14. The volume of a gas sample is enlarged. Why does the pressure the gas exerts decrease?

10.15. The temperature of a gas sample is raised. Why does the pressure the gas exerts increase?

10.16. A compressor pumps 50 liters of air at a pressure of 1 atm into an 8-liter tank. What is the absolute pressure (in atm) of the air in the tank?

10.17. How much air at a pressure of 1 atm can be stored in a 2-m³ tank which can safely withstand a pressure of 5×10^5 N/m²?

10.18. What is the Celsius equivalent of a temperature of 500 K?

10.19. What is the Kelvin equivalent of a temperature of 500 °C?

10.20. The *Rankine scale* is an absolute temperature scale that uses Fahrenheit degrees instead of Celsius degrees. (*a*) What is absolute zero on the Rankine scale? (*b*) Find formulas for converting Fahrenheit temperatures to Rankine temperatures and vice versa. (*c*) Express the freezing and boiling points of water on the Rankine scale.

10.21. A gas sample occupies 4 m^3 at an absolute pressure of 2×10^5 N/m^2 and a temperature of 320 K. Find its volume (a) at the same pressure and a temperature of 400 K, and (b) at the same temperature and a pressure of 4×10^4 N/m^2.

10.22. A gas sample occupies a volume of 1 m^3 at a temperature of 27 °C and a pressure of 1 atm. Find its volume (a) at 127 °C and 0.5 atm; (b) at 127 °C and 2 atm; (c) at −73 °C and 0.5 atm; and (d) at −73 °C and 2 atm.

10.23. Find the mass of the propane molecule, C_3H_8, and that of the glucose molecule, $C_6H_{12}O_6$.

10.24. A gas sample at 0 °C is heated until the average energy of its molecules doubles. What is its new temperature?

10.25. A gas sample at 0 °C is heated until the average velocity of its molecules doubles. What is its new temperature?

10.26. At room temperature oxygen molecules have an average velocity of about 1000 mi/hr. (a) What is the average velocity of hydrogen molecules, whose mass is 1/16 that of oxygen molecules, at this temperature? (b) What is the average velocity of sulfur dioxide molecules, whose mass is twice that of oxygen molecules, at this temperature?

10.27. Mercury is a gas at 500 °C. (a) What is the average energy of mercury atoms at this temperature? (b) What is the average velocity of mercury atoms at this temperature? The mass of a mercury atom is 3.3×10^{-25} kg.

Answers to Supplementary Problems

10.14. The pressure decreases partly because the gas molecules now must travel farther between impacts on the container walls and partly because these impacts are now distributed over a larger area.

10.15. The pressure increases partly because the gas molecules move faster than before and therefore strike the walls more often and partly because each impact yields a greater force than before.

10.16. 6.25 atm

10.17. 9.87 m^3

10.18. 227 °C

10.19. 773 K

10.20. −460 °F; $T_R = T_F + 460°$; $T_F = T_R - 460°$; 492 °R; 672 °R

10.21. 5 m^3; 20 m^3

10.22. 2.67 m^3; 0.67 m^3; 1.33 m^3; 0.33 m^3

10.23. 7.32×10^{-26} kg; 2.99×10^{-25} kg

10.24. 273 °C

10.25. 819 °C

10.26. 4000 mi/hr; 707 mi/hr

10.27. 1.6×10^{-20} J; 311 m/s

Chapter 11

Thermodynamics

MECHANICAL EQUIVALENT OF HEAT

It is possible to transform other types of energy into internal energy and vice versa. The ratio between an energy unit and a heat unit is called the *mechanical equivalent of heat* and has the values in the metric and British systems of

$$\mathcal{J} = 4185 \frac{J}{kcal}$$

$$\mathcal{J} = 778 \frac{ft\text{-}lb}{Btu}$$

HEAT ENGINES

To convert internal energy into mechanical energy is much more difficult than the reverse, and perfect efficiency is impossible. A *heat engine* is a device or system that can perform this conversion; the human body and the earth's atmosphere are heat engines, as well as gasoline and diesel motors, aircraft jet engines, and steam turbines.

All heat engines operate by absorbing heat from a reservoir of some kind at a high temperature, performing work, and then giving off heat to a reservoir of some kind at a lower temperature. According to the principle of conservation of energy, the work done in a complete cycle that returns the engine to its original state is equal to the difference between the heat absorbed and the heat given off; this statement constitutes the *first law of thermodynamics*.

A *refrigerator* is a heat engine that operates backwards to extract heat from a low-temperature reservoir and transfer it to a high-temperature reservoir. Because the natural tendency of heat is to flow from a hot region to a cold one, energy must be provided to a refrigerator to reverse the flow, and this energy adds to the heat exhausted by the refrigerator.

SECOND LAW OF THERMODYNAMICS

Internal energy resides in the kinetic energies of randomly moving atoms and molecules, whereas the output of a heat engine appears in the ordered motions of a piston or a wheel. Since all physical systems in the universe tend to go in the opposite direction, from order to disorder, no heat engine can completely convert heat into mechanical energy or, in general, into work. This fundamental principle leads to the *second law of thermodynamics*: It is impossible to construct a continuously-operating engine that takes heat from a source and performs an exactly equivalent amount of work.

Because some of the heat input to a heat engine must be wasted, and because heat flows from a hot reservoir to a cold one, every heat engine must have a low-temperature reservoir for exhaust heat to go to as well as a high-temperature reservoir from which the input heat is to come.

ENGINE EFFICIENCY

The efficiency of an ideal heat engine (often called a *Carnot engine*) in which there are no losses due to such practical difficulties as friction depends only upon the temperatures at which heat is absorbed and exhausted. If heat is absorbed at the absolute temperature T_1 and is given off at the absolute temperature T_2, the efficiency of such an engine is

$$\text{Efficiency} = 1 - \frac{T_2}{T_1}$$

The smaller the ratio between T_2 and T_1, the more efficient the engine. Because no reservoir can exist at a temperature of 0 K, which is absolute zero, no heat engine can be 100% efficient.

HEAT TRANSFER: CONDUCTION

The three mechanisms by which heat can be transferred from one place to another are conduction, convection, and radiation.

In *conduction*, heat is carried by means of collisions between rapidly moving molecules at the hot end of a body of matter and the slower molecules at the cold end. Some of the kinetic energy of the fast molecules passes to the slow molecules, and the result of successive collisions is a flow of heat through the body of matter. Solids, liquids, and gases all conduct heat. Conduction is poorest in gases because their molecules are relatively far apart and so interact less frequently than in the case of solids and liquids. Metals are the best conductors of heat because some of their electrons are able to move about relatively freely and can travel past many atoms between collisions.

CONVECTION

In *convection*, a volume of hot fluid (gas or liquid) moves from one region to another carrying internal energy with it. When a pan of water is heated on a stove, for instance, the hot water at the bottom expands slightly so that its density decreases, and the buoyancy of this water causes it to rise to the surface while colder, denser water descends to take its place at the bottom.

RADIATION

In *radiation*, energy is carried by the *electromagnetic waves* emitted by every object. Electromagnetic waves, of which light, radio waves, and X-rays are examples, travel at the velocity of light (3×10^8 m/s = 186,000 mi/s) and require no material medium for their passage. The higher the temperature of an object, the greater the rate at which it radiates energy; the rate is proportional to T^4, where T is the object's absolute temperature.

Solved Problems

11.1. Why is it impossible for a ship to use the internal energy of seawater to operate its engine?

　　There would be no suitable low-temperature reservoir to absorb the waste heat from the engine.

11.2. In an effort to cool a kitchen during the summer, the refrigerator door is left open and the kitchen's door and windows are closed. What will happen?

Since no refrigerator can be completely efficient, more heat is exhausted by the refrigerator into the kitchen than the heat it extracts from the kitchen. The net effect, then, is to increase the kitchen's temperature.

11.3. If all objects radiate electromagnetic energy, why do not the objects around us in everyday life grow colder and colder?

Every object also absorbs electromagnetic energy from its surroundings, and if both object and surroundings are at the same temperature, energy is emitted and absorbed at the same rate. When an object is at a higher temperature than its surroundings and heat is not supplied to it, it radiates more energy than it absorbs and cools down to the temperature of its surroundings.

11.4. When a pound of coal is burned, 14,000 Btu of heat is liberated. How many ft-lb of energy is this equivalent to?

$$E = \mathcal{J}Q = 778 \frac{\text{ft-lb}}{\text{Btu}} \times 14,000 \text{ Btu} = 1.09 \times 10^7 \text{ ft-lb}$$

11.5. An ice cube at 0 °C is dropped on the ground and melts to water at 0 °C. If all the kinetic energy of the ice went into melting it, from what height did it fall?

$$\mathcal{J} \times \text{mass of ice} \times \text{heat of fusion} = \text{initial potential energy of ice}$$

$$\mathcal{J}mL_f = mgh$$

$$h = \frac{\mathcal{J}L_f}{g} = \frac{4185 \text{ J/kcal} \times 80 \text{ kcal/kg}}{9.8 \text{ m/s}^2} = 3.4 \times 10^4 \text{ m}$$

11.6. Although the process described in Problem 11.5 is impossible from a practical point of view because some of the lost kinetic energy would go into heating the ground, it violates no physical principles. What about the reverse of this process, in which a puddle of water would rise into the air as it turns to ice, with the required energy coming from the heat of fusion?

The latter process would require the randomly-moving water molecules in the puddle to spontaneously assume more ordered motions that would enable forces to be exerted on the ground whose reaction forces would be sufficient to push the entire puddle upward as a unit. Since physical systems by themselves tend to go from order to disorder and not the reverse, which is the basis of the second law of thermodynamics, such an event would be exceedingly unlikely.

11.7. A 1-MW (10^6 W) generating plant has an overall efficiency of 40%. How much fuel oil whose heat of combustion is 11,000 kcal/kg does the plant burn each day?

In one day the plant produces

$$W = Pt = 10^6 \text{ W} \times 3600 \text{ s/hr} \times 24 \text{ hr/day} = 8.64 \times 10^{10} \text{ J}$$

of electric energy. Since its efficiency is 0.4 and Eff = work output/heat input, we have

$$\text{Heat input} = \frac{\text{work output}}{\text{Eff}} = \frac{8.64 \times 10^{10} \text{ J}}{0.4} = 2.16 \times 10^{11} \text{ J}$$

To convert the heat input to its equivalent in kcal we divide by 4185 J/kcal to get

$$\text{Heat input} = \frac{2.16 \times 10^{11} \text{ J}}{4.185 \times 10^3 \text{ J/kcal}} = 5.16 \times 10^7 \text{ kcal}$$

The fuel required to supply this amount of heat is

$$m = \frac{5.16 \times 10^7 \text{ kcal}}{1.1 \times 10^4 \text{ kcal/kg}} = 4.7 \times 10^3 \text{ kg}$$

11.8. A total of 0.8 kg of water at 20 °C is placed in a 1-kW electric kettle. How long a time is needed to raise the temperature of the water to 100 °C?

The required heat is

$$Q = mc\,\Delta T = 0.8 \text{ kg} \times 1 \text{ kcal/kg-°C} \times (100 \text{ °C} - 20 \text{ °C}) = 64 \text{ kcal}$$

The energy equivalent of this amount of heat is

$$E = \mathcal{J}Q = 4185 \text{ J/kcal} \times 64 \text{ kcal} = 2.68 \times 10^5 \text{ J}$$

Since $P = E/t$ and $P = 1$ kW $= 10^3$ J/s here, the time needed is

$$t = \frac{E}{P} = \frac{2.68 \times 10^5 \text{ J}}{10^3 \text{ J/s}} = 2.68 \times 10^2 \text{ s} = 268 \text{ s} = 4.5 \text{ min}$$

11.9. (a) Find the maximum possible efficiency of an engine that absorbs heat at a temperature of 327 °C and exhausts heat at a temperature of 127 °C. (b) What is the maximum amount of work (in joules) the engine can perform per kcal of heat input?

(a) Here $T_1 = 327$ °C $+ 273 = 600$ K and $T_2 = 127$ °C $+ 273 = 400$ K. Hence

$$\text{Eff} = 1 - \frac{T_2}{T_1} = 1 - \frac{400 \text{ K}}{600 \text{ K}} = \frac{1}{3} = 33\%$$

(b) Since 1 kcal $= 4185$ J,

$$\text{Work} = \text{Eff} \times \text{heat input} = \frac{1}{3} \times 4185 \text{ J} = 1395 \text{ J}$$

11.10. Three designs are proposed for an engine which is to operate between 500 K and 300 K. Design A is claimed to produce 3000 J of work per kcal of heat input, B is claimed to produce 2000 J, and C is claimed to produce 1000 J. Which design would you choose?

The efficiency of an ideal engine operating between $T_1 = 500$ K and $T_2 = 300$ K is

$$\text{Eff} = 1 - \frac{T_2}{T_1} = 1 - \frac{300 \text{ K}}{500 \text{ K}} = 0.40 = 40\%$$

Since 1 kcal $= 4185$ J, the claimed efficiencies of the proposed engines are

$$\text{Eff (A)} = \frac{\text{work output}}{\text{heat input}} = \frac{3000 \text{ J}}{4185 \text{ J}} = 0.72 = 72\%$$

$$\text{Eff (B)} = \frac{2000 \text{ J}}{4185 \text{ J}} = 0.48 = 48\%$$

$$\text{Eff (C)} = \frac{1000 \text{ J}}{4185 \text{ J}} = 0.24 = 24\%$$

Both A and B claim efficiencies greater than that of an ideal engine and hence could not possibly work as stated. Design C is therefore the only possible choice.

11.11. An engine is being planned which is to have an efficiency of 25% and which will absorb heat at a temperature of 267 °C. Find the maximum temperature at which it can exhaust heat.

The intake temperature is $T_1 = 267$ °C $+ 273 = 540$ K. We proceed as follows:

$$\text{Eff} = 1 - \frac{T_2}{T_1}, \quad \frac{T_2}{T_1} = 1 - \text{Eff}$$

$$T_2 = T_1(1 - \text{Eff}) = (540 \text{ K})(1 - 0.25) = 405 \text{ K}$$

The maximum exhaust temperature is therefore $T_2 = 405$ K $- 273 = 132$ °C.

Supplementary Problems

11.12. The first of Newton's laws of motion is a special case of the second law of motion. Is the first law of thermodynamics a special case of the second?

11.13. It is possible to convert mechanical energy to heat more efficiently than it is to convert heat to mechanical energy. What physical principle incorporates this observation?

11.14. Outdoors in the winter, why does a piece of metal feel colder than a piece of wood?

11.15. (a) How many kcal per hour are given off by a 100-W light bulb? (b) How many Btu per hour?

11.16. The conventional unit of refrigeration capacity is the "ton," which is defined as that rate of heat removal that can freeze 1 ton of water at 32 °F to ice at 32 °F in a day. How many hp is a refrigeration ton equivalent to? The heat of fusion of water is 144 Btu/lb and 1 hp = 550 ft-lb/s.

11.17. A typical gum drop contains 35 kcal of energy. If this energy were used to raise an 80-kg man above the ground, how high would he go?

11.18. A lead bullet traveling at 200 m/s strikes a tree and comes to a stop. If half the heat produced is retained by the bullet, by how much does its temperature increase? ($c_{lead} = 0.03$ kcal/kg-°C)

11.19. One and a half kg of water at 10 °C in a 300-g aluminum kettle is placed on a 2-kW electric hot plate. What is the temperature of the water after 3 min? ($c_{aluminum} = 0.22$ kcal/kg-°C)

11.20. A 2400-hp diesel locomotive burns 160 gal of fuel per hour. If the heat of combustion of the diesel oil used is 1.2×10^5 Btu/gal, find the efficiency of the engine.

11.21. Steam enters a turbine engine at 550 °C and emerges at 90 °C. The engine has an actual overall efficiency of 35%. What percentage of its ideal efficiency is this?

11.22. An engine absorbs 500 kcal of heat at 600 K and exhausts 300 kcal of heat at 300 K. (a) What is its efficiency? (b) If it were an ideal engine, what would its efficiency be and how much heat would it exhaust?

11.23. At what temperature must an ideal engine absorb heat if its efficiency is 33% and it exhausts heat at 60 °C?

Answers to Supplementary Problems

11.12. No. The first law of thermodynamics is the principle of conservation of energy. The second law states in essence that it is impossible to convert heat into work or into any other form of energy with perfect efficiency.

11.13. The second law of thermodynamics.

11.14. Metals are much better conductors of heat than wood and therefore conduct heat away from the hand more rapidly.

11.15.	86 kcal; 342 Btu	**11.20.**	32%
11.16.	4.7 hp	**11.21.**	63%
11.17.	187 m	**11.22.**	40%; 50%; 250 kcal
11.18.	80 °C	**11.23.**	224 °C
11.19.	65 °C		

Chapter 12

Electricity

ELECTRIC CHARGE

Electric charge, like mass, is one of the basic properties of certain of the elementary particles of which all matter is composed. There are two kinds of charge, *positive charge* and *negative charge*. The positive charge in ordinary matter is carried by *protons*, the negative charge by *electrons*. Charges of the same sign repel each other, charges of opposite sign attract each other.

The unit of charge is the *coulomb* (C). The charge of the proton is $+1.6 \times 10^{-19}$ C and the charge of the electron is -1.6×10^{-19} C. All charges in nature occur in multiples of $\pm e = \pm 1.6 \times 10^{-19}$ C.

According to the principle of *conservation of charge*, the net electric charge in an isolated system always remains constant. (Net charge means the total positive charge minus the total negative charge.) When matter is created from energy, equal amounts of positive and negative charge always come into being, and when matter is converted to energy, equal amounts of positive and negative charge disappear.

COULOMB'S LAW

The force one charge exerts on another is given by *Coulomb's law*:

$$\text{Electric force} = F = k\frac{Q_1 Q_2}{r^2}$$

where Q_1 and Q_2 are the magnitudes of the charges, r is the distance between them, and k is a constant whose value in free space is

$$k = 9.0 \times 10^9 \frac{\text{N-m}^2}{\text{C}^2}$$

The value of k in air is slightly greater.

ATOMIC STRUCTURE

An atom of any element consists of a small, positively charged *nucleus* with a number of electrons some distance away. The nucleus is composed of protons (charge $+e$, mass $= 1.673 \times 10^{-27}$ kg) and neutrons (uncharged, mass $= 1.675 \times 10^{-27}$ kg); the number of protons in the nucleus is normally equal to the number of electrons around it, so that the atom as a whole is electrically neutral. The forces between atoms that hold them together as solids and liquids are electrical in origin.

IONS

Under certain circumstances an atom may lose one or more electrons and become a *positive ion*, or it may gain one or more electrons and become a *negative ion*. Many solids consist of positive and negative ions rather than of atoms or molecules. An example is

67

ordinary table salt, which is made up of positive sodium ions (Na$^+$) and negative chlorine ions (Cl$^-$).   Solutions of such solids in water also contain ions.   Sparks, flames, and X-rays are among the influences that can ionize gases.   Ions of opposite sign in a gas come together soon after being formed and the excess electrons on the negative ions pass to the positive ones to form neutral molecules.   A gas can be maintained in an ionized state by passing an electric current through it (as in a neon sign) or by bombarding it with X-rays or ultraviolet light (as in the upper atmosphere of the earth, where the radiation comes from the sun).

ELECTRIC FIELD

An *electric field* is a region of space in which a charge would be acted upon by an electric force.   An electric field may be produced by one or more charges, and it may be uniform or it may vary in magnitude and/or direction from place to place.

If a charge Q at a certain point is acted upon by the force **F**, the electric field **E** at that point is defined as the ratio between **F** and Q:

$$\mathbf{E} = \frac{\mathbf{F}}{Q}$$

$$\text{Electric field} = \frac{\text{force}}{\text{charge}}$$

Electric field is a vector quantity whose direction is that of the force on a positive charge.   The unit of **E** is the newton/coulomb (N/C) or, more commonly, the equivalent volt/meter (V/m).

The advantage of knowing the electric field at some point is that we can at once establish the force on *any* charge Q placed there, which is

$$\mathbf{F} = Q\mathbf{E}$$

$$\text{Force} = \text{charge} \times \text{electric field}$$

ELECTRIC LINES OF FORCE

Lines of force are a means of describing a force field, such as an electric field, by using imaginary lines to indicate the direction and magnitude of the field.   The direction of an electric line of force at any point is the direction in which a positive charge would move if placed there, and lines of force are drawn close together where the field is strong and far apart where the field is weak.

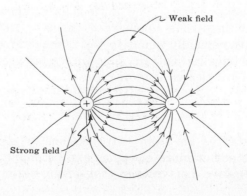

Fig. 12-1

POTENTIAL DIFFERENCE

The *potential difference* V between two points in an electric field is the amount of work needed to take a charge of 1 C from one of the points to the other.   Thus

$$V = \frac{W}{Q}$$

$$\text{Potential difference} = \frac{\text{work}}{\text{charge}}$$

The unit of potential difference is the *volt* (V):

$$1 \text{ volt} = 1 \frac{\text{joule}}{\text{coulomb}}$$

The potential difference between two points in a uniform electric field $\mathbf{E}$ is equal to the product of E and the distance s between the points in a direction parallel to $\mathbf{E}$:

$$V = Es$$

Since an electric field is usually produced by applying a potential difference between two metal plates s apart, this equation is most useful in the form

$$E = \frac{V}{s}$$

$$\text{Electric field} = \frac{\text{potential difference}}{\text{distance}}$$

A battery uses chemical reactions to produce a potential difference between its terminals; a generator uses electromagnetic induction (Chapter 15) for this purpose.

Solved Problems

12.1. Give some reasons in support of the idea that there are only two kinds of electric charge.

Experiments show the following: Every charge is either attracted by or repelled by a positive charge, and those charges that are attracted by a positive charge are repelled by a negative one and vice versa; the forces between all charges obey the same law; when matter is formed from energy, equal amounts of positive and negative charge come into being, and when matter is converted into energy, equal amounts of positive and negative charge disappear. Since no electrical phenomena are known that cannot be accounted for on the basis of two kinds of charge, there is no reason to believe any other kind of charge exists.

12.2. What are some of the similarities and differences between mass and electric charge?

Similarities.   Mass and charge are both fundamental properties which are possessed by most of the elementary particles of which all matter is composed.   The gravitational force between two masses and the electric force between two charges both follow laws of the same form:

$$F_{\text{grav}} = Gm_1 m_2/r^2 \quad \text{and} \quad F_{\text{elec}} = kQ_1 Q_2/r^2$$

Differences.   There is only a single kind of mass, and the gravitational forces between masses are always attractive; but there are two kinds of charge, and the electric forces between charges of the same kind are repulsive whereas those between charges of opposite kind are attractive. Mass is not conserved, since it can be converted into energy and energy can be converted into mass; but charge is conserved, and when matter is converted into energy and vice versa, equal amounts of opposite charges disappear or come into being so that the net charge of a system undergoing such an event is unchanged. Another difference is that electric forces between elementary particles are very much stronger than the gravitational forces between them.

12.3. How does a charged object (such as a hard rubber comb drawn through one's hair) attract an uncharged object (such as a piece of paper)?

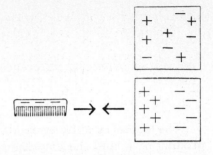

Let us say that the charged object has a negative charge. When it is brought near an uncharged object, negative charges in the latter are repelled and move as far away as they can while positive charges are attracted and move closer. As a result the part of the uncharged object near the charged one has a positive charge, and is attracted to it (Fig. 12-2). Since electric forces decrease with distance, the repulsive force of the negative charges that are farther away is weaker than the attractive forces of the positive charges.

Fig. 12-2

12.4. A rapidly moving proton can pass through a thin metal foil without stopping or being deflected by more than a small amount. To what aspect of atomic structure can this behavior be attributed?

Nearly all the volume occupied by an atom consists of empty space.

12.5. What is the magnitude and direction of the force on a charge of $+4 \times 10^{-9}$ C that is 5 cm from a charge of $+5 \times 10^{-8}$ C?

Since 5 cm $= 5 \times 10^{-2}$ m, we have from Coulomb's law

$$F = k\frac{Q_1 Q_2}{r^2} = \frac{9 \times 10^9 \text{ N-m}^2/\text{C}^2 \times 4 \times 10^{-9} \text{ C} \times 5 \times 10^{-8} \text{ C}}{(5 \times 10^{-2} \text{ m})^2} = 7.2 \times 10^{-4} \text{ N}$$

The force is directed away from the $+5 \times 10^{-8}$ C charge since both charges are positive.

12.6. A hydrogen atom consists of a proton (charge $+e$) and an electron (charge $-e$) that are an average of 5.3×10^{-11} m apart. Find the attractive force between them.

Since $e = 1.6 \times 10^{-19}$ C, we have from Coulomb's law

$$F = k\frac{Q_1 Q_2}{r^2} = \frac{9 \times 10^9 \text{ N-m}^2/\text{C}^2 \times (1.6 \times 10^{-19} \text{ C})^2}{(5.3 \times 10^{-11} \text{ m})^2} = 8.2 \times 10^{-8} \text{ N}$$

12.7. Two charges, one of $+5 \times 10^{-7}$ C and the other of -2×10^{-7} C, attract each other with a force of 100 N. How far apart are they?

From Coulomb's law we have

$$r = \sqrt{\frac{kQ_1 Q_2}{F}} = \sqrt{\frac{9 \times 10^9 \text{ N-m}^2/\text{C}^2 \times 5 \times 10^{-7} \text{ C} \times 2 \times 10^{-7} \text{ C}}{10^2 \text{ N}}}$$

$$= \sqrt{90 \times 10^{-7}} \text{ m} = \sqrt{9 \times 10^{-6}} \text{ m} = 3 \times 10^{-3} \text{ m} = 3 \text{ mm}$$

12.8. Two charges repel each other with a force of 10^{-5} N when they are 20 cm apart. (a) What is the force on each of them when they are 5 cm apart? (b) When they are 100 cm apart?

(a) Since F is proportional to $1/r^2$, the force increases when the charges are brought closer together to $(20/5)^2 = 16$ times what it was before, namely to 1.6×10^{-4} N.

(b) The force decreases when the charges are moved apart to $(20/100)^2 = 0.04$ times what it was before, namely to 4×10^{-7} N.

12.9. Under what circumstances, if any, is the gravitational attraction between two protons equal to their electrical repulsion?

Since the proton mass is 1.67×10^{-27} kg, the gravitational force between two protons that are a distance r apart is

$$F_{\text{grav}} = \frac{Gm_1m_2}{r^2} = \frac{6.67 \times 10^{-11} \text{ N-m}^2/\text{kg}^2 \times (1.67 \times 10^{-27} \text{ kg})^2}{r^2} = \frac{1.86 \times 10^{-64}}{r^2} \text{ N}$$

The electric force between the protons is

$$F_{\text{elec}} = \frac{kQ_1Q_2}{r^2} = \frac{9 \times 10^9 \text{ N-m}^2/\text{C}^2 \times (1.6 \times 10^{-19} \text{ C})^2}{r^2} = \frac{2.3 \times 10^{-28}}{r^2} \text{ N}$$

At every separation r, the electric force between the protons is greater than the gravitational force between them by a factor of more than 10^{36}; the forces are never equal.

12.10. The force holding the electron in a hydrogen atom to the proton (which is the atom's nucleus) is 8.2×10^{-8} N, as calculated in Problem 12.6. What is the electric field that acts upon the electron?

From the definition of electric field,

$$E = \frac{F}{Q} = \frac{F}{e} = \frac{8.2 \times 10^{-8} \text{ N}}{1.6 \times 10^{-19} \text{ C}} = 5.1 \times 10^{11} \text{ V/m}$$

12.11. The electric field in a certain neon sign is 5000 V/m. (a) What is the force this field exerts on a neon ion of mass 3.3×10^{-26} kg and charge $+e$? (b) What is the acceleration of the ion?

(a) The force on the neon ion is

$$F = QE = eE = 1.6 \times 10^{-19} \text{ C} \times 5 \times 10^3 \text{ V/m} = 8 \times 10^{-16} \text{ N}$$

(b) According to the second law of motion $F = ma$, and so here

$$a = \frac{F}{m} = \frac{8 \times 10^{-16} \text{ N}}{3.3 \times 10^{-26} \text{ kg}} = 2.4 \times 10^{10} \text{ m/s}^2$$

12.12. How strong an electric field is required to exert a force on a proton equal to its weight at sea level?

The electric force on the proton is $F = eE$ and its weight is mg.   Hence $eE = mg$ and

$$E = \frac{mg}{e} = \frac{1.67 \times 10^{-27} \text{ kg} \times 9.8 \text{ m/s}^2}{1.6 \times 10^{-19} \text{ C}} = 1.02 \times 10^{-7} \text{ V/m}$$

12.13. The potential difference between a certain thundercloud and the ground is 7×10^6 V. Find the energy dissipated when a charge of 50 C is transferred from the cloud to the ground in a lightning stroke.

$$W = QV = 50 \text{ C} \times 7 \times 10^6 \text{ V} = 3.5 \times 10^8 \text{ J}$$

12.14. A potential difference of 20 V is applied across two parallel metal plates and an electric field of 500 V/m is produced. How far apart are the plates?

Since $E = V/s$, here

$$s = \frac{V}{E} = \frac{20\ V}{500\ V/m} = 0.04\ m = 4\ cm$$

12.15. (a) What potential difference must be applied across two metal plates 15 cm apart if the electric field between them is to be 600 V/m? (b) What is the force on a charge of 10^{-10} C in this field? (c) How much kinetic energy will the charge have when it has moved through 5 cm in the field starting from rest?

(a) $$V = Es = 600\ V/m \times 0.15\ m = 90\ V$$

(b) $$F = QE = 10^{-10}\ C \times 600\ V/m = 6 \times 10^{-8}\ N$$

(c) Since the KE of the charge is equal to the work done on it by the electric field when it travels 0.05 m,

$$KE = W = Fs = 6 \times 10^{-8}\ N \times 0.05\ m = 3 \times 10^{-9}\ J$$

12.16. What potential difference must be applied to produce an electric field that can accelerate an electron to a velocity of 10^7 m/s?

The kinetic energy of such an electron is

$$KE = \tfrac{1}{2}mv^2 = \tfrac{1}{2} \times 9.1 \times 10^{-31}\ kg \times (10^7\ m/s)^2 = 4.6 \times 10^{-17}\ J$$

This KE is equal to the work W that must be done on the electron by the electric field and so, since $W = QV$ in general, we have here

$$V = \frac{W}{Q} = \frac{KE}{e} = \frac{4.6 \times 10^{-17}\ J}{1.6 \times 10^{-19}\ C} = 2.9 \times 10^2\ V = 290\ V$$

Supplementary Problems

12.17. The iron atom has 26 protons in its nucleus. How many electrons does this atom contain?

12.18. What information is provided by a sketch of the lines of force of an electric field?

12.19. Why is it impossible for the lines of force of an electric field to cross one another?

12.20. A rod with a charge of $+Q$ at one end and $-Q$ at the other is placed in a uniform electric field whose direction is parallel to the rod. How does the rod behave?

12.21. The rod of Problem 12.20 is placed in a uniform electric field whose direction is perpendicular to the rod. How does the rod behave?

12.22. A billion (10^9) electrons are added to a neutral pith ball. What is its charge?

12.23. What is the force between two $+1$-C charges located 1 m apart?

12.24. What is the magnitude and direction of the force on a charge of $+2 \times 10^{-7}$ C that is 0.3 m from a charge of -5×10^{-7} C?

12.25. Two electrons repel each other with a force of 10^{-8} N. How far apart are they?

12.26. Two charges attract each other with a force of 10^{-6} N when they are 1 cm apart. (*a*) How far apart should they be for the force between them to be 10^{-4} N? (*b*) 10^{-8} N?

12.27. How much force is exerted on a charge of 10^{-6} C by an electric field of 50 V/m?

12.28. An electron is present in an electric field of 10^4 V/m. (*a*) Find the force on the electron. (*b*) Find the electron's acceleration.

12.29. Two charges of $+10^{-6}$ C are located 1 cm apart. (*a*) What is the force on a charge of $+10^{-8}$ C halfway between them? (*b*) What is the force on a charge of -10^{-8} C at the same place? (*c*) What must be true of the strength of the electric field halfway between the two $+10^{-6}$-C charges?

12.30. A potential difference of 100 V is applied by a battery across a pair of metal plates 5 cm apart. (*a*) What is the electric field between the plates? (*b*) How much force does a charge of $+10^{-8}$ C experience in this field? (*c*) How much kinetic energy does this charge acquire when it goes from the positive plate to the negative plate?

12.31. A charge of -2×10^{-9} C in an electric field between two parallel metal plates 4 cm apart is acted upon by a force of 10^{-4} N. (*a*) What is the strength of the field? (*b*) What is the potential difference between the plates?

12.32. A proton is accelerated by a potential difference of 15,000 V. What is its kinetic energy?

12.33. A particle of charge 10^{-12} C starts to move from rest in an electric field of 500 V/m. (*a*) What is the force on the particle? (*b*) How much kinetic energy will it have when it has moved 1 cm in the field?

Answers to Supplementary Problems

12.17. Since all atoms are electrically neutral, the iron atom must have the same number of electrons as its nucleus has protons, namely 26 electrons.

12.18. Such a sketch shows how the magnitude and direction of the field vary in space. At a given point, the direction of the field is given by the direction of the nearest lines of force, and the relative magnitude of the field is indicated by how close together the lines of force are in the vicinity of the point.

12.19. By definition, a line of force represents the path a positively charged particle would follow in an electric field, and such a particle can travel in only one direction at any point.

12.20. The force exerted by the field on the $-Q$ charge is equal and opposite to the force exerted on the $+Q$ charge, so the rod does not move since the two forces have the same line of action.

12.21. The equal and opposite forces exerted on the charges now cause the rod to rotate until it is parallel to the electric field.

12.22. -1.6×10^{-10} C

12.23. 9×10^9 N; repulsive

12.24. 10^{-2} N directed toward the other charge

12.25. 1.5×10^{-10} m

12.26. 0.1 cm; 10 cm

12.27. 5×10^{-5} N

12.28. 1.6×10^{-15} N; 1.8×10^{15} m/s^2

12.29. 0; 0; $E = 0$

12.30. 2000 V; 2×10^{-5} N; 10^{-6} J

12.31. 5×10^4 V/m; 2000 V

12.32. 2.4×10^{-15} J

12.33. 5×10^{-10} N; 5×10^{-12} J

Chapter 13

Electric Current

ELECTRIC CURRENT

A flow of charge from one place to another constitutes an *electric current*. The direction of a current is conventionally considered to be that in which positive charge would have to move to produce the same effects as the actual current. Thus a current is always supposed to go from the positive terminal of a battery or generator to its negative terminal.

Electric currents in metal wires always consist of flows of electrons; such currents are assumed to occur in the direction opposite to that in which the electrons move. Since a positive charge going one way is for most purposes equivalent to a negative charge going the other way, this assumption makes no practical difference. Both positive and negative charges move when a current is present in a liquid or gaseous conductor.

If an amount of charge Q passes a given point in a conductor in the time interval t, the current in the conductor is

$$I = \frac{Q}{t}$$

$$\text{Electric current} = \frac{\text{charge}}{\text{time interval}}$$

The unit of electric current is the *ampere* (A), where

$$1 \text{ ampere} = 1 \frac{\text{coulomb}}{\text{second}}$$

OHM'S LAW

In order for a current to exist in a conductor, there must be a potential difference between its ends, just as a difference in height between source and outlet is necessary for a river current to exist. In the case of a metallic conductor, the current is proportional to the applied potential difference: doubling V causes I to double, tripling V causes I to triple, and so forth. This relationship is known as *Ohm's law* and is expressed in the form

$$I = \frac{V}{R}$$

$$\text{Electric current} = \frac{\text{potential difference}}{\text{resistance}}$$

The quantity R is a constant for a given conductor and is called its *resistance*. The unit of resistance is the *ohm* (Ω), where

$$1 \text{ ohm} = 1 \frac{\text{volt}}{\text{ampere}}$$

The greater the resistance of a conductor, the less the current when a certain potential difference is applied.

Ohm's law is not a physical principle but is an experimental relationship that most metals obey over a wide range of values of V and I.

RESISTORS IN SERIES

The equivalent resistance of a set of resistors depends upon the way in which they are connected as well as upon their values. If the resistors are joined in *series,* that is, consecutively (Fig. 13-1), the equivalent resistance R of the combination is the sum of the individual resistances.

$$R = R_1 + R_2 + R_3$$

Fig. 13-1

RESISTORS IN PARALLEL

In a *parallel* set of resistors, the corresponding terminals of the resistors are connected together (Fig. 13-2). The reciprocal of the equivalent resistance R of the combination, namely $1/R$, is the sum of the reciprocals of the individual resistances.

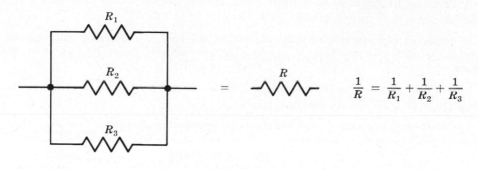

$$\frac{1}{R} = \frac{1}{R_1} + \frac{1}{R_2} + \frac{1}{R_3}$$

Fig. 13-2

If only two resistors are connected in parallel,

$$\frac{1}{R} = \frac{1}{R_1} + \frac{1}{R_2} = \frac{R_1 + R_2}{R_1 R_2} \quad \text{and so} \quad R = \frac{R_1 R_2}{R_1 + R_2}$$

ELECTRIC POWER

The rate at which work is done to maintain an electric current is given by the product of the current I and the potential difference V:

$$P = IV$$

Power = current × potential difference

When I is in amperes and V is in volts, P will be in watts.

If the conductor or device through which a current passes obeys Ohm's law, the power consumed may be expressed in the alternative forms

$$P = IV = I^2R = \frac{V^2}{R}$$

Solved Problems

13.1. Since electric current is a flow of charge, why are two wires rather than a single one used to carry current?

    If a single wire were used, charge of one sign or the other (depending upon the situation) would be permanently transferred from the source of current to the appliance at the far end of the wire. In a short time so much charge would have been transferred that the source would be unable to shift further charge against the repulsive force of the charge piled up at the appliance. Thus a single wire cannot carry a current continuously. The use of two wires, on the other hand, enables charge to be circulated from source to appliance and back, so that a continuous one-way flow of energy can take place.

13.2. Which solids are good electrical conductors and which are good insulators? How well do these substances conduct heat?

    All metals are good electrical conductors. All nonmetallic solids are good insulators, for instance glass, wood, plastics, rubber. In general, solids that are good conductors of electricity are also good conductors of heat, and solids that are good electrical insulators are poor conductors of heat. Metals are good conductors of heat and electricity because both are transferred through a metal by the freely moving electrons that are a characteristic feature of its structure.

13.3. To reduce the brightness of a light bulb, should an auxiliary resistance be connected in series with it or in parallel?

    In series, because in this way the current in the bulb is reduced; if the resistor were connected in parallel with the bulb, the current in the bulb and hence its brightness would not be affected.

13.4. Prove that $P = IV$ in an electric circuit.

    The work needed to take a charge Q through the potential difference V is $W = QV$. A current I involves the flow of the charge Q in the time t, so that $I = Q/t$ and $Q = It$. Hence

$$W = QV = IVt$$

Since power is the rate at which work is done,

$$P = \frac{W}{t} = \frac{IVt}{t} = IV$$

13.5. A wire carries a current of 1 A. How many electrons pass any point in the wire each second?

    The electron charge is of magnitude $e = 1.6 \times 10^{-19}$ C and so a current of 1 A = 1 C/s corresponds to a flow of

$$\frac{1 \text{ C/s}}{1.6 \times 10^{-19} \text{ C/electron}} = 6.3 \times 10^{18} \text{ electrons/s}$$

13.6. A 120-V toaster has a resistance of 12 Ω. What must be the minimum rating of the fuse in the electrical circuit to which the toaster is connected?

    The current in the toaster is

$$I = \frac{V}{R} = \frac{120 \text{ V}}{12 \text{ }\Omega} = 10 \text{ A}$$

so this must be the rating of the fuse.

13.7. A 120-V electric heater draws a current of 25 A. What is its resistance?

$$R = \frac{V}{I} = \frac{120 \text{ V}}{25 \text{ A}} = 4.8 \text{ }\Omega$$

13.8. It is desired to limit the current in a 50-Ω resistor to 10 A when it is connected to a 600-V power source. How should an auxiliary resistor be connected in the circuit and what should its resistance be?

For a current of 10 A, the total resistance in the circuit should be

$$R = \frac{V}{I} = \frac{600 \text{ V}}{10 \text{ A}} = 60 \ \Omega$$

Hence a 10-Ω resistor should be connected in series with the 50-Ω resistor to give a total of 60 Ω.

13.9. (*a*) What is the equivalent resistance of three 5-Ω resistors connected in series? (*b*) If a potential difference of 60 V is applied across the combination, what is the current in each resistor?

(*a*) $$R = R_1 + R_2 + R_3 = 5 \ \Omega + 5 \ \Omega + 5 \ \Omega = 15 \ \Omega$$

(*b*) The current in the entire circuit is

$$I = \frac{V}{R} = \frac{60 \text{ V}}{15 \ \Omega} = 4 \text{ A}$$

Since the resistors are in series, this current flows through each of them.

13.10. (*a*) What is the equivalent resistance of three 5-Ω resistors connected in parallel? (*b*) If a potential difference of 60 V is applied across the combination, what is the current in each resistor?

(*a*) $$\frac{1}{R} = \frac{1}{R_1} + \frac{1}{R_2} + \frac{1}{R_3} = \frac{1}{5 \ \Omega} + \frac{1}{5 \ \Omega} + \frac{1}{5 \ \Omega} = \frac{3}{5 \ \Omega}$$

$$R = \frac{5}{3} \ \Omega = 1.67 \ \Omega$$

(*b*) Since each resistor has a potential difference of 60 V across it, the current in each one is

$$I = \frac{V}{R} = \frac{60 \text{ V}}{5 \ \Omega} = 12 \text{ A}$$

13.11. The starting motor of a certain car develops 1 hp when it turns over the engine. How much current does it draw from a 12-V battery?

Here $P = 1 \text{ hp} = 746 \text{ W} = IV$, and so

$$I = \frac{P}{V} = \frac{746 \text{ W}}{12 \text{ V}} = 62 \text{ A}$$

13.12. A 12-V storage battery is charged by a current of 20 A for 1 hr. (*a*) How much power is required to charge the battery at this rate? (*b*) How much energy has been provided during the process?

(*a*) $$P = IV = 20 \text{ A} \times 12 \text{ V} = 240 \text{ W}$$

(*b*) $$W = Pt = 240 \text{ W} \times 3600 \text{ s} = 8.64 \times 10^5 \text{ J}$$

13.13. The 12-V battery of a certain car has a capacity of 80 A-hr, which means that it can furnish a current of 80 A for 1 hr, a current of 40 A for 2 hr, and so forth. (*a*) How much energy is stored in the battery? (*b*) If the car's lights require 60 W of power, how long can the battery keep them lit when the engine (and hence its generator) is not running?

(*a*) The 80 A-hr capacity of the battery is a way to express the amount of charge it can transfer from one of its terminals to the other. Here the amount of charge is

$$Q = 80 \text{ A-hr} \times 3600 \text{ s/hr} = 2.88 \times 10^5 \text{ A-s} = 2.88 \times 10^5 \text{ C}$$

and so the energy the battery can provide is

$$W = QV = 2.88 \times 10^5 \text{ C} \times 12 \text{ V} = 3.46 \times 10^6 \text{ J}$$

(b) Since $P = \dfrac{W}{t}$, $t = \dfrac{W}{P} = \dfrac{3.46 \times 10^6 \text{ J}}{60 \text{ W}} = 5.8 \times 10^4 \text{ s} = 16 \text{ hr}$

13.14. Two 240-Ω light bulbs are connected in series with a 120-V power source.   (a) What is the current in each bulb?   (b) How much power does each bulb dissipate?

(a) The equivalent resistance of the two bulbs is

$$R = R_1 + R_2 = 240 \ \Omega + 240 \ \Omega = 480 \ \Omega$$

The current in the circuit is therefore

$$I = \frac{V}{R} = \frac{120 \text{ V}}{480 \ \Omega} = 0.25 \text{ A}$$

and, since the bulbs are in series, this current passes through each of them.

(b) $P = I^2R = (0.25 \text{ A})^2 \times 240 \ \Omega = 15 \text{ W}$

13.15. Two 240-Ω light bulbs are connected in parallel to a 120-V power source.   (a) What is the current in each bulb?   (b) How much power does each bulb dissipate?

(a) The potential difference across each bulb is 120 V.   Hence the current in each bulb is

$$I = \frac{V}{R} = \frac{120 \text{ V}}{240 \ \Omega} = 0.5 \text{ A}$$

(b) $P = I^2R = (0.5 \text{ A})^2 \times 240 \ \Omega = 60 \text{ W}$

Supplementary Problems

13.16.   Bends in a pipe slow down the flow of water through it.   Do bends in a wire increase its electrical resistance?

13.17.   A number of light bulbs are to be connected to a single power outlet.   Will they provide more illumination if connected in series or in parallel?   Why?

13.18.   How many electrons pass through the filament of a 75-W, 120-V light bulb per second?

13.19.   Find the current in a 200-Ω resistor when the potential difference across it is 40 V.

13.20.   An electric water heater draws 10 A of current from a 240-V power line.   What is its resistance?

13.21.   It is desired to have a current of 20 A in a 5-Ω resistor when it is connected to an 80-V battery. Is there any way in which an auxiliary resistor can be connected in the circuit to increase the current in the 5-Ω resistor to this value?   If so, what should its resistance be?

13.22.   (a) Find the equivalent resistance of four 60-Ω resistors connected in series.   (b) If a potential difference of 12 V is applied across the combination, what is the current in each resistor?

13.23.   (a) Find the equivalent resistance of four 60-Ω resistors connected in parallel.   (b) If a potential difference of 12 V is applied across the combination, what is the current in each resistor?

13.24.   You have three 2-Ω resistors.   List the various resistances you can provide with them.

13.25. How much power is developed by an electric motor which draws a current of 4 A when operated at 240 V? How many horsepower is this?

13.26. What is the current in a 100-W light bulb when it is operated at 120 V?

13.27. A 32-V storage battery has a capacity of 10^6 J. How long can it supply a current of 5 A?

13.28. The 12-V battery of a car is required to be able to operate the 1.5-kW starting motor for a total of at least 10 min. (*a*) What should the minimum capacity of the battery be (in A-hr)? (*b*) How much energy is stored in such a battery?

13.29. A light bulb whose power is 100 W when operated at 240 V is instead connected to a 120-V source. (*a*) What is the current in the bulb? (*b*) How much power does it dissipate?

13.30. A 100-Ω resistor and a 200-Ω resistor are connected in series with a 40-V power source. (*a*) What is the current in each resistor? (*b*) How much power does each one dissipate?

13.31. A 100-Ω resistor and a 200-Ω resistor are connected in parallel with a 40-V power source. (*a*) What is the current in each resistor? (*b*) How much power does each one dissipate?

13.32. Currents of 5 A pass through two resistors, one of which has a potential difference of 100 V across it and the other of which has a potential difference of 300 V across it. (*a*) Compare the rates at which charge passes through each resistor. (*b*) Compare the rates at which energy is dissipated by each resistor.

Answers to Supplementary Problems

13.16. Bends in a wire have no effect on its electrical resistance because the electrons whose motion constitutes an electric current are extremely small and have very little mass and therefore can change direction readily.

13.17. In parallel, because this way each bulb has the maximum voltage across it and hence the maximum power dissipation.

13.18. 3.9×10^{18} electrons/s   **13.19.** 0.2 A   **13.20.** 24 Ω

13.21. There is no way in which an auxiliary resistor can be connected to increase the current in the 5-Ω resistor.

13.22. 240 Ω; 0.05 A   **13.27.** 6250 s = 1 hr 44 min 10 s

13.23. 15 Ω; 0.2 A   **13.28.** 21 A-hr; 9.0×10^5 J

13.24. 0.67 Ω; 1 Ω; 2 Ω; 3 Ω; 4 Ω; 6 Ω **13.29.** 0.21 A; 25 W

13.25. 960 W; 1.3 hp   **13.30.** 0.133 A; 0.133 A; 1.78 W; 3.55 W

13.26. 0.83 A   **13.31.** 0.4 A; 0.2 A; 16 W; 8 W

13.32. (*a*) Since the currents are the same, charge flows at the same rate through each resistor. (*b*) Since $P = IV$, energy is dissipated by the second resistor three times as fast as by the first resistor.

Chapter 14

Magnetism

THE NATURE OF MAGNETISM

Two electric charges at rest exert forces on each other according to Coulomb's law. When the charges are in motion, the forces are different, and it is customary to attribute the differences to *magnetic forces* that occur between moving charges in addition to the electric forces between them. In this interpretation, the total force on a charge Q at a certain time and place can be divided into two parts: an electric force that depends only upon the value of Q, and a magnetic force that depends upon the velocity v of the charge as well as upon Q.

In reality there is only a single interaction between charges, the *electromagnetic interaction*. The theory of relativity provides the link between electric and magnetic forces: just as the mass of an object moving with respect to an observer is greater than when it is at rest, so the electric force between two charges appears altered to an observer when the charges are moving with respect to him. Magnetism is not distinct from electricity in the way that, for example, gravitation is. Despite the unity of the electromagnetic interaction, however, it is convenient for many purposes to treat electric and magnetic effects separately.

MAGNETIC FIELD

A *magnetic field* **B** is present wherever a magnetic force acts on a moving charge. The direction of **B** at a certain place is that along which a charge can move without experiencing a magnetic force; along any other direction the charge would be acted upon by such a force. The magnitude of **B** is equal numerically to the force on a charge of 1 C moving at 1 m/s perpendicular to **B**.

The unit of magnetic field is the *tesla* (T), where

$$1 \text{ tesla} = 1 \frac{\text{newton}}{\text{ampere-meter}}$$

The tesla is sometimes called the *weber/m²*. The *gauss*, equal to 10^{-4} T, is another unit of magnetic field sometimes used.

MAGNETIC FIELD OF A STRAIGHT CURRENT

The magnetic field a distance s from a long, straight current I has the magnitude

$$B = K \frac{I}{s}$$

where K is a constant whose value is

$$K = 2.00 \times 10^{-7} \text{ N/A}^2$$

The lines of force of the field are in the form of concentric circles around the current. To find the direction of **B**, place the thumb of the right hand in the direction of the current; the curled fingers of that hand then point in the direction of **B** (Fig. 14-1).

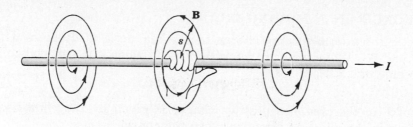

Fig. 14-1

MAGNETIC FIELD OF A CURRENT LOOP

The magnetic field at the center of a current loop whose radius is r has the magnitude

$$B = \pi K \frac{I}{r}$$

The lines of force of **B** are perpendicular to the plane of the loop, as shown in Fig. 14-2(a). To find the direction of **B**, grasp the loop so the curled fingers of the right hand point in the direction of the current; the thumb of that hand then points in the direction of **B** [Fig. 14-2(b)].

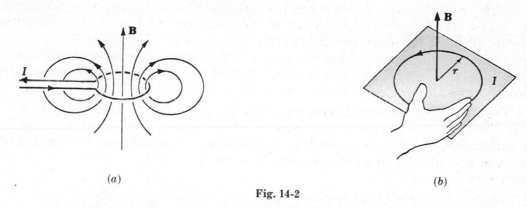

(a)   (b)

Fig. 14-2

A *solenoid* is a coil consisting of many loops of wire. If the turns are close together and the solenoid is long compared with its diameter, the magnetic field inside it is uniform and parallel to the axis with the magnitude

$$B = 2\pi K \frac{N}{L} I$$

In this formula N is the number of turns, L is the length of the solenoid, and I is the current. The direction of **B** is as shown in Fig. 14-3.

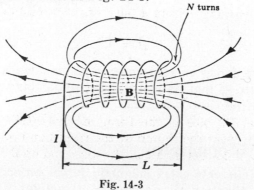

Fig. 14-3

MAGNETIC FORCE ON A MOVING CHARGE

The magnetic force on a moving charge Q in a magnetic field varies with the relative directions of $\mathbf{v}$ and $\mathbf{B}$. When $\mathbf{v}$ is parallel to $\mathbf{B}$, $F = 0$; when $\mathbf{v}$ is perpendicular to $\mathbf{B}$, F has its maximum value of

$$F = QvB \quad (\mathbf{v} \perp \mathbf{B})$$

The direction of $\mathbf{F}$ in the case of a positive charge is given by the right-hand rule shown in Fig. 14-4; $\mathbf{F}$ is in the opposite direction when the charge is negative.

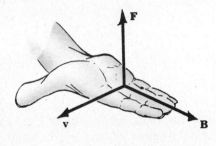
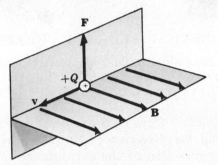

Fig. 14-4

MAGNETIC FORCE ON A CURRENT

Since a current consists of moving charges, it follows that a current-carrying wire will experience no force when parallel to a magnetic field $\mathbf{B}$ and maximum force when perpendicular to $\mathbf{B}$. In the latter case, F has the value

$$F = ILB \quad (\mathbf{I} \perp \mathbf{B})$$

where I is the current and L is the length of wire in the magnetic field. The direction of the force is as shown in Fig. 14-5.

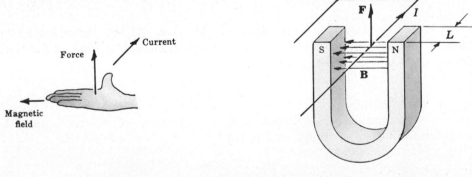

Fig. 14-5

Owing to the different forces exerted on each of its sides, a current loop in a magnetic field always tends to rotate so that its plane is perpendicular to $\mathbf{B}$. This is the effect that underlies the operation of all electric motors.

FORCE BETWEEN TWO CURRENTS

Two parallel electric currents exert magnetic forces on each other (Fig. 14-6). If the currents are in the same direction, the forces are attractive; if the currents are in opposite directions, the forces are repulsive.   The force per unit length F/L on each current depends upon the currents I_1 and I_2 and their separation s:

$$\frac{F}{L} = K\frac{I_1 I_2}{s}$$

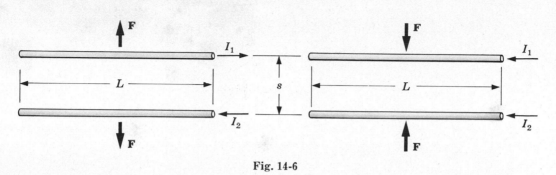

Fig. 14-6

FERROMAGNETISM

The magnetic field produced by a current is altered by the presence of a substance of any kind.   Usually the change, which may be an increase or a decrease in **B**, is very small, but in certain cases there is an increase in **B** by hundreds or thousands of times.   Substances that have the latter effect are called *ferromagnetic*; iron and iron alloys are familiar examples.   An *electromagnet* is a solenoid with a ferromagnetic core to increase its magnetic field.

Ferromagnetism is a consequence of the magnetic properties of the electrons all atoms contain.   An electron behaves in some respects as though it is a spinning charged sphere, and is therefore magnetically equivalent to a tiny current loop.   In most substances the magnetic fields of the atomic electrons cancel each other out, but in ferromagnetic substances the cancellation is not complete and each atom has a certain magnetic field of its own. The atomic magnetic fields align themselves with an external magnetic field to produce a much stronger total **B**.   When the external field is removed, the atomic magnetic fields may remain aligned to produce a *permanent magnet*.   The field of a bar magnet has the same form as that of a solenoid because both fields are due to parallel current loops (Fig. 14-7).

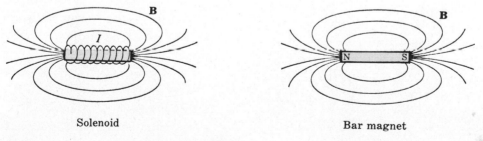

Solenoid   Bar magnet

Fig. 14-7

THE EARTH'S MAGNETIC FIELD

The earth has a magnetic field which arises from electric currents in its liquid iron core. The field is like that which would be produced by a current loop centered a few hundred miles from the earth's center whose plane is tilted by 11° from the plane of the equator (Fig. 14-8).   The *geomagnetic poles* are the points where the magnetic axis passes through the earth's surface.   The magnitude of the earth's magnetic field varies from place to place; a typical sea-level value is 3×10^{-5} T.

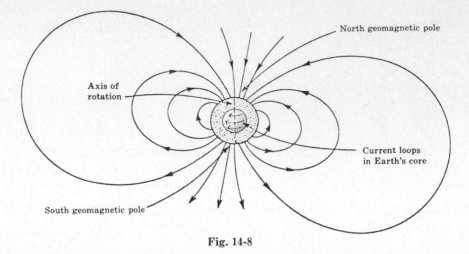

Fig. 14-8

Solved Problems

14.1. In what ways are electric and magnetic fields similar? In what ways are they different?

Similarities. Both fields originate in electric charges, and both fields can exert forces on electric charges.

Differences. All electric charges give rise to electric fields, but only a charge in motion relative to an observer gives rise to a magnetic field. Electric fields exert forces on all charges, but magnetic fields only exert forces on moving charges.

14.2. A positive charge is moving vertically upward when it enters a magnetic field directed to the north. In what direction is the force on the charge?

To apply the right-hand rule here, the fingers of the right hand are pointed north and the thumb of that hand is pointed upward. The palm of the hand faces west, which is therefore the direction of the force on the charge.

14.3. A stream of protons is moving parallel to a stream of electrons. Do the streams tend to come together or to move apart?

The electric force between the streams is attractive but the magnetic force is repulsive. Which of the forces is stronger depends upon how fast the particles are moving.

14.4. The ends of a bar magnet are traditionally called its "poles," with the end that tends to point north called the "north pole" and the end that tends to point south called the

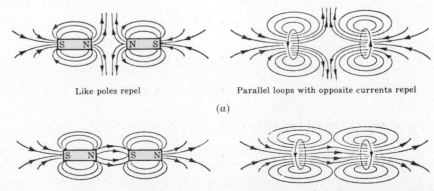

Like poles repel   Parallel loops with opposite currents repel

(a)

Unlike poles attract   Parallel loops with similar currents attract

(b)

Fig. 14-9

"south pole." It is observed that like poles of nearby magnets repel each other and that unlike poles attract. Explain this behavior in terms of the interaction of current loops.

Bar magnets with like poles facing each other are equivalent to parallel current loops whose currents are in opposite directions [Fig. 14-9(a)]. Such loops repel. Bar magnets with opposite poles facing each other are equivalent to parallel current loops whose currents are in the same direction [Fig. 14-9(b)]. Such loops attract.

14.5. How does a permanent magnet attract an unmagnetized iron object?

The presence of the magnet induces the atomic magnets in the object to line up with its field (Fig. 14-10), and the attraction of opposite poles then produces a net force on the object.

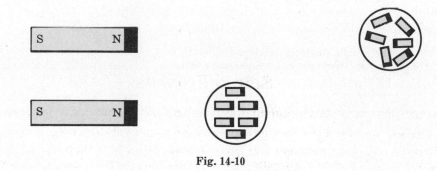

Fig. 14-10

14.6. A proton is moving in a uniform magnetic field. Describe the path of the proton if its initial direction is (a) parallel to the field, (b) perpendicular to the field, and (c) at an intermediate angle to the field.

(a) There is no magnetic force on the proton so it continues to move in a straight line [Fig. 14-11(a)].

(b) The force on the proton is perpendicular to its velocity **v** and also perpendicular to **B**, hence it moves in a circle [Fig. 14-11(b)].

(c) The proton moves in a helical path [Fig. 14-11(c)] because the component of **v** parallel to **B** is not changed while the component of **v** perpendicular to **B** leads to an inward force as in (b).

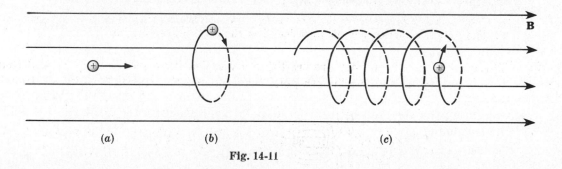

(a)   (b)   (c)

Fig. 14-11

14.7. A charged particle moving perpendicular to a uniform magnetic field follows a circular path. Find the radius of the circle.

The magnetic force QvB on the particle provides the centripetal force mv^2/r that keeps it moving in a circle of radius r.   Hence

$$F_{\text{magnetic}} = F_{\text{centripetal}}$$

$$QvB = \frac{mv^2}{r}$$

$$r = \frac{mv}{QB}$$

The radius is directly proportional to the particle's momentum mv and inversely proportional to its charge Q and the magnetic field B.

14.8. A cable 5 m above the ground carries a current of 100 A from east to west. Find the direction and magnitude of the magnetic field on the ground directly beneath the cable. (Neglect the earth's magnetic field.)

From the right-hand rule, the direction of the field is south. The magnitude of the field is

$$B = K\frac{I}{s} = 2 \times 10^{-7} \text{ N/A}^2 \times \frac{100 \text{ A}}{5 \text{ m}} = 4 \times 10^{-6} \text{ T}$$

14.9. Two parallel wires 10 cm apart carry currents in the same direction of 8 A. What is the magnetic field halfway between them?

The magnetic field halfway between the wires is 0 because the fields of the currents are opposite in direction and have the same magnitude there.

14.10. Two parallel wires 10 cm apart carry currents in opposite directions of 8 A. What is the magnetic field halfway between them?

Here the magnetic fields of the two currents are in the same direction and hence add together. Since $s = 5 \text{ cm} = 5 \times 10^{-2}$ m halfway between the wires, the field of each current there is

$$B = K\frac{I}{s} = \frac{2 \times 10^{-7} \text{ N/A}^2 \times 8 \text{ A}}{5 \times 10^{-2} \text{ m}} = 3.2 \times 10^{-5} \text{ T}$$

and the total field is $2B = 6.4 \times 10^{-5}$ T.

14.11. A 100-turn flat circular coil has a radius of 5 cm. Find the magnetic field at the center of the coil when the current is 4 A.

Each turn of the coil acts as a separate loop in contributing to the total magnetic field. If there are N turns, the result is a field N times stronger than each turn produces by itself. Hence

$$B = \frac{\pi K N I}{r} = \frac{\pi \times 2 \times 10^{-7} \text{ N/A}^2 \times 100 \times 4 \text{ A}}{5 \times 10^{-2} \text{ m}} = 5.03 \times 10^{-3} \text{ T}$$

14.12. A solenoid 0.2 m long has 1000 turns of wire and is oriented with its axis parallel to the earth's magnetic field at a place where the latter is 2.5×10^{-5} T. What should the current in the solenoid be in order that its field exactly cancel the earth's field inside the solenoid?

The magnetic field inside a solenoid is

$$B = 2\pi K \frac{N}{L} I$$

Here $N = 10^3$, $L = 0.2$ m, and $B = 2.5 \times 10^{-5}$ T, so

$$I = \frac{BL}{2\pi KN} = \frac{2.5 \times 10^{-5} \text{ T} \times 0.2 \text{ m}}{2\pi \times 2 \times 10^{-7} \text{ N/A}^2 \times 10^3} = 0.004 \text{ A}$$

14.13. In a certain electric motor wires that carry a current of 5 A are perpendicular to a magnetic field of 0.8 T. What is the force on each cm of these wires?

$$F = ILB = 5 \text{ A} \times 0.01 \text{ m} \times 0.8 \text{ T} = 0.04 \text{ N}$$

14.14. The wires that supply current to a 120-V, 2-kW electric heater are 2 mm apart. What is the force per meter between the wires?

Since $P = IV$ the current in the wires is

$$I = I_1 = I_2 = \frac{P}{V} = \frac{2000 \text{ W}}{120 \text{ V}} = 16.7 \text{ A}$$

Since $s = 2$ mm $= 2 \times 10^{-3}$ m, the force between the wires is

$$\frac{F}{L} = \frac{KI_1I_2}{s} = \frac{2 \times 10^{-7} \text{ N/A}^2 \times (16.7 \text{ A})^2}{2 \times 10^{-3} \text{ m}} = 0.028 \text{ N/m}$$

The currents are in opposite directions, so the force is repulsive.

Supplementary Problems

14.15. An observer is able to measure electric, magnetic, and gravitational fields. Which of these does he detect when (a) a proton moves past him, and (b) he moves past a proton?

14.16. A beam of electrons that are moving slowly at first are accelerated to higher and higher velocities. What happens to the diameter of the beam as this happens?

14.17. An electric current is flowing south along a power line. What is the direction of the magnetic field above it? Below it?

14.18. A charged particle moves through a magnetic field perpendicular to **B**. Is the particle's energy affected? Is its momentum?

14.19. In a sketch of magnetic lines of force, what is the significance of lines of force that are closer together in a particular region than they are elsewhere?

14.20. A negative charge is moving west when it enters a magnetic field directed vertically downward. In what direction is the force on the charge?

14.21. A wire carrying a current is placed in a magnetic field **B**. (a) Under what circumstances, if any, will the force on the wire be 0? (b) Under what circumstances will the force on the wire be a maximum?

14.22. Under what circumstances, if any, does a current-carrying wire loop not tend to rotate in a magnetic field?

14.23. The magnetic field 5 cm from a certain straight wire is 10^{-4} T. Find the current in the wire.

14.24. How far away from a compass should a wire carrying a 1-A current be located if its magnetic field at the compass is not to exceed 1 percent of the earth's magnetic field, which is typically 3×10^{-5} T?

14.25. Two parallel wires 20 cm apart carry currents in the same direction of 5 A. Find the magnetic field between the wires 5 cm from one of them and 15 cm from the other.

14.26. What should the current be in a wire loop 1 cm in radius if the magnetic field in the center of the loop is to be 0.01 T?

14.27. What is the magnetic field inside a solenoid wound with 20 turns/cm when the current in it is 5 A?

14.28. An electron in a television picture tube travels at 3×10^7 m/s. Does the earth's gravitational field or its magnetic field exert the greater force on the electron? Assume **v** is perpendicular to **B**.

14.29.   (a) A wire 1 m long is perpendicular to a magnetic field of 0.01 T. What is the force on the wire when it carries a current of 10 A? (b) What is the force on the wire when it is parallel to the magnetic field?

14.30.   A horizontal wire 10 cm long whose mass is 1 g is to be supported magnetically against the force of gravity. The current in the wire is 10 A and goes from north to south. (a) What should be the direction of the magnetic field? (b) What should its magnitude be?

14.31.   The starting motor of a certain car is connected to the battery by a pair of cables that are 8 mm apart for a distance of 50 cm. Find the force between the cables when the current in them is 100 A.

Answers to Supplementary Problems

14.15.   All three fields are detected in both cases.

14.16.   At first the mutual electric repulsion of the electrons causes the beam diameter to increase, but as they go faster the magnetic attraction becomes more significant and the beam diameter decreases.

14.17.   west; east

14.18.   The particle's energy is not changed since the magnetic force on it is perpendicular to its direction of motion and so no work is done on it by the field. The particle's direction changes, however, and hence its momentum, which is a vector quantity, also changes.

14.19.   The closer together the lines of force are drawn in a particular region, the stronger the field is there.

14.20.   To the north.

14.21.   When the wire is parallel to **B**; when it is perpendicular to **B**.

14.22.   Such a loop does not tend to rotate when its plane is perpendicular to the direction of the magnetic field.

14.23.   25 A

14.24.   67 cm

14.25.   Since the fields are in opposite directions,
$$B \;=\; 2 \times 10^{-5} \text{ T} - 0.67 \times 10^{-5} \text{ T} \;=\; 1.33 \times 10^{-5} \text{ T}$$

14.26.   159 A

14.27. 1.26×10^{-2} T

14.28.   The magnetic force evB is more than 10^{13} times greater than the gravitational force mg.

14.29.   0.1 N; 0

14.30.   The direction of the field should be toward the west in order that the force on the wire be upward; 9.8×10^{-3} T

14.31.   0.125 N; the force is repulsive

Chapter 15

Electromagnetic Induction

ELECTROMAGNETIC INDUCTION

A current is produced in a conductor whenever it cuts across magnetic lines of force. If the motion is parallel to the lines of force, there is no effect. It is not even necessary for there to be relative motion of a wire and a source of magnetic field, since a magnetic field whose strength is changing has moving lines of force associated with it and a current will be induced in a wire that is in the path of these moving lines of force.

Electromagnetic induction originates in the force a magnetic field exerts on a moving charge. When a wire moves across a magnetic field, the electrons it contains experience sidewise forces which push them along the wire to produce a current.

THE GENERATOR

In a *generator,* a coil of wire is rotated in a magnetic field to produce an electric current. In this way mechanical energy is converted into electrical energy.

When a wire loop is rotated about a diameter in a magnetic field, each half of the loop moves up and then down through the field. As a result the current induced in the loop reverses its direction twice per complete turn. Such a constantly changing current is an *alternating current,* and all generators produce alternating current. If direct current is required, either the output of the generator can be switched back and forth automatically as the coil turns, or a device called a *rectifier* can be used which permits current to pass through it only in one direction. Both approaches are widely used.

ALTERNATING CURRENT

The *frequency* of an alternating current is the number of complete back-and-forth cycles it goes through each second. The unit of frequency is the cycle/s, which has been given the name *hertz* (Hz). Thus a current that undergoes 60 cycles/s has a frequency of 60 Hz.

Because an alternating current varies continuously (Fig. 15-1), its maximum value $\pm I_{max}$ does not indicate its ability to do work or to produce heat as the magnitude of a direct current does. Instead it is customary to refer to the *effective current*

$$I = \frac{I_{max}}{\sqrt{2}} = 0.707\, I_{max}$$

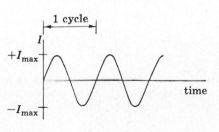

Fig. 15-1

which is such that a direct current of this value does as much work or produces as much heat as the alternating current whose maximum value is $\pm I_{max}$. Similarly the effective potential difference in an alternating-current circuit is

$$V = \frac{V_{max}}{\sqrt{2}} = 0.707\, V_{max}$$

89

THE TRANSFORMER

A *transformer* consists of two coils of wire, usually wound on an iron core. When an alternating current is passed through one of the windings, the changing magnetic field it gives rise to induces an alternating current in the other winding. The potential difference per turn is the same in both primary and secondary windings, so the ratio of turns in the windings determines the ratio of voltages across them:

$$\frac{V_1}{V_2} = \frac{N_1}{N_2}$$

$$\frac{\text{Primary voltage}}{\text{Secondary voltage}} = \frac{\text{primary turns}}{\text{secondary turns}}$$

Since the power $I_1 V_1$ going into a transformer must equal the power $I_2 V_2$ going out, where I_1 and I_2 are the primary and secondary currents respectively, the ratio of currents is inversely proportional to the ratio of turns:

$$\frac{I_1}{I_2} = \frac{N_2}{N_1}$$

$$\frac{\text{Primary current}}{\text{Secondary current}} = \frac{\text{secondary turns}}{\text{primary turns}}$$

Solved Problems

15.1. According to *Lenz's law*, an induced current is always in such a direction that its own magnetic field tends to oppose the effect that brought it about. For example, when a bar magnet is passed through a wire loop, a current is produced in the loop whose own magnetic field is opposite to the field of the magnet. How is Lenz's law related to the principle of conservation of energy?

If the induced current were in such a direction that its own magnetic field would add to the field of external origin, this new field would increase the rate of change of the total field, and more and more current would be induced in the loop even if the external field were to stop changing. But energy is associated with every current, and to have an induced current increasing by itself with no external agency to provide additional energy means violating the principle of conservation of energy.

15.2. What would happen if the primary winding of a transformer were connected to a battery?

When the connection is made, there will be a momentary current in the secondary winding as the current in the primary builds up to its final value. Afterward, since the primary current will be constant and hence its magnetic field will not change, there will be no current in the secondary.

15.3. Alternating current with a maximum value of 10 A is passed through a 20-Ω resistor. At what rate does the resistor dissipate energy?

The effective current is

$$I = 0.707\, I_{\max} = 0.707 \times 10 \text{ A} = 7.07 \text{ A}$$

and so the power dissipated is

$$P = I^2 R = (7.07 \text{ A})^2 \times 20\ \Omega = 100 \text{ W}$$

15.4. Alternating current is in wide use chiefly because its voltage can be so easily changed by a transformer. Since $P = IV$, the higher the voltage, the lower the current, and vice versa. In transmitting electrical energy through long distances, a small current is desirable in order to minimize energy loss to heat, which is equal to I^2R where R is the resistance of the transmission line. On the other hand, both the generation and final use of electrical energy is best accomplished at moderate potential differences. Hence electricity is typically generated at 10,000 V or so, stepped up by transformers at the power station to 500,000 V or even more for transmission, and near the point of consumption other transformers reduce the potential difference to 240 V or 120 V. To verify the advantage of high-voltage transmission, find the rate of energy loss to heat when a 5-Ω cable is used to transmit 1 kW of electricity at 100 V and at 100,000 V.

Since $P = IV$, the currents in the cable are respectively

$$I_A = \frac{P}{V_A} = \frac{1000 \text{ W}}{100 \text{ V}} = 10 \text{ A}$$

$$I_B = \frac{P}{V_B} = \frac{1000 \text{ W}}{100,000 \text{ V}} = 0.01 \text{ A}$$

The rates of heat production per kW are respectively

$$I_A^2 R = (10 \text{ A})^2 \times 5 \text{ }\Omega = 500 \text{ W}$$

$$I_B^2 R = (0.01 \text{ A})^2 \times 5 \text{ }\Omega = 0.0005 \text{ W} = 5 \times 10^{-4} \text{ W}$$

Transmission at 100 V therefore means 10^6 — a million — times more energy lost as heat than does transmission at 100,000 V.

15.5. A transformer has 100 turns in its primary winding and 500 turns in its secondary winding. If the primary voltage and current are respectively 120 V and 3 A, what are the secondary voltage and current?

$$V_2 = \frac{N_2}{N_1}V_1 = \frac{500 \text{ turns}}{100 \text{ turns}} \times 120 \text{ V} = 600 \text{ V}$$

$$I_2 = \frac{N_1}{N_2}I_1 = \frac{100 \text{ turns}}{500 \text{ turns}} \times 3 \text{ A} = 0.6 \text{ A}$$

15.6. A transformer rated at a maximum power of 10 kW is used to connect a 5000-V transmission line to a 240-V circuit. (*a*) What is the ratio of turns in the windings of the transformer? (*b*) What is the maximum current in the 240-V circuit?

(*a*)
$$\frac{N_1}{N_2} = \frac{V_1}{V_2} = \frac{5000 \text{ V}}{240 \text{ V}} = 20.8$$

(*b*) Since $P = IV$, here
$$I_2 = \frac{P}{V_2} = \frac{10,000 \text{ W}}{240 \text{ V}} = 41.7 \text{ A}$$

15.7. A transformer connected to a 120-V AC power line has 200 turns in its primary winding and 50 turns in its secondary winding. The secondary is connected to a 100-Ω light bulb. How much current is drawn from the 120-V power line?

The voltage across the secondary is

$$V_2 = \frac{N_2}{N_1}V_1 = \frac{50 \text{ turns}}{200 \text{ turns}} \times 120 \text{ V} = 30 \text{ V}$$

and so the current in the secondary circuit is

$$I_2 = \frac{V_2}{R} = \frac{30 \text{ V}}{100 \text{ }\Omega} = 0.3 \text{ A}$$

Hence the current in the primary circuit is

$$I_1 = \frac{N_2}{N_1}I_2 = \frac{50 \text{ turns}}{200 \text{ turns}} \times 0.3 \text{ A} = 0.075 \text{ A}$$

Supplementary Problems

15.8. Why is it harder to rotate the shaft of a generator when it is connected to an external circuit than when it is not?

15.9. What is it whose action on the secondary winding of a transformer causes an alternating potential difference to occur across its ends even though there is no connection between the primary and secondary windings?

15.10. How many times per second does a 60-Hz alternating current reverse its direction?

15.11. Find the maximum voltage in a 120-V AC power line.

15.12. A transformer has 50 turns in its primary winding and 100 turns in its secondary. (a) If a 60-Hz, 3-A current passes through the primary winding, what is the nature and magnitude of the current in the secondary? (b) If a 3-A direct current passes through the primary winding, what is the nature and magnitude of the current in the secondary?

15.13. The transformer in an electric welding machine draws 3 A from a 240-V AC power line and delivers 400 A. What is the potential difference across the secondary of the transformer?

15.14. A 240-V, 400-W electric mixer is connected to a 120-V power line through a transformer. (a) What is the ratio of turns in the transformer? (b) How much current is drawn from the power line?

Answers to Supplementary Problems

15.8. When the generator is connected to an external circuit, turning its shaft provides power to the circuit, and work must be done to provide this power.

15.9. The changing magnetic field produced by an alternating current in the primary winding.

15.10. 120 times per second

15.11. 170 V

15.12. 60 Hz, 1.5 A; no current

15.13. 1.8 V

15.14. 2 : 1; 3.3 A

Chapter 16

Waves

WAVES

A *wave* is, in general, a disturbance that moves through a medium. A wave carries energy, but there is no transport of matter. In a *periodic wave*, pulses of the same kind follow one another in regular succession.

In a *transverse wave*, the particles of the medium move back and forth perpendicular to the direction of the wave. Waves that travel down a stretched string when one end is shaken are transverse (Fig. 16-1).

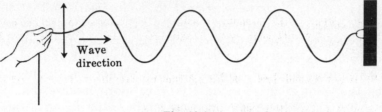

Fig. 16-1

In a *longitudinal wave*, the particles of the medium move back and forth in the same direction as the wave. Waves that travel down a coil spring when one end is pulled out are longitudinal (Fig. 16-2).

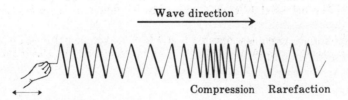

Fig. 16-2

Water waves are a combination of longitudinal and transverse waves. Each particle near the surface moves in a circular orbit, so that a succession of crests and troughs occurs. As in all types of wave motion, there is no net motion of matter from one place to another.

FREQUENCY AND WAVELENGTH

The *frequency f* of a periodic wave is the number of waves that pass a given point per second. The unit of frequency is the *hertz* (Hz), where

$$1 \text{ Hz} = 1 \text{ cycle/s}$$

Multiples of the hertz are the kilohertz (kHz) and megahertz (mHz), equal respectively to 10^3 Hz and 10^6 Hz.

The wavelength λ (Greek letter *lambda*) of a periodic wave is the distance between adjacent wave crests. The frequency and wavelength are related to the wave velocity by the formula

$$v = f\lambda$$

Wave velocity = frequency × wavelength

SOUND

Sound waves are longitudinal waves in which alternate regions of compression and rarefaction move away from a source. Sound waves can travel through solids, liquids, and gases. The velocity of sound is a constant for a given material at a given pressure and temperature; in air at 1 atm pressure and 0 °C it is 331 m/s = 1086 ft/s.

The *intensity* of a sound wave (or any other kind of wave) is the rate at which it transports energy per unit area perpendicular to its direction of motion. The response of the human ear to sound intensity is not proportional to the intensity, so that doubling the actual intensity of a certain sound does not lead to the sensation of a sound twice as loud but only of one that is slightly louder than the original. For this reason the *decibel* (db) scale is used for sound intensity. An intensity of 10^{-12} W/m², which is just audible, is given the value 0 db; a sound 10 times more intense is given the value 10 db; a sound 10^2 times more intense than 0 db is given the value 20 db; a sound 10^3 times more intense than 0 db is given the value 30 db; and so forth. Normal conversation might be 60 db, city traffic noise might be 90 db, and a jet aircraft might produce 140 db (which produces damage to the ear) at a distance of 100 ft.

ELECTROMAGNETIC WAVES

Electromagnetic waves consist of coupled electric and magnetic fields that vary periodically as they move through space. The electric and magnetic fields are perpendicular to each other and to the direction in which the waves travel (Fig. 16-3), so the waves are transverse, and the variations in **E** and **B** occur simultaneously. Electromagnetic waves transport energy and require no material medium for their passage. Radio waves, light waves, X-rays, and gamma rays are examples of electromagnetic waves, and differ only in frequency. The color of light waves depends upon their frequency, with red light having the lowest visible frequencies and violet light the highest. White light contains light waves of all frequencies.

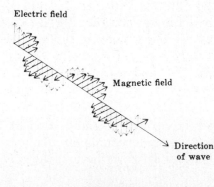

Fig. 16-3

Electromagnetic waves are generated by accelerated electric charges, usually electrons. Electrons oscillating back and forth in an antenna give off radio waves, for instance, and accelerated electrons in atoms give off light waves.

In free space all electromagnetic waves have the *velocity of light*, which is

Velocity of light = c = 3.00×10^8 m/s = 186,000 mi/s

DOPPLER EFFECT

When there is relative motion between a source of waves and an observer of them, the frequency perceived by the observer is not the same as that produced by the source. Thus the frequency of a fire engine siren appears higher when it is moving toward us than when it is at rest, and lower when it is moving away from us. Such frequency changes constitute the *Doppler effect*.

In sound, the Doppler effect obeys the formula

$$f_l = f_s \left(\frac{v + v_l}{v - v_s} \right)$$

where f_l = frequency found by observer

f_s = frequency produced by source

v = velocity of sound

v_l = velocity of observer (+ if toward source, − if away from source)

v_s = velocity of source (+ if toward observer, − if away from observer)

In light, the Doppler effect obeys the formula

$$f = f_s \sqrt{\frac{1 + v/c}{1 - v/c}}$$

where f = frequency found by observer

f_s = frequency produced by source

v = relative velocity between source and observer
(+ if approaching, − if receding)

c = velocity of light

REFLECTION OF LIGHT

When light is reflected from a smooth, plane surface, the angle of reflection equals the angle of incidence (Fig. 16-4). The image of an object in a plane mirror has the same size and shape as the object but with left and right reversed; the image is the same distance behind the mirror that the object is in front of it.

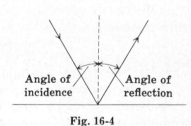

Fig. 16-4

REFRACTION OF LIGHT

When light passes obliquely from one medium to another in which its velocity is different, its direction is changed (Fig. 16-5). The greater the ratio between the two velocities, the greater the deflection. If the light goes from the medium of high velocity to the one of low velocity, it is bent toward the normal to the surface; if the light goes the other way, it is bent away from the normal. Light moving along the normal is not deflected.

The *index of refraction* of a transparent medium is the ratio between the velocity of light in free space and its velocity in the medium:

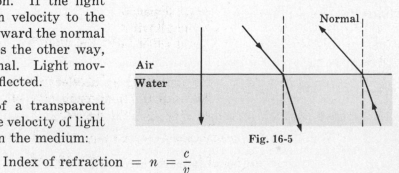

Fig. 16-5

$$\text{Index of refraction} = n = \frac{c}{v}$$

The greater the index of refraction, the more is light deflected on entering a medium from air.

In general the index of refraction of a medium increases with increasing frequency of the light. For this reason a beam of white light is separated into its component frequencies, each of which produces the sensation of a particular color, when it passes through an object whose sides are not parallel, for instance a glass prism. The resulting band of color is called a *spectrum*.

INTERFERENCE

Interference occurs when waves of the same nature from different sources meet at the same place. In *constructive interference* the waves are in phase ("in step") and reinforce each other; in *destructive interference* the waves are out of phase and partially or completely cancel each other. All types of waves exhibit interference under appropriate circumstances.

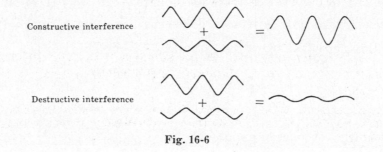

Fig. 16-6

DIFFRACTION

The ability of a wave to bend around the edge of an obstacle is called *diffraction*. Diffraction limits the sharpness of the images produced by optical systems; the larger the diameter of a lens or mirror and the shorter the wavelength of the light, the less significant diffraction is.

POLARIZATION

A *polarized* beam of light is one in which the electric fields of the waves are all in the same direction. If the electric fields are in random directions (though always perpendicular to the direction of propagation), the beam is *unpolarized*.

Solved Problems

16.1. Why can light waves travel through a vacuum whereas sound waves cannot?

Light waves consist of coupled fluctuations in electric and magnetic fields and hence require no material medium for their passage. Sound waves, on the other hand, are pressure fluctuations and cannot occur without a material medium to transmit them.

16.2. When is it appropriate to think of light as consisting of waves and when as consisting of rays?

When paths or path differences are involved whose lengths are comparable with the wavelengths found in light, the wave nature of light is significant and must be taken into account. Thus diffraction and interference can be understood only on a wave basis. When paths are involved that are many wavelengths long and neither diffraction nor interference occurs, as in reflection and refraction, it is more convenient to consider light as consisting of rays.

16.3. When a light beam enters one medium from another, which, if any, of the following quantities never changes? The direction of the beam, its velocity, its frequency, its wavelength.

> Only the frequency never changes.

16.4. Why does an object submerged in water appear closer to the surface than it actually is?

> As Fig. 16-7 shows, light leaving the object is bent away from the normal to the water-air surface as it leaves the water. Since an observer interprets what he sees in terms of the straight-line propagation of light, the object seems closer to the surface than its true depth.

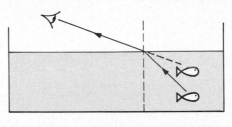

Fig. 16-7

16.5. When two light beams interfere, the result is a pattern of bright and dark lines. What becomes of the energy of the light waves whose destructive interference leads to the dark lines?

> The missing energy is found in the bright lines, whose brightness is greater than the simple addition of the two light beams would produce in the absence of interference. The total energy remains the same.

16.6. When two trains of waves meet on the surface of a body of water, the resulting interference pattern is obvious. However, when the light beams from two flashlights overlap on a screen, there is no evidence of an interference pattern. Why not? Is there any way in which the interference of light can be demonstrated?

> There are two reasons why such an experiment does not yield a conspicuous interference pattern. First, the wavelengths found in light are so short that such a pattern would be on an extremely small scale. Second and more important, all sources of light (except lasers) emit light waves as short trains of random phase and not as continuous trains. The interference that occurs between light beams from two independent sources is therefore averaged out during all but the briefest of observation times and cannot be seen by eye or recorded on photographic film. Such light sources are said to be *incoherent.*
>
> To exhibit an interference pattern in light, sources must be used whose waves have fixed phase relationships during the observation period. The waves from one source can be in step with those from the other when they are produced, or out of step, or something in between, but the essential thing is that the relationship be constant. Such sources are *coherent.* Three ways to construct coherent sources are:
>
> 1. Pass light from a single source (such as an illuminated slit or a narrow filament) through two or more other slits. The waves that emerge from the latter slits are necessarily coordinated and can interfere to produce a visible pattern.
>
> 2. Combine a direct light beam from a source with an indirect beam from the same source produced by refraction or reflection. This is how the interference patterns produced by thin oil films floating on water are caused.
>
> 3. Coordinate the radiating atoms in each individual source so that the radiating atoms always emit wave trains in step with one another. This is done in the laser.

16.7. Light is scattered by small obstacles or irregularities in a medium in its path. The higher the frequency of the light, the more readily scattering occurs. Use this fact to explain why the sky is blue.

When we look at the sky in daytime, what we see is light from the sun that has been scattered by irregularities in the upper atmosphere so that it seems to come from all around the earth. Because blue light has the highest frequency in the visible spectrum, it is scattered to the greatest extent, hence skylight is blue. At sunrise or sunset, light from the sun must travel a long distance through the atmosphere, and the sun appears red at such times because more blue light is scattered from its direct beam than when it is overhead.

16.8. As a phonograph record turns, a certain groove passes the needle at 25 cm/s. If the wiggles in the groove are 0.1 mm apart, what is the frequency of the sound that results?

Here the wavelength of the wiggles is $\lambda = 0.1$ mm $= 10^{-4}$ m, so they pass the needle at the rate of

$$f = \frac{v}{\lambda} = \frac{0.25 \text{ m/s}}{10^{-4} \text{ m}} = 2500 \text{ Hz}$$

This is therefore the frequency of the sound waves that are produced by the electronic system of the record player.

16.9. The velocity of sound in seawater is 5020 ft/s. Find the wavelength in seawater of a sound wave whose frequency is 256 Hz.

$$\lambda = \frac{v}{f} = \frac{5020 \text{ ft/s}}{256 \text{ Hz}} = 19.6 \text{ ft}$$

16.10. An anchored boat is observed to rise and fall once every 4 s as waves whose crests are 30 m apart pass by it. What is the velocity of these waves?

The frequency of the waves is $f = 1/(4 \text{ s}) = 0.25$ Hz and so their velocity is

$$v = f\lambda = 0.25 \text{ Hz} \times 30 \text{ m} = 7.5 \text{ m/s}$$

16.11. A marine radar operates at a wavelength of 3.2 cm. What is the frequency of the radar waves?

Radar waves are electromagnetic and hence travel with the velocity of light c. Therefore

$$f = \frac{c}{\lambda} = \frac{3 \times 10^8 \text{ m/s}}{3.2 \times 10^{-2} \text{ m}} = 9.4 \times 10^9 \text{ Hz}$$

16.12. The index of refraction of diamond is 2.42. What is the velocity of light in diamond?

Since $n = c/v$, here

$$v = \frac{c}{n} = \frac{3 \times 10^8 \text{ m/s}}{2.42} = 1.24 \times 10^8 \text{ m/s}$$

16.13. The frequency of a police car siren is 1000 Hz. If the car approaches a stationary person at 100 ft/s (68 mi/hr), what frequency does he hear?

Here $f_s = 1000$ Hz, $v = 1086$ ft/s, $v_l = 0$, and $v_s = +100$ ft/s (since the source is moving toward the observer). Hence

$$f_l = f_s\left(\frac{v + v_l}{v - v_s}\right) = 1000 \text{ Hz} \times \frac{1086 \text{ ft/s}}{(1086 - 100) \text{ ft/s}} = 1101 \text{ Hz}$$

16.14. The police car of Problem 16.13 is chasing another car whose velocity is 60 ft/s (41 mi/hr). What frequency does a person in the other car hear?

Here $f_s = 1000$ Hz, $v = 1086$ ft/s, $v_l = -60$ ft/s (since the observer is headed away from the source of sound), and $v_s = +100$ ft/s (since the source is moving toward the observer). Hence

$$f_l = f_s\left(\frac{v + v_l}{v - v_s}\right) = 1000 \text{ Hz} \times \frac{(1086 - 60) \text{ ft/s}}{(1086 - 100) \text{ ft/s}} = 1041 \text{ Hz}$$

16.15. When a certain train is in a station, its whistle has a frequency of 800 Hz, and when it moves away, the frequency to a listener in the station drops to 750 Hz. What is the train's velocity in m/s?

Here $v_l = 0$. We begin by solving the Doppler effect formula for the unknown velocity v_s:

$$f_l = f_s\left(\frac{v + v_l}{v - v_s}\right) = f_s\left(\frac{v}{v - v_s}\right)$$

$$\frac{f_l}{f_s} = \frac{v}{v - v_s}, \qquad v - v_s = v\left(\frac{f_s}{f_l}\right)$$

$$v_s = v - v\left(\frac{f_s}{f_l}\right) = v\left(1 - \frac{f_s}{f_l}\right)$$

Now we substitute $v = 331$ m/s, $f_s = 800$ Hz, and $f_l = 750$ Hz:

$$v_s = 331 \text{ m/s} \times \left(1 - \frac{800 \text{ Hz}}{750 \text{ Hz}}\right) = 331 \text{ m/s} \times (-0.067) = -22 \text{ m/s}$$

The minus sign means that the train is moving away from the listener.

16.16. A distant galaxy of stars is moving away from the earth at one-tenth of the velocity of light. Does light emitted at specific frequencies by the galaxy (these are the "spectral lines" discussed in Chapter 20) appear shifted to the red or to the violet end of the spectrum? By what percentage?

Motion away from an observer leads to a shift to lower frequencies, which means toward the red end of the spectrum. Since $v = -0.1\,c$ (the minus sign is needed because the relative motion is away from the observer), $v/c = -0.1$ and the observed frequencies are

$$\frac{f}{f_s} = \sqrt{\frac{1 + v/c}{1 - v/c}} = \sqrt{\frac{1 - 0.1}{1 + 0.1}} = \sqrt{0.82} = 0.90 = 90\%$$

of their normal values.

Supplementary Problems

16.17. What quantity is carried by all types of waves from their source to the place where they are eventually absorbed?

16.18. What is the relationship between the direction of an electromagnetic wave and the directions of its electric and magnetic fields?

16.19. Why is a beam of white light that passes perpendicularly though a flat pane of glass not dispersed into a spectrum?

16.20. Is it possible for the index of refraction of a substance to be less than 1?

16.21. The image formed by an optical system of any kind is called "real" if light rays actually pass through it and "virtual" if they do not. Which of the following produce real images and which produce virtual images? A plane mirror, a pair of binoculars, a camera, a slide projector.

16.22. How can the Doppler effect be used to determine how rapidly the sun is rotating?

16.23. Why do radio waves diffract readily around buildings whereas light waves do not?

16.24. A woman 5 ft 6 in. tall wishes to buy a mirror in which she can see herself at full length. What is the minimum height of such a mirror? Where should the mirror be placed?

16.25. What is the frequency of radio waves whose wavelength is 20 m?

16.26. A certain radio station transmits at a frequency of 1050 kHz. What is the wavelength of these waves?

16.27. The velocity of sound waves in air at sea level is 331 m/s. Find the wavelength in air of a sound wave whose frequency is 440 Hz.

16.28. A tuning fork vibrating at 600 Hz is immersed in a tank of water, and the resulting sound waves in the water are found to have a wavelength of 8.2 ft. What is the velocity of sound in the water?

16.29. The velocity of light in ice is 2.3×10^8 m/s. What is its index of refraction?

16.30. A fire engine with a 600-Hz siren goes past a stationary observer at 50 ft/s. What frequency does the observer hear (*a*) when the fire engine is approaching him? (*b*) When it is directly opposite him? (*c*) When it is moving away from him?

16.31. A concertgoer hurries down the aisle to his seat at 3 m/s. What frequency does he hear when the note A (440 Hz) is being played?

16.32. A person standing in the street hears a marching band in the distance. If all the notes appear 0.5% lower in frequency than their correct values, find the velocity in ft/s at which the band is marching and whether it is approaching or receding.

16.33. A certain galaxy is receding from the earth at a velocity of 1.5×10^7 m/s. When the galaxy emits light of frequency 8.0×10^{14} Hz, what frequency does the light have when it arrives at the earth?

Answers to Supplementary Problems

16.17. Energy

16.18. The electric and magnetic fields are perpendicular to the direction of the wave and to each other.

16.19. Light incident perpendicularly on a surface is not deflected, so light of the various frequencies in white light stays together despite their different velocities in the glass.

16.20. No, because this would mean a velocity of light greater than its velocity in free space *c*, which is not possible.

16.21. *Real*: camera, slide projector. *Virtual*: plane mirror, binoculars.

16.22. As the sun rotates, one side of its rim is moving toward us and the other side away from us. The light emitted from these regions will have its characteristic frequencies respectively increased and decreased through the Doppler effect, and from the amounts of these changes the rate of rotation can be calculated.

16.23. The wavelengths of visible light are very short relative to the size of a building so their diffraction is imperceptible. The wavelengths of radio waves are more nearly comparable with the size of a building.

16.24. Since the angle of reflection equals the angle of incidence, the mirror should be half her height, or 2 ft 9 in., and placed so its top is level with the middle of her forehead (Fig. 16-8). The distance between the woman and the mirror does not matter.

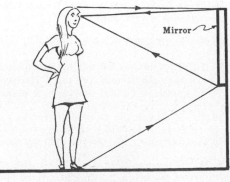

Fig. 16-8

16.25. 1.5×10^7 Hz

16.26. 286 m

16.27. 0.75 m

16.28. 4920 ft/s

16.29. 1.3

16.30. 629 Hz; 600 Hz; 574 Hz

16.31. 444 Hz

16.32. 5.5 ft/s; receding

16.33. 7.6×10^{14} Hz

Chapter 17

Quantum Physics

QUANTUM THEORY OF LIGHT

Certain features of the behavior of light can be explained only on the basis that light consists of individual *quanta* or *photons*. The energy of a photon of light whose frequency is f is

$$\text{Quantum energy} = E = hf$$

where h is *Planck's constant*:

$$\text{Planck's constant} = h = 6.63 \times 10^{-34} \text{ joule-second}$$

A photon has most of the properties associated with particles — it is localized in space and possesses energy and momentum — but it has no mass. Photons travel with the velocity of light.

The electromagnetic and quantum theories of light complement each other: under some circumstances light exhibits a wave character, under other circumstances it exhibits a particle character. Both are aspects of the same basic phenomenon.

X-RAYS

X-rays are high-frequency electromagnetic waves produced when fast electrons impinge on a target. If the electrons are accelerated through a potential difference of V, each electron has the energy $KE = eV$. If all of this energy goes into creating an X-ray photon,

$$eV = hf$$

$$\text{Electron kinetic energy} = \text{X-ray photon energy}$$

and the frequency of the X-rays is $f = eV/h$.

THE ELECTRON VOLT

A common energy unit in atomic and quantum physics is the *electron volt* (eV), which is defined as the energy an electron gains when it moves through a potential difference of 1 volt. Hence

$$1 \text{ eV} = 1.60 \times 10^{-19} \text{ J}$$

Multiples of the eV are the *keV*, *MeV*, and *GeV*, where

$$1 \text{ keV} = 10^3 \text{ eV} \qquad 1 \text{ MeV} = 10^6 \text{ eV} \qquad 1 \text{ GeV} = 10^9 \text{ eV}$$

MATTER WAVES

Under certain conditions moving bodies exhibit wave properties. The quantity whose variations constitute the *matter waves* (or *de Broglie waves*) of a moving body is known as its *wave function* ψ (Greek letter *psi*). The likelihood of finding the body at a particular time and place is proportional to the value of ψ^2 at that time and place. A large value of ψ^2 signifies a high probability of finding the body, a small value of ψ^2 signifies a low probability of finding the body.

101

The matter waves associated with a moving body are in the form of a group of waves that travels with the same velocity as the body.

The wavelength of the matter waves of a body of mass m and velocity v is

$$\text{de Broglie wavelength} = \lambda = \frac{h}{mv}$$

UNCERTAINTY PRINCIPLE

A consequence of the wave nature of moving bodies is the *uncertainty principle*: it is impossible to determine both the exact position and the exact momentum of a body at the same time. If Δx is the uncertainty in position and Δmv is the uncertainty in momentum, then

$$\Delta x\,\Delta mv \geq \frac{h}{2\pi}$$

where the symbol $\geq$ means "equal to or greater than." Since Planck's constant h is so small, the uncertainty principle is significant only for very small bodies such as elementary particles.

Solved Problems

17.1. The human eye can respond to as few as three photons of light. If the light is yellow ($f = 5 \times 10^{14}$ Hz), how much energy does this represent?

The energy of each photon is

$$E = hf = 6.63 \times 10^{-34}\text{ J-s} \times 5 \times 10^{14}\text{ Hz} = 3.3 \times 10^{-19}\text{ J}$$

The total energy is $3E = 10^{-18}$ J.

17.2. The average wavelength of the light emitted by a certain 100-W light bulb is 5.5×10^{-7} m. How many photons per second does the light bulb emit?

The frequency of the light is

$$f = \frac{c}{\lambda} = \frac{3 \times 10^8\text{ m/s}}{5.5 \times 10^{-7}\text{ m}} = 5.5 \times 10^{14}\text{ Hz}$$

and the energy of each photon is

$$E = hf = 6.63 \times 10^{-34}\text{ J-s} \times 5.5 \times 10^{14}\text{ Hz} = 3.6 \times 10^{-19}\text{ J}$$

Since 100 W = 100 J/s, the number of photons emitted per second is

$$\frac{100\text{ J/s}}{3.6 \times 10^{-19}\text{ J/photon}} = 2.8 \times 10^{20}\text{ photons/s}$$

17.3. What is the kinetic energy in eV of an electron whose velocity is 10^7 m/s?

$$\text{KE} = \tfrac{1}{2}mv^2 = \tfrac{1}{2} \times 9.1 \times 10^{-31}\text{ kg} \times (10^7\text{ m/s})^2 = 4.55 \times 10^{-17}\text{ J}$$

Since 1 eV = 1.6×10^{-19} J,

$$\text{KE} = \frac{4.55 \times 10^{-17}\text{ J}}{1.6 \times 10^{-19}\text{ J/eV}} = 284\text{ eV}$$

17.4. A proton ($m = 1.67 \times 10^{-27}$ kg) is accelerated through a potential difference of 200 V. (*a*) What is its kinetic energy in eV? (*b*) What is its velocity?

(*a*) Since the charge on the proton is $+e$, its kinetic energy is 200 eV.

(*b*) $$KE = 200 \text{ eV} \times 1.6 \times 10^{-19} \text{ J/eV} = 3.2 \times 10^{-17} \text{ J}$$

Since $KE = \frac{1}{2}mv^2$,

$$v = \sqrt{\frac{2KE}{m}} = \sqrt{\frac{2 \times 3.2 \times 10^{-17} \text{ J}}{1.67 \times 10^{-27} \text{ kg}}} = 2 \times 10^5 \text{ m/s}$$

17.5. An unstable elementary particle known as the muon has a rest energy of 106 MeV. What is its rest mass? How many times the electron mass is this?

First we convert 106 MeV to joules:

$$E = 106 \times 10^6 \text{ eV} \times 1.6 \times 10^{-19} \text{ J/eV} = 1.7 \times 10^{-11} \text{ J}$$

Since $E = mc^2$, the muon rest mass is

$$m = \frac{E}{c^2} = \frac{1.7 \times 10^{-11} \text{ J}}{(3 \times 10^8 \text{ m/s})^2} = 1.88 \times 10^{-28} \text{ kg}$$

The electron rest mass is $m_e = 9.1 \times 10^{-31}$ kg, and so

$$m = \frac{1.88 \times 10^{-28} \text{ kg}}{9.1 \times 10^{-31} \text{ kg}/m_e} = 207 \, m_e$$

17.6. Every chemical reaction that involves the absorption or release of energy (typically several eV per molecular change) necessarily involves a corresponding increase or decrease in the masses of the reacting atoms or molecules. To see why such mass changes are undetectable, find the change in mass associated with the absorption or release of 1 eV and compare this with the mass of the H_2O (water) molecule, which is 3×10^{-26} kg.

Since 1 eV $= 1.6 \times 10^{-19}$ J and $E = mc^2$, the mass change is

$$\Delta m = \frac{\Delta E}{c^2} = \frac{1.6 \times 10^{-19} \text{ J}}{(3 \times 10^8 \text{ m/s})^2} = 1.8 \times 10^{-36} \text{ kg}$$

Relative to the mass of the H_2O molecule this mass change is

$$\frac{\Delta m}{m} = \frac{1.8 \times 10^{-36} \text{ kg}}{3 \times 10^{-26} \text{ kg}} = 6 \times 10^{-9}$$

or six-billionths the mass of the H_2O molecule.

17.7. In the *photoelectric effect,* light directed at the surfaces of certain metals causes electrons to be emitted. In the case of potassium, 2 eV of work must be done to remove an electron from the surface. (*a*) If light of wavelength 5×10^{-7} m falls on a potassium surface, what is the maximum energy of the photoelectrons that emerge? (*b*) If light of wavelength 4×10^{-7} m falls on the same surface, will the photoelectrons have more or less energy?

(*a*) Since $c = f\lambda$, the frequency of the light is

$$f = \frac{c}{\lambda} = \frac{3 \times 10^8 \text{ m/s}}{5 \times 10^{-7} \text{ m}} = 6 \times 10^{14} \text{ Hz}$$

The energy of each photon is therefore

$$E = hf = 6.63 \times 10^{-34} \text{ J-s} \times 6 \times 10^{14} \text{ Hz} = 3.98 \times 10^{-19} \text{ J}$$

Since 1 eV $= 1.6 \times 10^{-19}$ J,

$$E = \frac{3.98 \times 10^{-19} \text{ J}}{1.6 \times 10^{-19} \text{ J/eV}} = 2.49 \text{ eV}$$

This is the maximum energy that can be given to an electron by a photon of this light. Because 2 eV is needed to remove an electron, the maximum energy of the photoelectrons in this situation is 0.49 eV.

(b) A shorter wavelength means a higher frequency and hence more energy to be imparted to the photoelectrons.

17.8. In a certain television picture tube, electrons are accelerated through a potential difference of 10,000 V. Find the frequency of the X-rays that are emitted when these electrons strike the screen.

Since $hf = eV$, here we have

$$f = \frac{eV}{h} = \frac{1.6 \times 10^{-19}\text{ C} \times 10^4\text{ V}}{6.63 \times 10^{-34}\text{ J-s}} = 2.4 \times 10^{18}\text{ Hz}$$

17.9. An X-ray tube emits X-rays whose wavelength is 2×10^{-11} m. What is the operating voltage of the tube?

The frequency of the X-rays is

$$f = \frac{c}{\lambda} = \frac{3 \times 10^8\text{ m/s}}{2 \times 10^{-11}\text{ m}} = 1.5 \times 10^{19}\text{ Hz}$$

Since $hf = eV$,

$$V = \frac{hf}{e} = \frac{6.63 \times 10^{-34}\text{ J-s} \times 1.5 \times 10^{19}\text{ Hz}}{1.6 \times 10^{-19}\text{ C}} = 62{,}000\text{ V}$$

17.10. What is the de Broglie wavelength of a 1000-kg car whose velocity is 20 m/s? Would you expect the wave properties of such a car to be noticeable?

$$\lambda = \frac{h}{mv} = \frac{6.63 \times 10^{-34}\text{ J-s}}{10^3\text{ kg} \times 20\text{ m/s}} = 3.3 \times 10^{-38}\text{ m}$$

The wave properties of such a car would be impossible to observe.

17.11. An electron microscope uses a beam of fast electrons which are focused by electric and magnetic fields to produce an enlarged image of a thin specimen on a screen or photographic plate. Find the resolving power of an electron microscope that uses 15-keV electrons by assuming that this is equal to the electron wavelength.

The kinetic energy of the electrons is

$$\text{KE} = 1.5 \times 10^4\text{ eV} \times 1.6 \times 10^{-19}\text{ J/eV} = 2.4 \times 10^{-15}\text{ J}$$

Since $\text{KE} = \frac{1}{2}mv^2$, the velocity of the electrons is

$$v = \sqrt{\frac{2\text{KE}}{m}} = \sqrt{\frac{2 \times 2.4 \times 10^{-15}\text{ J}}{9.1 \times 10^{-31}\text{ kg}}} = 7.26 \times 10^7\text{ m/s}$$

The electron wavelength is therefore

$$\lambda = \frac{h}{mv} = \frac{6.63 \times 10^{-34}\text{ J-s}}{9.1 \times 10^{-31}\text{ kg} \times 7.26 \times 10^7\text{ m/s}} = 1.0 \times 10^{-11}\text{ m}$$

17.12. The atoms in a solid would possess a certain minimum "zero-point" kinetic energy even at 0 K, although molecules in an ideal gas would have no kinetic energy at 0 K. Use the uncertainty principle to account for this difference.

Each atom in a solid is limited to a certain definite region of space — otherwise the assembly of atoms would not be a solid. The uncertainty in the position of each atom is therefore finite, and its momentum and hence energy cannot be zero. There is no restriction on the position of a molecule in an ideal gas, and so the uncertainty in its position is effectively infinite and its momentum and hence energy can be zero.

17.13. The position of a certain electron is determined at a certain time with an uncertainty of 10^{-9} m. (a) Find the uncertainty in the electron's momentum. (b) Find the uncertainty in the electron's velocity and, from this, the uncertainty in its position 1 s after the original measurement was made.

(a)
$$\Delta mv = \frac{h}{2\pi \Delta x} = \frac{6.63 \times 10^{-34} \text{ J-s}}{2\pi \times 10^{-9} \text{ m}} = 1.06 \times 10^{-25} \text{ kg-m/s}$$

(b)
$$\Delta v = \frac{\Delta mv}{m} = \frac{1.06 \times 10^{-25} \text{ kg-m/s}}{9.1 \times 10^{-31} \text{ kg}} = 1.2 \times 10^{5} \text{ m/s}$$

Hence the uncertainty in the electron's position 1 s later will be

$$\Delta x' = t\,\Delta v = 1 \text{ s} \times 1.2 \times 10^{5} \text{ m/s} = 1.2 \times 10^{5} \text{ m}$$

which is 120 km!

17.14. Most atoms are a little over 10^{-10} m in radius. (a) Find the uncertainty in the momentum of an electron whose position is known to within 10^{-10} m. (b) Find the corresponding uncertainty in the electron's kinetic energy. What is the significance of this figure?

(a) The momentum uncertainty is

$$\Delta mv = \frac{h}{2\pi \Delta x} = \frac{6.63 \times 10^{-34} \text{ J-s}}{2\pi \times 10^{-10} \text{ m}} = 1.06 \times 10^{-24} \text{ kg-m/s}$$

(b) Since $\text{KE} = \frac{1}{2}mv^2 = \frac{1}{2m}(mv)^2$,

$$\Delta\text{KE} = \frac{1}{2m}(\Delta mv)^2 = \frac{(1.06 \times 10^{-24} \text{ kg-m/s})^2}{2 \times 9.11 \times 10^{-31} \text{ kg}} = 6.1 \times 10^{-19} \text{ J} = 3.8 \text{ eV}$$

Electrons in atoms must have greater kinetic energies than this, as in fact they do.

Supplementary Problems

17.15. The energy of a light beam is carried by separate photons, yet we do not perceive light as a series of tiny flashes. Why not?

17.16. If Planck's constant were equal to 6.63 J-s instead of 6.63×10^{-34} J-s, would quantum phenomena be more or less conspicuous in everyday life than they are now?

17.17. Why is the uncertainty principle only significant for such extremely small particles as electrons and protons even though it applies to objects of all sizes?

17.18. Light from the sun arrives at the earth at the rate of about 1400 W/m² of area perpendicular to the direction of the light. Assuming that sunlight consists exclusively of light of wavelength 6×10^{-7} m, find the number of photons per second that fall on each m² of the earth's surface directly facing the sun.

17.19. How many photons per second are emitted by a 50-kW radio transmitter that operates at a frequency of 1200 kHz?

17.20. What is the operating voltage of an X-ray tube that produces X-rays of frequency 10^{19} Hz?

17.21. Find the wavelength of the X-rays produced by a 50,000-V X-ray machine.

17.22. What is the kinetic energy in eV of a proton $(m = 1.67 \times 10^{-27}$ kg) whose velocity is 5×10^6 m/s?

17.23. Find the rest energy of the proton in MeV.

17.24. Find the velocity of a 50-eV electron.

17.25. The de Broglie wavelength of an electron is 1.66×10^{-10} m. Find its energy in eV.

17.26. Find the de Broglie wavelength of a 10-g rifle bullet traveling at the velocity of sound, 331 m/s.

17.27. The mass of an electron moving at 75% of the velocity of light is 1.5 times its rest mass. Find the de Broglie wavelength of such an electron.

17.28. The work needed to remove an electron from the surface of sodium is 2.3 eV. Find the maximum wavelength of light that will cause photoelectrons to be emitted from sodium. (See Problem 17.7.)

17.29. Photoelectrons are emitted by a copper surface only when light whose frequency is 1.1×10^{15} Hz or more is directed at it. What is the maximum energy of the photoelectrons when light of frequency 1.5×10^{15} Hz is directed at the surface?

17.30. An experiment is planned in which the momentum of an electron is to be measured to within $\pm 10^{-29}$ kg-m/s while its position is known to within $\pm 10^{-10}$ m. Do you think the experiment will succeed?

17.31. A proton is confined to a region 10^{-9} m across. (*a*) Find the uncertainty in its momentum. (*b*) Find the minimum energy it can have.

Answers to Supplementary Problems

17.15. Even a weak light involves many photons per second. Visual responses persist for a short time, so successive photons give the impression of a continuous transfer of energy.

17.16. More conspicuous.

17.17. The uncertainties in the position and momentum of an object much larger than an elementary particle are so small compared with its dimensions and momentum as to be undetectable.

17.18. 4.2×10^{21} photons/m^2-s   **17.24.** 4.2×10^6 m/s

17.19. 6.3×10^{31} photons/s   **17.25.** 55 eV

17.20. 41,400 V   **17.26.** 2×10^{-34} m

17.21. 2.5×10^{-11} m   **17.27.** 2.16×10^{-12} m

17.22. 1.3×10^5 eV $= 0.13$ MeV   **17.28.** 5.4×10^{-7} m

17.23. 938 MeV   **17.29.** 1.7 eV

17.20. No, because the uncertainty principle limits $\Delta mv\, \Delta x$ to a minimum of 1.05×10^{-34} J-s.

17.31. 1.05×10^{-25} kg-m/s; 3.3×10^{-24} J $= 0.21$ MeV

Chapter 18

The Nucleus

NUCLEAR STRUCTURE

The nucleus of an atom is composed of protons and neutrons whose masses are respectively

$$m_p = 1.673 \times 10^{-27} \text{ kg} = 1.007277 \text{ u}$$
$$m_n = 1.675 \times 10^{-27} \text{ kg} = 1.008665 \text{ u}$$

The proton has a charge of $+e$ and the neutron is uncharged. The *atomic number* of an element is the number of protons in the nucleus of one of its atoms. Protons and neutrons are jointly called *nucleons*.

Although all the atoms of an element have the same number of protons in their nuclei, the number of neutrons may be different. Each variety of nucleus found in a given element is called an *isotope* of the element. Symbols for isotopes follow the pattern

$$_Z^A X$$

where X = chemical symbol of element

Z = atomic number of element = number of protons in nucleus

A = mass number of isotope = number of protons + neutrons in nucleus

BINDING ENERGY

The mass of an atom is always less than the sum of the masses of the neutrons, protons, and electrons of which it is composed. The energy equivalent of the missing mass is called the *binding energy* of the nucleus; the greater its binding energy, the more stable the nucleus. The mass defect Δm of a nucleus with Z protons and N neutrons may be found from its atomic mass m by using the formula

$$\Delta m = (Zm_{\text{H}} + Nm_n) - m$$

where m_{H}, the mass of the hydrogen atom, is

$$m_{\text{H}} = 1.007825 \text{ u}$$

To find the binding energy in MeV, the usual unit, ΔM can be multiplied by the conversion factor 931 MeV/u.

FUNDAMENTAL FORCES

The force between nucleons that holds an atomic nucleus together despite the repulsive electrical forces its protons exert on each other is the result of what is known as the *strong interaction*. This is a fundamental interaction in the same sense as the gravitational and electromagnetic interactions are: none can be explained in terms of any of the others. The strong interaction has only a very short range, unlike the gravitational and electromagnetic interactions, and is only effective within nuclei.

There is another interaction involving nuclei called the *weak interaction* which is responsible for beta decays (Chapter 19). Recent evidence indicates that the weak interaction may be electromagnetic in origin and not a fundamental interaction as had hitherto been believed.

NUCLEAR REACTIONS

Nuclei can be transformed into others of a different kind by interaction with each other. Since nuclei are all positively charged, a high-energy collision is necessary between two nuclei if they are to get close enough together to react. Because it has no charge, a neutron can initiate a nuclear reaction even if it is moving slowly. In any nuclear reaction, the total number of neutrons and the total number of protons in the products must be equal to the corresponding total numbers in the reactants.

FISSION AND FUSION

Nuclei of intermediate size have the highest binding energies per nucleon and therefore are more stable than lighter and heavier nuclei. If a heavy nucleus is split into two smaller ones, the greater binding energy of the latter means that energy will be liberated. This process is called *nuclear fission*. Certain very large nuclei, such as $^{235}_{92}$U, undergo fission when they absorb a neutron; since the products of the fission include several neutrons as well as two daughter nuclei, a *chain reaction* can be established in an assembly of a suitable fissionable isotope. If uncontrolled, the result is an atomic bomb; if controlled so that the rate at which fission events occur is constant, the result is a nuclear reactor that can serve as an energy source for generating electricity or for ship propulsion.

In *nuclear fusion*, two light nuclei combine to form a heavier one whose binding energy per nucleon is greater. The difference in binding energies is liberated in the process. To bring about a fusion reaction, the initial nuclei must be moving rapidly when they collide to overcome their electrical repulsion. Nuclear fusion is the source of energy in the sun and stars, where the high temperatures in the interiors mean that nuclei there have sufficiently high velocities and the high pressures mean that nuclear collisions occur frequently. In the operation of a hydrogen bomb, a fission bomb is first detonated to produce the high temperature and pressure necessary for fusion reactions to occur. The problem in constructing a fusion reactor for controlled energy production is to contain a sufficiently hot and dense mixture of suitable isotopes for long enough to yield a net energy output.

Solved Problems

18.1. Only the gravitational interaction influences the motion of the earth about the sun. Why are the other fundamental interactions not involved as well?

The strong and weak interactions are too limited in range to affect planetary motion. As for the electromagnetic interaction, because like charges repel and unlike charges attract, it is very difficult to separate neutral matter into large-scale assemblies of opposite charge, hence all astronomical bodies are electrically neutral.

18.2. The largest stable nucleus is that of the bismuth isotope $^{209}_{83}$Bi. Why are larger nuclei unstable?

The range of the strong interaction, which provides the attractive forces that hold nucleons together, is quite short, whereas the electric repulsive forces that act between protons have unlimited range. Hence beyond a certain size the repulsive forces become comparable with the attractive ones and such nuclei are unstable.

18.3. State the number of protons and neutrons in each of the following nuclei:

$$^{6}_{3}\text{Li}, \ ^{12}_{6}\text{C}, \ ^{36}_{16}\text{S}, \ ^{137}_{56}\text{Ba}$$

A nucleus designated $^{A}_{Z}X$ contains Z protons and $A - Z$ neutrons. Accordingly the numbers of protons and neutrons in the given nuclei are as follows:

$$^{6}_{3}\text{Li: } 3 \text{ protons, } 3 \text{ neutrons}$$

$$^{12}_{6}\text{C: } 6 \text{ protons, } 6 \text{ neutrons}$$

$$^{36}_{16}\text{S: } 16 \text{ protons, } 20 \text{ neutrons}$$

$$^{137}_{56}\text{Ba: } 56 \text{ protons, } 81 \text{ neutrons}$$

18.4. Ordinary chlorine is a mixture of 75.53% of the $^{35}_{17}\text{Cl}$ isotope and 24.47% of the $^{37}_{17}\text{Cl}$ isotope. The atomic masses of these isotopes are respectively 34.969 u and 36.966 u. Find the atomic mass of ordinary chlorine.

The procedure is to multiply the mass of each isotope by the proportion of the whole it represents, and then to add the results together. Thus we obtain

$$0.7553 \times 34.969 \text{ u} + 0.2447 \times 36.966 \text{ u} = 35.458 \text{ u}$$

which is the atomic mass of ordinary chlorine.

18.5. The atomic mass of $^{16}_{8}\text{O}$ is 15.9949 u. (*a*) What is its binding energy? (*b*) What is its binding energy per nucleon?

(*a*) $^{16}_{8}\text{O}$ contains 8 protons and 8 neutrons in its nucleus. The mass of 8 H atoms is $8m_{\text{H}} = 8 \times 1.007825 \text{ u} = 8.0626 \text{ u}$ and the mass of 8 neutrons is $8m_n = 8 \times 1.008665 \text{ u} = 8.0693 \text{ u}$. Hence the mass deficit in $^{16}_{8}\text{O}$ is

$$\Delta m = (8.0626 + 8.0693) \text{ u} - 15.9949 \text{ u} = 0.1370 \text{ u}$$

and the binding energy is, since 1 u = 931 MeV,

$$\Delta E = 0.1370 \text{ u} \times 931 \text{ MeV/u} = 127.5 \text{ MeV}$$

(*b*) There are 16 nucleons in $^{16}_{8}\text{O}$, so the binding energy per nucleon is 127.6 MeV/16 nucleons = 7.97 MeV/nucleon.

18.6. The binding energy of $^{20}_{10}\text{Ne}$ is 160.6 MeV. Find its atomic mass.

$^{20}_{10}\text{Ne}$ contains 10 protons and 10 neutrons in its nucleus. The mass of 10 H atoms and 10 neutrons is

$$m_0 = 10.07825 \text{ u} + 10.08665 \text{ u} = 20.1649 \text{ u}$$

The mass equivalent of 160.6 MeV is

$$\Delta m = \frac{160.6 \text{ MeV}}{931 \text{ MeV/u}} = 0.1725 \text{ u}$$

and so the mass of the $^{20}_{10}\text{Ne}$ atom is

$$m = m_0 - \Delta m = 20.1649 \text{ u} - 0.1725 \text{ u} = 19.9924 \text{ u}$$

18.7. Complete the following nuclear reactions:

$$^{6}_{3}\text{Li} + ^{2}_{1}\text{H} \longrightarrow ^{4}_{2}\text{He} + \text{?}$$

$$^{35}_{17}\text{Cl} + \text{?} \longrightarrow ^{32}_{16}\text{S} + ^{4}_{2}\text{He}$$

$$^{9}_{4}\text{Be} + ^{4}_{2}\text{He} \longrightarrow ^{1}_{0}n + \text{?}$$

In each of these reactions, the number of protons and the number of neutrons must be the same on both sides of the equation. Hence the complete reactions must be as follows:

$$^6_3\text{Li} + {}^2_1\text{H} \longrightarrow {}^4_2\text{He} + {}^4_2\text{He}$$

$$^{35}_{17}\text{Cl} + {}^1_1\text{H} \longrightarrow {}^{32}_{16}\text{S} + {}^4_2\text{He}$$

$$^9_4\text{Be} + {}^4_2\text{He} \longrightarrow {}^1_0n + {}^{12}_6\text{C}$$

18.8. In a typical fission reaction, a $^{235}_{92}\text{U}$ nucleus absorbs a neutron and splits into a $^{140}_{54}\text{Xe}$ nucleus and a $^{94}_{38}\text{Sr}$ nucleus. How many neutrons are liberated in this process?

In order that the total numbers of protons and neutrons be the same before and after the fission reaction, two neutrons must be liberated. Hence the reaction is

$$^{235}_{92}\text{U} + {}^1_0n \longrightarrow {}^{140}_{54}\text{Xe} + {}^{94}_{38}\text{Sr} + {}^1_0n + {}^1_0n + \Delta E$$

In this case ΔE is about 200 MeV.

18.9. The atomic masses of $^{16}_8\text{O}$, $^{15}_8\text{O}$, and $^{15}_7\text{N}$ are respectively 15.9949 u, 15.0030 u, and 15.0001 u. (a) How much energy is required to remove one proton from $^{16}_8\text{O}$? (b) How much energy is required to remove one neutron from $^{16}_8\text{O}$? (c) Why are these figures so different?

(a) The reaction is $^{16}_8\text{O} + \Delta E \longrightarrow {}^{15}_7\text{N} + {}^1_1\text{H}$. The mass deficiency Δm is

$$\Delta m = m(^{15}_7\text{N}) + m(^1_1\text{H}) - m(^{16}_8\text{O}) = 15.0001 \text{ u} + 1.0078 \text{ u} - 15.9949 \text{ u} = 0.0130 \text{ u}$$

Hence $\Delta E = 0.0130 \text{ u} \times 931 \text{ MeV/u} = 12.1 \text{ MeV}$.

(b) The reaction is $^{16}_8\text{O} + \Delta E \longrightarrow {}^{15}_8\text{O} + {}^1_0n$. The mass deficiency Δm is

$$\Delta m = m(^{15}_8\text{O}) + m(^1_0n) - m(^{16}_8\text{O}) = 15.0030 \text{ u} + 1.0087 \text{ u} - 15.9949 \text{ u} = 0.0168 \text{ u}$$

Hence $\Delta E = 0.0168 \text{ u} \times 931 \text{ MeV/u} = 15.6 \text{ MeV}$.

(c) Less energy is needed to remove a proton than a neutron because of the repulsive electric force exerted on the proton by the other protons in the nucleus. Only attractive forces are exerted on a neutron within a nucleus, except when it is extremely close to another nucleon.

18.10. When $^{235}_{92}\text{U}$ undergoes fission, about 0.1% of the original mass is released as energy. (a) How much energy is released when 1 kg of $^{235}_{92}\text{U}$ undergoes fission? (b) How much $^{235}_{92}\text{U}$ must undergo fission per day in a nuclear reactor that provides energy to a 100-megawatt (10^8-W) electric power plant? Assume perfect efficiency. (c) When coal is burned, about 7800 kcal/kg of heat is liberated. How many kg of coal would be consumed per day by a conventional coal-fired 100-MW electric power plant?

(a) $$E = mc^2 = 0.001 \text{ kg} \times (3 \times 10^8 \text{ m/s})^2 = 9 \times 10^{13} \text{ J}$$

(b) Energy = power $\times$ time, and so here

$$E = Pt = 10^8 \text{ W} \times 3600 \text{ s/hr} \times 24 \text{ hr/day} = 8.64 \times 10^{12} \text{ J/day}$$

Hence the mass of $^{235}_{92}\text{U}$ required is

$$\frac{8.64 \times 10^{12} \text{ J/day}}{9 \times 10^{13} \text{ J/kg}} = 9.6 \times 10^{-2} \text{ kg/day} = 96 \text{ g/day}$$

(c) The energy liberated per kg of coal burned is

$$7800 \text{ kcal/kg} \times 4185 \text{ J/kcal} = 3.26 \times 10^7 \text{ J}$$

Hence the mass of coal required is

$$\frac{8.64 \times 10^{12} \text{ J/day}}{3.26 \times 10^7 \text{ J/kg}} \; = \; 2.65 \times 10^5 \text{ kg/day}$$

which is 265 metric tons.

18.11. In the sun and most other stars the principal energy-liberating process is the conversion of hydrogen into helium in a series of nuclear fusion reactions in the course of which *positrons* (positively charged electrons) are emitted. (a) Write the equation for the overall process in which four protons form a helium nucleus. (b) How much energy is liberated in each such process? The masses of $^1_1$H, $^4_2$He, and the electron are respectively 1.007825 u, 4.002603 u, and 0.000549 u.

(a) Two positrons must be given off in order that charge be conserved. Hence the overall process is

$$^1_1\text{H} + ^1_1\text{H} + ^1_1\text{H} + ^1_1\text{H} \; \longrightarrow \; ^4_2\text{He} + e^+ + e^+$$

(b) Since a helium atom has only two electrons around its nucleus, two electrons as well as two positrons are lost when each helium atom is formed. The mass change is therefore

$$\Delta m \; = \; 4m_H - (m_{He} + 4m_e) \; = \; 4 \times 1.007825 \text{ u} - (4.002603 \text{ u} + 4 \times 0.000549 \text{ u})$$

$$= \; 0.026501 \text{ u}$$

and the energy liberated is $0.026501 \text{ u} \times 931 \text{ MeV/u} = 24.7 \text{ MeV}$.

Supplementary Problems

18.12. (a) Which of the fundamental interactions has the least significance in nuclear physics? (b) Which two apparently are related?

18.13. In experiments involving nuclear fusion, magnetic fields rather than solid containers are used to confine atomic nuclei that are to react. Why?

18.14. What are similarities and differences between nuclear fission and nuclear fusion?

18.15. What parts of its structure are chiefly responsible for an atom's mass and for its chemical behavior?

18.16. State the numbers of protons and neutrons in each of the following nuclei: $^{15}_7$N, $^{35}_{17}$Cl, $^{64}_{30}$Zn, $^{200}_{80}$Hg.

18.17. Ordinary boron is a mixture of 20% of the $^{10}_5$B isotope and 80% of the $^{11}_5$B isotope. The atomic masses of these isotopes are respectively 10.013 u and 11.009 u. Find the atomic mass of ordinary boron.

18.18. The atomic mass of $^3_2$He is 3.01603. (a) What is its binding energy? (b) What is its binding energy per nucleon?

18.19. The atomic mass of $^{35}_{17}$Cl is 34.96885 u. (a) What is its binding energy? (b) What is its binding energy per nucleon?

18.20. The binding energy of $^{42}_{20}$Ca is 361.7 MeV. Find its atomic mass.

18.21. Deuterium is the hydrogen isotope $^2_1$H. A gamma-ray photon with at least 2.22 MeV of energy is able to break a deuterium nucleus apart into a proton and a neutron. Find the atomic mass of deuterium.

18.22. Complete the following nuclear reactions:

$$^{14}_{7}\text{N} + ^{4}_{2}\text{He} \longrightarrow ^{1}_{1}\text{H} + \text{?}$$

$$^{11}_{5}\text{B} + ^{1}_{1}\text{H} \longrightarrow ^{11}_{6}\text{C} + \text{?}$$

$$^{6}_{3}\text{Li} + \text{?} \longrightarrow ^{7}_{4}\text{Be} + ^{1}_{0}n$$

18.23. When $^{235}_{92}\text{U}$ undergoes fission, about 0.1% of the original mass is released as energy. (a) How much energy is released by an atomic bomb that contains 10 kg of $^{235}_{92}\text{U}$? (b) When a ton of TNT is exploded, about 4×10^9 J is released. How many tons of TNT are equivalent in destructive power to the above bomb?

18.24. In some stars three $^{4}_{2}\text{He}$ nuclei fuse together in sequence to form a $^{12}_{6}\text{C}$ nucleus ($m = 12.000000$ u). How much energy is liberated each time this happens?

Answers to Supplementary Problems

18.12. the gravitational interaction; the weak and electromagnetic interactions

18.13. In such experiments, the nuclei form a gas at very high temperature that is called a *plasma*. A plasma would be cooled upon contact with a solid container, and atoms of the container would also be dislodged and enter the plasma where they might affect the reaction unfavorably. It is not likely that the container would actually melt, since the total internal energy of the plasma, as distinguished from its temperature, is not very great.

18.14. In fission, a large nucleus splits into smaller ones; in fusion, two small nuclei join to form a larger one. In both processes, the products of the reaction have less mass than the original nucleus or nuclei, with the missing mass being released as energy.

18.15. The number of protons and neutrons in its nucleus determines the mass of an atom, and the number of electrons in the electron cloud surrounding the nucleus governs its chemical behavior.

18.16. 7 p, 8 n; 17 p, 18 n; 30 p, 34 n; 80 p, 120 n

18.17. 10.81 u

18.18. 7.71 MeV; 2.57 MeV

18.19. 298 MeV; 8.5 MeV

18.20. 41.9586 u

18.21. 2.01411 u

18.22. $^{17}_{8}\text{O}$; $^{1}_{0}n$; $^{2}_{1}\text{H}$

18.23. 9×10^{14} J; 2.25×10^5 tons

18.24. 7.27 MeV

Chapter 19

Radioactivity and Elementary Particles

RADIOACTIVE DECAY

Certain nuclei are unstable and undergo *radioactive decay*. Three kinds of radiation may be emitted:

1. *Alpha particles,* which are $_2^4$He nuclei;
2. *Beta particles,* which are electrons;
3. *Gamma rays,* which are high-energy photons whose frequencies are higher than those of X-rays.

A nucleus may be unstable because it is so large that the short-range forces holding its nucleons together are unable to balance the long-range electric repulsive forces its protons exert upon each other. Such a nucleus can achieve smaller size and hence more stability by emitting an alpha particle, which reduces its mass number by 4 and its atomic number by 2. Another source of instability is too great or too small a number of neutrons for the number of protons present.   In the former case a neutron can become a proton within the nucleus and an electron is emitted, which constitutes beta decay:

$$n^0 \longrightarrow p^+ + e^-$$

In the latter case a proton becomes a neutron and a positron (positively charged electron) is emitted:

$$p^+ \longrightarrow n^0 + e^+$$

Gamma decay, in which a nucleus emits one or more photons, occurs when a nucleus has excess energy, which often happens following alpha or beta decay.

HALF-LIFE

A nucleus subject to radioactive decay always has a certain definite probability of decay during any time interval.   The *half-life* of a radioactive isotope is the time required for half of any initial quantity of it to decay.   If an isotope has a half-life of, say, 5 hr and we start with 100 g of it, after 5 hr, 50 g will be left undecayed; after 10 hr, 25 g will be left undecayed; after 15 hr, 12.5 g will be left undecayed; and so on.

ELEMENTARY PARTICLES

The protons, neutrons, and electrons of ordinary matter are called *elementary particles* because they do not consist of smaller particles as, for instance, atoms do.   A great many other elementary particles are known, nearly all of which are highly unstable and decay soon after coming into being.

Elementary particles other than the photon fall into two main classes: *hadrons,* which are subject to the strong nuclear interaction, and *leptons,* which are not.   Electrons are leptons, nucleons are hadrons.   Hadrons are further divided into *mesons,* whose masses are less than that of the proton, and *baryons,* whose masses are greater.

ANTIPARTICLES

Most elementary particles have *antiparticles*. The antiparticle of a particle has the same rest mass but its electric charge is opposite in sign. Thus the positron (e^+) is the antiparticle of the electron (e^-), and the antiproton (p^-) is the antiparticle of the proton (p^+). The neutron, too, has an antiparticle because, although uncharged, it has other properties (such as angular momentum) which are different in the two forms. The photon is an example of a particle which has no antiparticle.

If a particle and its antiparticle come together, they are *annihilated*: the two disappear simultaneously with their mass being converted into energy. Thus the annihilation of a positron and an electron yields two gamma-ray photons; two photons must carry off the energy in order that momentum be conserved.

The reverse of annihilation is also possible. In *pair production*, a particle and its antiparticle are created from energy when a gamma-ray photon (or other particle) of sufficient energy passes near a nucleus.

Antimatter composed of antiprotons, antineutrons, and positrons would behave exactly like ordinary matter and it is entirely possible that parts of the universe consist of antimatter rather than of ordinary matter. If matter and antimatter come together, of course, both will be annihilated with the evolution of much energy.

Solved Problems

19.1. What happens to the atomic number and mass number of a nucleus that emits (a) an electron? (b) A positron? (c) An alpha particle?

(a) Z increases by 1, A is unchanged. (b) Z decreases by 1, A is unchanged. (c) Z decreases by 2, A decreases by 4.

19.2. How many successive alpha decays occur in the decay of the thorium isotope $^{228}_{90}$Th into the lead isotope $^{212}_{82}$Pb?

Each alpha decay means a reduction of 2 in atomic number and of 4 in mass number. Here Z decreases by 8 and A by 16, which means that 4 alpha particles are emitted.

19.3. The oxygen isotopes $^{14}_{8}$O and $^{19}_{8}$O are both unstable and undergo beta decay. Which would you think emits an electron and which a positron?

$^{19}_{8}$O has five more neutrons than $^{14}_{8}$O for the same number of protons. Hence $^{19}_{8}$O emits an electron, thereby converting one of its neutrons into a proton, and $^{14}_{8}$O emits a positron, thereby converting one of its protons into a neutron.

19.4. Tritium is the hydrogen isotope $^3_1$H whose nucleus contains two neutrons and a proton. Tritium is beta-radioactive and emits an electron. (a) What does tritium become after beta decay? (b) The half-life of tritium is 12.5 years. How much of a 1-g sample will remain undecayed after 25 years?

(a) In the beta decay of a nucleus, one of its neutrons becomes a proton. Since the atomic number 2 corresponds to helium, the beta decay of $^3_1$H is given by

$$^3_1\text{H} \longrightarrow \ ^3_2\text{He} + e^-$$

and the new atom is $^3_2$He.

(b) Twenty-five years is two half-lives of tritium, and so $\frac{1}{2} \times \frac{1}{2} \times 1$ g $= \frac{1}{4}$ g of tritium remains undecayed.

19.5. The half-life of the sodium isotope $^{24}_{11}$Na against beta decay is 15 hr.   How long does it take for 7/8 of a sample of this isotope to decay?

After 7/8 has decayed, 1/8 is left, and $\frac{1}{8} = \frac{1}{2} \times \frac{1}{2} \times \frac{1}{2}$ which is 3 half-lives.   Hence the answer is 3×15 hr $= 45$ hr.

19.6. The carbon isotope $^{14}_{6}$C (called "radiocarbon") is beta-radioactive with a half-life of 5600 years.   Radiocarbon is produced in the earth's atmosphere by the action of cosmic rays on nitrogen atoms, and the carbon dioxide of the atmosphere contains a small proportion of radiocarbon as a result.   All plants and animals therefore contain a certain amount of radiocarbon along with the stable isotope $^{12}_{6}$C. When a living thing dies, it stops taking in radiocarbon, and the radiocarbon it already contains decays steadily.   By measuring the ratio between the $^{14}_{6}$C and $^{12}_{6}$C contents of the remains of an animal or plant and comparing it with the ratio of these isotopes in living organisms, the time that has passed since the death of the animal or plant can be found.   (a) How old is a piece of wood from an ancient dwelling if its relative radiocarbon content is $\frac{1}{4}$ that of a modern specimen?   (b) If it is $\frac{1}{16}$ that of a modern specimen?

(a) Since $\frac{1}{4} = \frac{1}{2} \times \frac{1}{2}$, the specimen is two half-lives old, which is 11,200 years old.

(b) Since $\frac{1}{16} = \frac{1}{2} \times \frac{1}{2} \times \frac{1}{2} \times \frac{1}{2}$, the specimen is four half-lives old, which is 22,400 years old.

19.7. The rate at which a sample of a radioactive substance decays is called its *activity*. The unit of activity is the *curie*, where 1 curie $= 3.7 \times 10^{10}$ decays/s.   If a luminous watch dial contains 5 microcuries of the radium isotope $^{226}_{88}$Ra, how many decays per second occur in it?   (This isotope emits alpha particles which cause flashes of light when they strike a special material the isotope is mixed with.)

A microcurie is 10^{-6} curies, and so the activity of the watch dial is

$$3.7 \times 10^{10} \, \frac{\text{decays/s}}{\text{curie}} \times 5 \times 10^{-6} \, \text{curie} \; = \; 1.85 \times 10^5 \, \text{decays/s}$$

19.8. The *neutrino* is a massless lepton whose properties are discussed in most physical science texts.   What are the similarities and differences between the photon and the neutrino?

Both particles have no rest mass but can possess energy, linear momentum, and angular momentum.   The photon is associated with the electromagnetic interaction; the neutrino is associated with the weak interaction.   There is only one kind of photon, which has no antiparticle; there are two kinds of neutrino, each with an antiparticle.

19.9. Neutrinos are able to travel immense distances through matter before interacting — the equivalent of 10^{18} m of iron on the average. Why?

Neutrinos can interact with anything else only via the weak interaction, which is very much feebler than the electromagnetic or strong interactions.

19.10. (a) What is the minimum energy a gamma-ray photon must have if it is to materialize into a neutron-antineutron pair?   (b) Is this more or less than the minimum energy needed to form a proton-antiproton pair?

(a) The neutron mass is 1.008665 u, so the energy equivalent of a neutron-antineutron pair is

$$E \; = \; 2 \times 1.008665 \text{ u} \times 931 \text{ MeV/u} \; = \; 1878 \text{ MeV}$$

(b) Since the neutron mass is greater than the proton mass, more energy is needed to form a neutron-antineutron pair than to form a proton-antiproton pair.

19.11. If baryons are assigned the *baryon number* $B = +1$ and antiprotons are assigned the baryon number $B = -1$, it is found that in all known processes the total value of B remains constant. How does the conservation of B allow the neutron to beta decay into a proton in free space, whereas the proton cannot decay at all in free space?

When a neutron undergoes beta decay, $B = +1$ both before and after the decay since $B = +1$ for both neutron and proton. Energy is also conserved in this decay since the neutron mass is greater than the proton mass. However, since the proton is the lightest baryon, it cannot decay in free space into any other baryon, such as a neutron, and still conserve energy. (Inside a nucleus this can occur, because energy is available from the other nucleons.) The conservation of B means that the proton cannot decay into a meson or lepton, for which $B = 0$. Hence there is no way for the proton to decay in free space, where it is stable.

Supplementary Problems

19.12. Radium spontaneously decays into the elements helium and radon. Why is radium itself considered an element and not simply a chemical compound of helium and radon?

19.13. What happens to the atomic number and mass number of a nucleus that emits a gamma-ray photon? What happens to its mass?

19.14. The uranium isotope $^{238}_{92}\text{U}$ decays into a stable lead isotope through the successive emission of 8 alpha particles and 6 electrons. What is the symbol of the lead isotope?

19.15. The half-life of $^{238}_{92}\text{U}$ against alpha decay is 4.5×10^9 years. How long does it take for 7/8 of a sample of this isotope to decay? For 15/16 to decay?

19.16. The half-life against beta decay of the strontium isotope $^{90}_{38}\text{Sr}$ is 28 years. (a) What does $^{90}_{38}\text{Sr}$ become after beta decay? (b) What percentage of a sample of $^{90}_{38}\text{Sr}$ will remain undecayed after 112 years?

19.17. How can a positron be distinguished from an electron in a device such as a bubble chamber in which the paths of charged particles are made visible?

19.18. Which of the four fundamental interactions is the proton subject to? The neutron? The electron?

19.19. A neutron in free space decays into a proton, an electron, and an antineutrino. Into what does an antineutron decay?

19.20. What is the minimum binding energy a neutron must contribute to a nucleus of which it is a part in order that the neutron not undergo beta decay inside the nucleus? How does this figure compare with the observed binding energies per nucleon in stable nuclei?

19.21. A particle called the *neutral pion* (symbol π^0) has a rest mass of 264 electron masses. The π^0 has an average lifetime of 9×10^{-17} s and decays into two gamma-ray photons of equal energy. (a) What is the energy of each photon? (b) Why must the photons have the same energy?

19.22. A positron with a kinetic energy of 0.5 MeV collides head on with an electron whose kinetic energy is also 0.5 MeV. Find the energy and frequency of each of the resulting photons.

19.23. The average lifetime of many elementary particles is about 10^{-23} s, and they move at nearly the velocity of light after being created in high-energy collisions between other elementary particles. Compare the distance such a particle can travel before it decays with nuclear diameters, which are typically about 10^{-14} m.

Answers to Supplementary Problems

19.12. Helium and radon cannot be combined to form radium, nor can radium be broken down into helium and radon by chemical means.

19.13. Z and A are unchanged, but the actual mass decreases in proportion to the energy lost.

19.14. $^{206}_{82}\text{Pb}$

19.15. 1.35×10^{10} yr; 1.8×10^{10} yr

19.16. $^{90}_{39}\text{Y}$; 6.25%

19.17. By applying a magnetic field to the chamber and observing which way the particle's path is deviated.

19.18. The proton and neutron are subject to all four interactions, the electron to all except the strong interaction.

19.19. An antiproton, a positron, and a neutrino.

19.20. 0.78 MeV, which is less than the observed binding energies per nucleon in stable nuclei.

19.21. 68 MeV; in order that linear momentum be conserved

19.22. 1.01 MeV; 2.4×10^{20} Hz

19.23. Such a particle travels about 3×10^{-15} m before decaying, which is less than a nuclear diameter.

Chapter 20

Theory of the Atom

BOHR MODEL OF THE HYDROGEN ATOM

The hydrogen atom consists of a single electron and a single proton, which is the nucleus. In the classical model of this atom, the electron is imagined to circle the proton in an orbit such that the electrical attraction of the proton provides the required centripetal force. The flaw in this model is that, according to electromagnetic theory, because the electron is accelerated it must radiate electromagnetic waves and thus it will lose energy until it spirals into the proton.

In the Bohr model of the hydrogen atom, it is postulated that stable orbits exist in which the angular momentum of the electron is a multiple of $h/2\pi$, that is, that the angular momentum is $nh/2\pi$ where $n = 1, 2, 3, \ldots$. This postulate is equivalent to requiring that each orbit be a whole number of de Broglie wavelengths in circumference. If n is the *quantum number* of an orbit, then the orbit radius is

$$r_n = n^2 r_1, \quad n = 1, 2, 3, \ldots$$

where r_1, the radius of the smallest orbit, is 5.3×10^{-11} m.

ENERGY LEVELS

The total energy (kinetic energy plus electric potential energy) of a hydrogen atom whose electron is in the nth orbit is given by

$$E_n = \frac{E_1}{n^2}, \quad n = 1, 2, 3, \ldots$$

where $E_1 = -13.6 \text{ eV} = -2.18 \times 10^{-18}$ J. The permitted energies of an atom are called its *energy levels*. The energy levels are all negative, which means that the electron does not have enough energy to escape from the proton. The lowest level, corresponding to $n = 1$, is the *ground state*; higher levels are *excited states*. As n increases, E_n approaches zero; when $E_n = 0$, the electron is no longer bound to the proton and the atom breaks up. The work needed to remove an electron from an atom in its ground state is called the *ionization energy*; it is 13.6 eV in the case of hydrogen.

ATOMIC SPECTRA

When a gas or vapor is excited by the passage of an electric current, light is given off which consists of certain specific wavelengths only. Every element has a characteristic *emission line spectrum*. The wavelengths in this spectrum fall into definite series whose member wavelengths are related by simple formulas.

When white light is passed through a cool gas or vapor, light of certain specific wavelengths is absorbed. The wavelengths in the resulting *absorption line spectrum* correspond to a number of the wavelengths in the emission spectrum of that element.

Line spectra owe their origin to the presence of energy levels in atoms. An atom in an excited state can remain there only a brief time (normally about 10^{-8} s) before dropping to a lower state. The difference in energy appears as a photon of frequency f, where

$$E_{\text{initial}} - E_{\text{final}} = hf$$

The accompanying energy level diagram shows the possible transitions in the hydrogen atom that are responsible for its emission line spectrum. The larger the energy difference between initial and final energy levels, as indicated by the lengths of the arrows, the higher the frequency of the photon that is emitted. The names of the various spectral series in hydrogen are indicated.

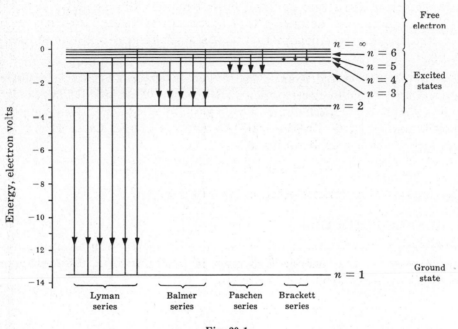

Fig. 20-1

An absorption spectrum is produced by transitions in the opposite direction, from the ground state to excited states. Light of frequencies that correspond to the various energy differences is absorbed by atoms illuminated by light whose spectrum is continuous (that is, which contains all frequencies). These atoms then reradiate light as they fall to their ground states, but the reradiation occurs in random directions and so is much fainter in the direction of the original beam.

QUANTUM THEORY OF THE ATOM

In the quantum theory of the atom, no compromise is made with mechanical analogies and instead an entirely probabilistic concept is developed. This theory holds for many-electron atoms as well as for the hydrogen atom. Four quantum numbers turn out to be needed to describe the physical state of an atomic electron, in place of the single quantum number of the Bohr model. These are as follows:

Name	Symbol	Possible values	Quantity determined
Principal	n	$1, 2, 3, \ldots$	Electron energy
Orbital	ℓ	$0, 1, 2, \ldots, n-1$	Magnitude of angular momentum
Magnetic	m_ℓ	$-\ell, \ldots, 0, \ldots, +\ell$	Direction of angular momentum
Spin magnetic	m_s	$-\frac{1}{2}, +\frac{1}{2}$	Direction of electron spin

The possible energies of the electron are chiefly determined by n and only to a smaller extent by ℓ and m_ℓ. For the hydrogen atom, $E_n = E_1/n^2$ in the quantum theory as in the Bohr theory.

Every electron behaves in certain respects as though it is a spinning charged sphere. The amount of spin is the same for every electron, but there are two possible directions in which the angular momentum vector can point in a magnetic field, "up" ($m_s = +\frac{1}{2}$) and "down" ($m_s = -\frac{1}{2}$).

ATOMIC ORBITALS

The distribution in space of the *probability density* ψ^2 of an atomic electron depends upon its quantum numbers n, ℓ, and m_ℓ. As mentioned in Chapter 17, the larger the value of ψ^2 in a certain place at a certain time, the greater the likelihood of finding the electron there. The quantum theory of the atom enables ψ^2 to be calculated for any combination of n, ℓ, and m_ℓ values for any atom. The region in space where ψ^2 is appreciable for a given quantum state is called an *orbital*. The ℓ value of an orbital is usually designated by a letter according to this scheme:

$$\ell = 0: \ s \qquad \ell = 1: \ p \qquad \ell = 2: \ d \qquad \ell = 3: \ f \qquad \ell = 4: \ g$$

(Higher values of ℓ are designated in alphabetical order, so that $\ell = 5$ is h and so forth.) Thus a $3p$ orbital corresponds to $n = 3$, $\ell = 1$.

THE EXCLUSION PRINCIPLE

According to the *Pauli exclusion principle*, no two electrons in an atom can exist in the same quantum state. Thus each electron must be described by a different set of quantum numbers n, ℓ, m_ℓ, m_s.

Solved Problems

20.1. In the Bohr model of the hydrogen atom, the radius of the electron's orbit is 5.3×10^{-11} m in the ground state. What aspect of the quantum-theoretical model might correspond to this relationship?

   The maximum probability of finding the electron occurs at this distance from the nucleus.

20.2. How many electrons are able to share an orbital in an atom?

   Since an orbital is characterized by a given n, ℓ, and m_ℓ the exclusion principle permits two electrons to occupy each orbital in an atom, one with $m_s = +\frac{1}{2}$ and the other with $m_s = -\frac{1}{2}$.

20.3. Describe two mechanisms by which the atoms of a gas can be excited so that they emit light whose frequencies make up the characteristic line spectrum of the element involved.

   (a) One mechanism is a collision with another atom, as a result of which some of their kinetic energy becomes excitation energy within one or both of the atoms. The excited atoms then lose this energy by emitting one or more photons. In an electric discharge in a gas, for instance in a neon sign or a sodium-vapor highway lamp, an electric field accelerates electrons and ions to velocities sufficient for atomic excitation.

(b) Another mechanism is the absorption by an atom of a photon of light for which hf is just right to raise the atom from its ground state to one of its excited states. If white light is directed at the gas, photons of those energies that correspond to such transitions will be absorbed. When the energy is reradiated, all of the possible transitions from the highest excited state reached will show up in the emitted light.

20.4. Find the energies of the first three excited states of the hydrogen atom.

Since $E_n = E_1/n^2$ and $E_1 = -13.6$ eV, we have

$$E_2 = \frac{E_1}{2^2} = \frac{-13.6 \text{ eV}}{4} = -3.4 \text{ eV}$$

$$E_3 = \frac{E_1}{3^2} = \frac{-13.6 \text{ eV}}{9} = -1.51 \text{ eV}$$

$$E_4 = \frac{E_1}{4^2} = \frac{-13.6 \text{ eV}}{16} = -0.85 \text{ eV}$$

20.5. Find the velocity of the electron in a ground-state hydrogen atom according to the Bohr model.

In the Bohr model, the de Broglie wavelength $\lambda = h/mv$ of the electron in the $n = 1$ state is equal to the orbit circumference of $2\pi r_1$. Hence

$$\lambda = \frac{h}{mv} = 2\pi r_1$$

$$v = \frac{h}{2\pi m r_1} = \frac{6.63 \times 10^{-34} \text{ J-s}}{2\pi \times 9.1 \times 10^{-31} \text{ kg} \times 5.3 \times 10^{-11} \text{ m}} = 2.2 \times 10^6 \text{ m/s}$$

20.6. To what temperature must a hydrogen sample be heated in order that the average molecular energy equal the binding energy of the hydrogen atom?

The binding (or ionization) energy of the hydrogen atom is 13.6 eV $= 2.18 \times 10^{-18}$ J. Since the average molecular energy in a gas whose absolute temperature is T is equal to $\frac{3}{2}kT$, here

$$\frac{3}{2} kT = E$$

$$T = \frac{2E}{3k} = \frac{2 \times 2.18 \times 10^{-18} \text{ J}}{3 \times 1.38 \times 10^{-23} \text{ J/K}} = 1.05 \times 10^5 \text{ K}$$

20.7. A proton and an electron come together to form a hydrogen atom in its ground state. Under the assumption that a single photon is emitted in this process, what is its frequency?

The energy of a hydrogen atom in its ground state is -13.6 eV $= -2.13 \times 10^{-18}$ J. Hence 2.18×10^{-18} J of energy must be given off when the atom is being formed, and if a single photon is emitted, its frequency is found from $E = hf$ to be

$$f = \frac{E}{h} = \frac{2.18 \times 10^{-18} \text{ J}}{6.63 \times 10^{-34} \text{ J-s}} = 3.3 \times 10^{15} \text{ Hz}$$

20.8. Which of the spectral series of hydrogen contains the shortest wavelengths? Why?

The Lyman series, since the transitions that lead to it represent the greatest energy differences and hence the highest frequencies.

20.9. Of the following transitions between energy levels in a hydrogen atom, which one involves (*a*) the emission of the photon of highest frequency? (*b*) The emission of the photon of lowest frequency? (*c*) The absorption of the photon of highest frequency? (*d*) The absorption of the photon of lowest frequency? The transitions are $n = 1$ to $n = 2$; $n = 2$ to $n = 1$; $n = 2$ to $n = 6$; and $n = 6$ to $n = 2$.

In general, photon emission occurs during a transition to a state of lower n, and photon absorption occurs during a transition to a state of higher n. By inspecting the energy level diagram of hydrogen, we see that the energy difference between the $n = 1$ and $n = 2$ levels is greater than that between the $n = 2$ and $n = 6$ levels. Hence the answers are as follows: (*a*) $n = 2$ to $n = 1$; (*b*) $n = 6$ to $n = 2$; (*c*) $n = 1$ to $n = 2$; (*d*) $n = 2$ to $n = 6$.

20.10. A sample of hydrogen gas is bombarded by a beam of electrons. How much energy must the electrons have if the first line of the Balmer spectral series, corresponding to a transition from the $n = 3$ state to the $n = 2$ state, is to be radiated?

The energy of the $n = 3$ state in hydrogen is

$$E_3 = \frac{E_1}{3^2} = \frac{-13.6 \text{ eV}}{9} = -1.5 \text{ eV}$$

The difference in energy between the ground ($n = 1$) state and the $n = 3$ state is $13.6 \text{ eV} - 1.5 \text{ eV} = 12.1 \text{ eV}$, so the energy needed by the bombarding electrons is 12.1 eV. The reason the energy difference between the $n = 1$ and $n = 3$ states is involved is that the hydrogen atoms are initially in the $n = 1$ state.

20.11. What are the quantum numbers that characterize the atomic state in which an electron has the lowest energy?

$$n = 1, \ \ell = 0, \ m_\ell = 0, \ m_s = \pm\tfrac{1}{2}$$

20.12. What are the possible values of the orbital and magnetic quantum numbers of an atomic electron whose principal quantum number is $n = 3$?

Since ℓ ranges from 0 up to $n - 1$ for an electron of principal quantum number n and m_ℓ ranges from $-\ell$ through 0 to $+\ell$ for an electron of orbital quantum number ℓ, we have here:

$$\ell = 0 \qquad m_\ell = 0$$

$$\ell = 1 \quad \begin{cases} m_\ell = +1 \\ m_\ell = 0 \\ m_\ell = -1 \end{cases}$$

$$\ell = 2 \quad \begin{cases} m_\ell = +2 \\ m_\ell = +1 \\ m_\ell = 0 \\ m_\ell = -1 \\ m_\ell = -2 \end{cases}$$

Supplementary Problems

20.13. In the Bohr model of the hydrogen atom, what condition must be obeyed by the electron if it is to move in its orbit indefinitely without radiating energy?

20.14. How does the ionization energy of an atom compare with the binding energy per nucleon of its nucleus?

20.15. What is the nature of the spectrum found in (a) light from the hot filament of a light bulb; (b) light from a neon sign; (c) light from a light bulb that has passed through cool neon gas?

20.16. Radiation with a continuous spectrum is passed through a container of hydrogen gas whose atoms are all in their ground states. What spectral series will appear in the resulting absorption spectrum?

20.17. Electrons of 13.0 eV energy are used to bombard a sample of hydrogen gas. Lines of what spectral series will be emitted?

20.18. How is the size of an orbital related to its principal quantum number?

20.19. Which quantum number is not involved in describing an atomic orbital?

20.20. The earth's mass is 6×10^{24} kg, the radius of its orbit around the sun is 1.5×10^{11} m, and its orbital velocity is 3×10^4 m/s. (a) Find the de Broglie wavelength of the earth. (b) Find the quantum number of the earth's orbit. (c) Do you think quantum considerations play an important part in the earth's orbital motion?

20.21. How much energy is needed to remove the electron from a hydrogen atom when it is in the $n = 4$ state?

20.22. How does the average energy per molecule in a gas at 20 °C compare with the energy needed to raise a hydrogen atom from its ground state to its lowest excited state?

20.23. Find the velocity of the electron in the $n = 5$ state of a hydrogen atom according to the Bohr model. Is this more or less than the velocity of the electron when it is in the $n = 1$ state?

20.24. To what transitions do the spectral lines of the highest and lowest frequencies in the Balmer series of hydrogen correspond?

20.25. How many m_ℓ values are possible for a $4f$ electron?

20.26. Verify that there are n^2 possible orbitals for every value of the principal quantum number n.

Answers to Supplementary Problems

20.13. The orbit must be one de Broglie wavelength in circumference; or, equivalently, the orbital angular momentum of the electron must be equal to $h/2\pi$.

20.14. The ionization energy is hundreds of thousands of times smaller than the binding energy per nucleon.

20.15. (a) continuous emission spectrum; (b) emission line spectrum; (c) absorption line spectrum

20.16. The Lyman series.

20.17. The highest energy level reached by the hydrogen atoms will be $n = 4$, so lines of the Paschen, Balmer, and Lyman series will be present.

20.18. The larger the value of n, the larger the size of the orbital.

20.19. The spin magnetic quantum number m_s.

20.20. (a) 3.68×10^{-63} m; (b) 2.56×10^{74}; (c) no

20.21. 0.85 eV

20.22. $KE_{av} = 6.07 \times 10^{-21}$ J; the energy difference between the $n = 1$ and $n = 2$ states is 1.64×10^{-18} J, which is 270 times greater.

20.23. 8.8×10^4 m/s; less

20.24. Highest frequency, $n = \infty$ to $n = 2$; lowest frequency, $n = 3$ to $n = 2$.

20.25. Seven, since $\ell = 3$ corresponds to an f state. They are $m_\ell = 0, +1, -1, +2, -2, +3, -3$.

Chapter 21

The Periodic Law

THE PERIODIC TABLE

According to the *periodic law,* when the elements are listed in order of increasing atomic number, elements with similar chemical and physical properties recur at regular intervals. A *periodic table* is a way of arranging the elements to exhibit these regularities. A simple periodic table is given in Appendix C.

Elements with similar properties form the *groups* that appear as vertical columns in the periodic table. Thus Group I consists of hydrogen plus the alkali metals, all of which are extremely active chemically and whose atoms tend to lose an electron when forming a compound. Group VII consists of the halogens, which are volatile, active nonmetals whose atoms tend to gain an electron when forming a compound. Group VIII (sometimes called Group 0) consists of the inert gases, elements so inactive that they almost never form compounds with other elements.

The horizontal rows in the periodic table are called *periods.* Across each period is a more or less steady transition from an active metal through less active metals and weakly active nonmetals to highly active nonmetals and finally to an inert gas. Within each column there are also regular changes in properties, but they are far less conspicuous than those in each period. For example, increasing atomic number in the alkali metals is accompanied by greater chemical activity, while the reverse is true in the halogens. In general, an active element is one whose compounds are very stable and require much energy to decompose; an inactive element usually forms unstable compounds that are readily decomposed.

A series of *transition elements* appears in each period after the third between the Group II and Group III elements. The transition elements are metals with considerable chemical resemblance to one another but no pronounced resemblance to the elements in the other groups. Fifteen of the transition elements in Period 6 are virtually indistinguishable in their properties and are known as the *lanthanide* (or *rare earth*) elements. A similar group of closely related metals, the *actinide* elements, is found in Period 7.

ATOMIC STRUCTURE

The two basic rules that govern the electron structures of many-electron atoms are:

1. An atom is stable when all its electrons are in quantum states of the lowest energy possible.

2. Only one electron can occupy each quantum state of an atom; this is the exclusion principle.

Electrons with the same principal quantum number n in an atom are usually about the same distance from the nucleus and have similar energies. Such electrons are said to occupy the same *shell.* The higher the value of n, the greater the energy. Electrons in a given shell that have different orbital quantum numbers ℓ have different probability-density distributions and therefore somewhat different energies; those with the same value of ℓ are said to occupy the same *subshell* and have very nearly the same energy. The higher the value of ℓ in a given subshell, the greater the energy.

124

As described in Chapter 20, a subshell is identified by its principal quantum number n followed by a letter corresponding to its orbital quantum number ℓ. A superscript after the letter indicates the number of electrons in that subshell in a particular atom. Thus the electron configuration of sodium is written $1s^2 2s^2 2p^6 3s^1$, which means that this atom contains two $1s$ electrons ($n = 1$, $\ell = 0$), two $2s$ electrons ($n = 2$, $\ell = 0$), six $2p$ electrons ($n = 2$, $\ell = 1$), and one $3s$ electron ($n = 3$, $\ell = 0$). The sequence in which electron subshells are filled in atoms corresponds to the order of increasing electron energy:

$$1s,\ 2s,\ 2p,\ 3s,\ 3p,\ 4s,\ 3d,\ 4p,\ 5s,\ 4d,\ 5p,\ 6s,\ 4f,\ 5d,\ 6p,\ 7s,\ 6d$$

The sequence is not always in order of principal quantum number n because orbitals of high ℓ are relatively far from the nucleus and the corresponding energy levels may be higher than those for low-ℓ subshells of the next shell. Thus a $4s$ electron is, on the average, closer to the nucleus than a $3d$ electron, and the $4s$ subshell is filled before the $3d$ subshell.

EXPLAINING THE PERIODIC TABLE: THE INERT GASES

An atomic shell or subshell that contains the maximum number of electrons permitted by the exclusion principle is said to be *closed*. A closed s subshell holds two electrons, a closed p subshell holds six electrons, a closed d subshell holds ten electrons, and so on.

The electrons in a closed shell are all tightly bound to the atom, since the positive nuclear charge that holds them in place is large compared with the negative charge of the electrons in inner shells. For instance, each $n = 2$ electron in neon ($Z = 10$) is acted on by the $+10e$ nuclear charge minus the effect of the two electrons in the inner $n = 1$ shell, so the effective charge acting on it is $+8e$ (Fig. 21-1). In addition, an atom that consists only of closed shells or sub-

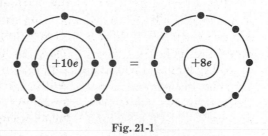

Fig. 21-1

shells has a uniform distribution of electron probability density ψ^2, so that it does not attract other electrons nor are any of its own electrons easily removed. An atom of this kind should exhibit little if any chemical activity, and the inert gases of Group VIII, whose electron structures contain only closed shells or subshells, indeed behave in this way.

THE ALKALI METALS

An atom that has only a single electron in its outer shell tends to lose this electron. There are two reasons for this. First, the outer electron is shielded by electrons in inner shells from all but $+e$ of the nuclear charge. Second, the outer electron is relatively far from the nucleus, and the electric force between charges varies as $1/r^2$. The elements in Group I of the periodic table, namely hydrogen and the alkali metals, all have structures with single outer electrons, and the ease with which these electrons can be detached accounts for their chemical behavior.

THE HALOGENS

Atoms whose outer shells lack a single electron of being closed readily pick up such an electron because of the strong attraction of the nuclear charge, which is only partly shielded by the inner electrons. In fluorine, for instance, the nuclear charge of $+9e$ is shielded by only two electrons in the inner $n = 1$ shell, so the effective charge felt by an outer electron is $+7e$. The resulting attractive force is sufficient to overcome the repulsive force exerted by the other electrons, and the fluorine atom reacts chemically by acquiring an electron. The elements in Group VII of the periodic table, namely the halogens, have similar structures and similar chemical behavior.

Solved Problems

21.1. The transition elements in any period have the same or nearly the same outer electron shells and add electrons successively to inner shells. How does this bear upon their chemical similarity?

The outermost electron shell of an atom determines its chemical behavior, hence the similarity of the outermost shells of the transition elements means that their chemical behavior must also be similar.

21.2. What is the effective nuclear charge that acts on each electron in the outer shell of the chlorine ($Z = 17$) atom?

The chlorine atom has closed $n = 1$ and $n = 2$ shells with, respectively, two and eight electrons in each. Hence the $n = 3$ shell contains seven electrons that are shielded by the ten inner ones, and the effective charge acting on these $n = 3$ electrons is accordingly $+7e$ (Fig. 21-2).

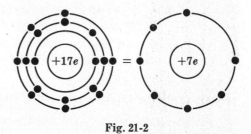

Fig. 21-2

21.3. Why is a chlorine atom able to pick up only a single electron when it reacts chemically whereas an oxygen atom can pick up two electrons?

A chlorine atom lacks one electron of a closed outer shell. When it picks up an electron, which it can do easily because the other electrons in the outer shell do not shield the nuclear charge, this shell is closed and any further electrons would have to go into the next shell. However, an electron in the next shell would find a net charge of $-e$ inside its orbital, not a net charge of $+7e$ as the electrons in the previous shell do, and hence would be repelled rather than held to the atom. Since an oxygen atom lacks two electrons of a closed outer shell, it can accommodate two additional electrons in this shell; further electrons would be repelled as in the case of a Cl^- ion.

21.4. Why does a closed d subshell contain ten electrons?

The electrons in a d subshell have the orbital quantum number $\ell = 2$. The possible values of the magnetic quantum number m_ℓ when $\ell = 2$ are $m_\ell = -2, -1, 0, +1, +2$, which is a total of five. An electron with a given value of m_ℓ can have $m_s = \pm\frac{1}{2}$, so the total number of permitted quantum states in a d subshell is twice five or ten.

21.5. What is the electron configuration of strontium, $Z = 38$?

We begin by noting that an s subshell can contain at most 2 electrons, a p subshell can contain at most 6 electrons, and a d subshell can contain at most 10 electrons. Now we write down the various subshells in the order in which they are filled, keeping track of the total number of electrons until it reaches 38:

Subshell	Electrons in subshell	Configuration	Total electrons
$1s$	2	$1s^2$	2
$2s$	2	$2s^2$	4
$2p$	6	$2p^6$	10
$3s$	2	$3s^2$	12
$3p$	6	$3p^6$	18
$4s$	2	$4s^2$	20
$3d$	10	$3d^{10}$	30
$4p$	6	$4p^6$	36
$5s$	2	$5s^2$	38

Although the $4s$ subshell is filled before the $3d$ subshell, it is customary to list closed subshells in the order of their principal quantum number n. Thus the electron configuration of strontium is $1s^22s^22p^63s^23p^63d^{10}4s^24p^65s^2$.

21.6. An atom has the electron configuration $1s^22s^22p^53s^1$. (a) What kind of atom is this? (b) Is it in its ground state or in an excited state? (c) If it is in an excited state, what is the ground state?

(a) There are $2+2+5+1 = 10$ electrons, hence $Z = 10$ and the atom is a neon atom.

(b) The $2p$ subshell can contain a total of 6 electrons. Since there are only 5 in this subshell here and one electron in the $3s$ subshell whose energy is greater, the atom must be in an excited state.

(c) In the ground state of an atom the subshells of lowest energy are occupied first. Hence the ground-state configuration of neon is $1s^22s^22p^6$.

21.7. Is it possible for an atom to have the electron configuration $1s^22s^32p^4$?

An s subshell can contain a maximum of two electrons, hence no atom can have this configuration.

21.8. What is the electron configuration of the lithium ion Li^+ in its ground state?

The superscript plus sign means that the atom has a net charge of $+e$, hence one electron is missing from its normal three. Such an ion in its ground state has the same configuration as a ground-state atom with two electrons, namely $1s^2$.

21.9. What is the electron configuration of the oxygen ion O^{--} in its ground state?

The two superscript minus signs mean that the atom has a net charge of $-2e$, hence that it has two electrons more than its normal eight. Such an ion in its ground state has the same configuration as a ground-state atom with 10 electrons, namely $1s^22s^22p^6$.

21.10. The ionization energy of an atom is the energy required to detach one of its outer electrons. Figure 21-3 shows how the ionization energies of the elements vary with atomic number. Account for the main features of the graph.

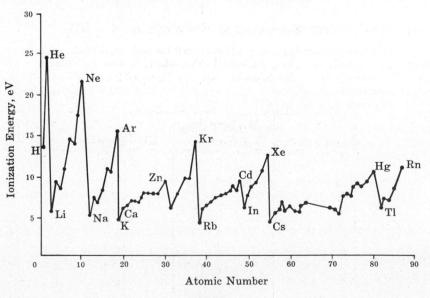

Fig. 21-3

The atoms of the alkali metals have the lowest ionization energies because each has a single *s* electron in its outer shell. The electrons in the inner shells partially shield the outer electron from the full nuclear charge of $+Ze$ so that the effective charge holding the outer electron to the atom is just $+e$ rather than $+Ze$. Relatively little work is needed to detach such an electron and the alkali metals form positive ions readily. The larger an atom, the farther its outer electrons are from the nucleus and the weaker is the force on them; this is why the ionization energy generally decreases going down any group in the periodic table. The increase in ionization energy from left to right across any period is accounted for by the increase in nuclear charge while the number of inner shielding electrons stays constant. Thus the inert gases have the highest ionization energies in their respective periods.

21.11. What relationship would you expect between the chemical activity of a metal and its ionization energy? How is this borne out in the alkali metals?

The lower the ionization energy of a metal, the more easily one of its electrons can be detached in a chemical reaction and hence the more active it is. In the alkali metals, both ionization energy and chemical activity increase with atomic number.

21.12. Why do the alkali metals have the largest atoms in each period?

Each alkali metal atom has a single electron in its outer shell, hence this electron is shielded by the inner electrons from all but $+e$ of the nuclear charge. The attractive force on the outer electron is therefore weak, and the repulsion of the inner electrons keeps this electron relatively far from the nucleus and so results in a large effective size for the atom.

21.13. Why does the effective size of an atom decrease more or less steadily from left to right across each period?

Across each period, electrons are usually added one by one to the outer shells of the atoms. These electrons are about the same distance from the nucleus and cannot shield each other from its increasing positive charge, whose force acts to pull the electrons inward and thus to reduce progressively the size of the atom.

Supplementary Problems

21.14. Are the majority of elements metals or nonmetals?

21.15. What is true in general of the chemical properties of the elements in each vertical column of the periodic table? Of the elements in each horizontal row?

21.16. How many electrons are present in the outermost shells of the atoms of Group II elements? Are these elements metals or nonmetals?

21.17. What is the effective nuclear charge that acts on each electron in the outer shell of the calcium ($Z = 20$) atom? Would you think that such an electron is relatively easy or relatively hard to detach from the atom?

21.18. What is the effective nuclear charge that acts on each electron in the outer shell of the sulfur ($Z = 16$) atom? Would you think that such an electron is relatively easy or relatively hard to detach from the atom?

21.19. How many electrons are there in a closed f ($\ell = 3$) subshell?

21.20. What is the ground-state electron configuration of sulfur, $Z = 16$?

21.21. What is the ground-state electron configuration of calcium, $Z = 20$?

21.22. Give the electron configuration of oxygen in its excited state of lowest energy.

21.23. Of the following electron configurations, (a) which represent atoms in their ground states? (b) Which represent atoms in excited states? (c) Which are impossible? State the names and symbols of the elements that correspond to the atoms of (a) and (b) and give the ground-state configurations of the atoms of (b).

$$1s^2 2s$$
$$1s^2 2p$$
$$1s^2 2s^2 3s$$
$$1s^2 2s^2 2p 3s$$
$$1s^2 2s^2 2p^8 3s^2$$
$$1s^2 2s^2 2p^6 3s^2 3p^5$$

21.24. What are the electron configurations of the following negative ions in their ground states? F^-, S^{--}, N^{---}.

21.25. What are the electron configurations of the following positive ions in their ground states? H^+, Na^+, Ca^{++}.

21.26. In each of the following pairs of atoms, which would you expect to be larger? Li and F; Li and Na; F and Cl; Na and Si.

21.27. Would you expect a Na atom or a Na^+ ion to be larger? A Cl atom or a Cl^- ion?

21.28. In each of the following pairs of atoms, which would you expect to have the greater chemical activity? Li and Be; Li and Na; Cl and Br; Br and Kr.

Answers to Supplementary Problems

21.14. Most elements are metals, as the sketch of the periodic table, Fig. 21-4, shows.

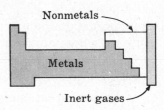

Fig. 21-4

21.15. they are similar; they are different

21.16. two electrons; metals

21.17. $+2e$; relatively easy

21.18. $+6e$; relatively hard

21.19. 14 electrons

21.20. $1s^2 2s^2 2p^6 3s^2 3p^4$

21.21. $1s^2 2s^2 2p^6 3s^2 3p^6 4s^2$

21.22. $1s^2 2s^2 2p^3 3s$

21.23. (a) $1s^2 2s$, lithium, Li
$1s^2 2s^2 2p^6 3s^2 3p^5$, chlorine, Cl

  (b) $1s^2 2p$, lithium, Li, $1s^2 2s$
$1s^2 2s^2 3s$, boron, B, $1s^2 2s^2 2p$
$1s^2 2s^2 2p 3s$, carbon, C, $1s^2 2s^2 2p^2$

  (c) $1s^2 2s^2 2p^8 3s^2$ is impossible because the maximum number of electrons in a p subshell is 6.

21.24. $1s^2 2s^2 2p^6$; $1s^2 2s^2 2p^6 3s^2 3p^6$; $1s^2 2s^2 2p^6$

21.25. H^+ is a bare proton with no electrons around it; $1s^2 2s^2 2p^6$; $1s^2 2s^2 2p^6 3s^2 3p^6$

21.26. Li; Na; Cl; Na

21.27. Na atom; Cl^- ion

21.28. Li; Li; Cl; Br

Chapter 22

Chemical Bonding

CHEMICAL BONDS

When a compound is formed, atoms of the elements present are linked together by *chemical bonds*. It is customary to classify chemical bonds as being *ionic* or *covalent,* although actual bonds are often intermediate between the two extremes. In an ionic bond, one or more electrons from one atom are transferred to another atom, and the resulting positive and negative ions then attract each other. In a covalent bond, one or more pairs of electrons are shared by two adjacent atoms.

A molecule is a group of atoms that are held together tightly enough by covalent bonds to behave as a single particle. A molecule always has a definite composition and structure, and has little tendency to gain or lose atoms. Ionic bonds usually result in crystalline solids, not in molecules; such solids consist of aggregates of positive and negative ions in a stable arrangement characteristic of the compound involved. Some crystalline solids are covalent rather than ionic, as discussed below.

THE IONIC BOND

Ionic bonds occur between metal atoms, which tend to lose their outermost electrons, and nonmetal atoms, which tend to gain electrons to complete their outermost shells. Thus sodium, whose atoms have one $n = 3$ electron, can combine with chlorine, whose atoms have seven $n = 3$ electrons and thus can accommodate another electron, by the transfer of electrons from Na atoms to Cl atoms to produce Na^+ and Cl^- ions.

The ionization energy (see Problem 21.10) of a metal atom is a measure of how easily one of its outermost electrons can be detached; the lower the ionization energy, the more readily a metal atom can contribute an electron to an ionic bond and hence the more chemically active it is. Thus potassium, with an ionization energy of 4.3 eV, is more active than sodium, whose ionization energy is 5.1 eV.

The *electron affinity* of a nonmetal atom is the energy released when an electron is added to it; this is the same as the work needed to remove the additional electron. The greater the electron affinity, the more securely the additional electron is held, and the more chemically active the atom is. Thus chlorine, with an electron affinity of 3.6 eV, is more active than oxygen, whose electron affinity is 1.5 eV. Since Group I and II elements have the lowest ionization energies and Group VI and VII elements have the highest electron affinities, the combination of elements from each end of the periodic table usually produces an ionic compound. Compounds consisting entirely of nonmetals are covalent.

Ionic compounds are not limited to combinations of equal numbers of atoms of two elements, as in NaCl. Magnesium atoms, for instance, have two electrons in their outer shells and both of these electrons can be transferred to other atoms in forming an ionic bond; an example is magnesium bromide, $MgBr_2$, in which there are two Br^- ions for each Mg^{++} ion. Also, some groups of atoms, such as NO_3 and CO_3, act as units when they participate in ionic bonds, in the former case picking up one electron to become an NO_3^- ion and in the latter case picking up two electrons to become a CO_3^{--} ion.

THE COVALENT BOND

The wave nature of a moving electron is what permits it to be shared by two atoms instead of belonging to one of them exclusively. The probability density ψ^2 of such an electron does not have a sharp boundary but instead gradually decreases away from its "parent" atom; if another atom is close enough, the electron can shift across to it for a while, then return to first atom, and so forth. The region where ψ^2 is large for a shared electron is called a *molecular orbital* and, as in the case of atomic orbitals, two electrons of opposite spin can occupy it. An occupied molecular orbital that is concentrated in the region between two atoms is equivalent to a negative electric charge between two positive ions and thus acts to hold them together; it is a *bonding orbital*. Whenever they are available, such an orbital is occupied by two electrons, usually one contributed by each atom but sometimes both by one of them. A molecular orbital which is concentrated outside the atoms acts to pull them apart; it is an *antibonding orbital* and does not lead to a stable configuration.

A molecular orbital may be regarded as being composed of the atomic orbitals of the outer electrons of the atoms involved. Only s ($l = 0$) atomic orbitals are spherically symmetric. Other atomic orbitals have lobes of high ψ^2 in particular directions, and when an atom with such orbitals forms covalent bonds with other atoms, a definite geometrical arrangement is the result. Thus the water molecule H_2O is "bent" with an angle of 104.5° between the two O—H covalent bonds, and the ammonia molecule NH_3 is pyramidal in shape with an angle of 107.5° between the N—H covalent bonds.

In covalent bonds between dissimilar atoms, the shared electron pair usually tends to be closer to one of the atoms. One part of the molecule therefore is negatively charged and another part positively charged. Such molecules are called *polar covalent*.

MULTIPLE BONDS

More than one pair of electrons may be shared by two atoms. For instance, there are two covalent bonds in the oxygen molecule O_2, which can be represented as O=O, and three covalent bonds in the nitrogen molecule N_2, which can be represented as N≡N. Carbon atoms, with four electrons in their outer shells, can participate in four covalent bonds; an example is the ethylene molecule C_2H_4, whose structure is

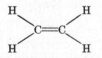

The chemistry of carbon is the subject of Chapter 32.

BONDING IN SOLIDS

Most solids are crystalline, with the ions, atoms, or molecules of which they consist being arranged in a regular pattern. Four kinds of bonds are found in crystals: ionic, covalent, metallic, and van der Waals.

A crystal of ordinary salt, NaCl, is an example of an ionic solid, with Na$^+$ and Cl$^-$ ions in alternate positions in a simple lattice (Fig. 22-1).

An example of a covalent solid is diamond, each of whose carbon atoms is joined by covalent bonds to four other carbon atoms in a structure that is repeated throughout the crystal (Fig. 22-2).

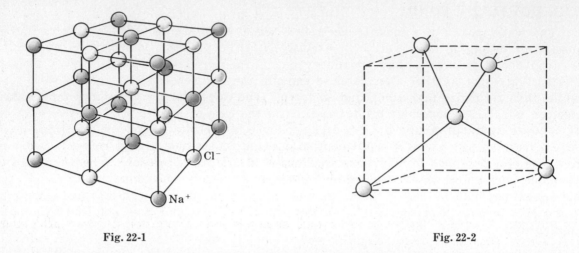

Fig. 22-1   Fig. 22-2

Both ionic and covalent solids are hard and have high melting points, which are reflections of the strength of the bonds. Ionic solids are much more common than covalent ones.

In a metal, the outermost electrons of each atom are shared by the entire assembly, so that a "gas" or "sea" of electrons moves relatively freely throughout it. The interaction between this electron sea and the positive metal ions leads to a cohesive force, much as in the case of the shared electrons in a covalent bond but on a larger scale. The presence of the free electrons accounts for such typical properties of metals as their opacity, surface luster, and high electrical and heat conductivities.

MOLECULAR CRYSTALS

All molecules, and even inert-gas atoms such as those of helium, exhibit weak, short-range attractions for one another due to *van der Waals forces*. These forces are responsible for the condensation of gases into liquids and the freezing of liquids into solids even in the absence of ionic, covalent, or metallic bonds between the atoms or molecules involved. Such familiar aspects of the behavior of matter as friction, viscosity, and adhesion arise from van der Waals forces.

Van der Waals forces are in essence due to the attraction of polar molecules for each other, with the positive end of one molecule being pulled to the negative end of the other. Not all molecules are permanently polar, of course, but the electrons of a nonpolar molecule are in constant motion and at each instant their distribution is not symmetric. When two nonpolar molecules are close together, these momentary charge asymmetries tend to shift together, with the positive part of one molecule always near the negative part of the other even though the locations of these parts are always changing. Van der Waals forces are quite weak, and substances composed of whole molecules, such as water, usually have low melting and boiling points and little mechanical strength in the solid state.

Solved Problems

22.1. When an ionic bond is formed, what keeps the ions of opposite sign from coming so close together that their electron structures overlap?

There are two reasons why such an overlap does not occur. First, if it were to happen, the positively charged nuclei of the two ions would no longer be electrically shielded by their electron structures and would repel each other directly. Second, the Pauli exclusion principle states that no two electrons in the same atomic system can exist in the same quantum state. If the electron structures of two ions overlap, they constitute a single atomic system, and in consequence some of the electrons must occupy quantum states of higher energy than their normal ones. Because atomic systems are stable only in configurations of lowest energy, such an overlap cannot take place.

22.2. Why do only the outermost electrons of an atom usually participate in bonding?

Inner electrons are much more tightly held to a nucleus both because they are closer to it and because they are shielded by fewer intervening electrons; hence they are unable either to transfer to another atom in an ionic bond or to be shared with another atom in a covalent bond. Also, the repulsive interatomic forces discussed in Problem 22.1 become dominant while the inner shells of the atoms involved are still relatively far apart.

22.3. Would you expect the inert gas atoms as a rule to have high or low ionization energies? High or low electron affinities?

The inert gas atoms have filled outer shells, hence the effective nuclear charge on their outer electrons is a maximum in their respective periods and the ionization energies are accordingly large. An additional electron would have to go to a new shell where the nuclear charge would be entirely shielded by the other electrons, hence the electron affinity of an inert gas atom is low.

22.4. What are the ions present in the crystal structure of MgO?

Since magnesium is in Group II of the periodic table, an Mg atom combines chemically by giving up the two electrons in its outer shell to form an Mg^{++} ion. An oxygen atom, on the other hand, combines chemically by picking up two electrons to complete its outer shell, so that an O atom tends to become an O^{--} ion. Thus MgO consists of Mg^{++} and O^{--} ions.

22.5. The melting points of NaF, NaCl, NaBr, and NaI are respectively 990 °C, 801 °C, 740 °C, and 660 °C. Why do the melting points vary in this way?

The sizes of the halogen atoms increase in the order F ($Z = 9$), Cl ($Z = 17$), Br ($Z = 35$), and I ($Z = 53$). The larger an atom, the less strongly its outer electrons are held since they are at greater distances from the nucleus. Hence the abilities of F, Cl, Br, and I atoms to hold electrons decrease in that order, and the ionic bonds they form with Na are weaker in the same order. The weaker the bonds in a solid are, the more readily its ordered structure can be disrupted into the disordered structure of a liquid.

22.6. In acetylene, H—C ⟨?⟩ C—H, how many covalent bonds are present between the carbon atoms?

A carbon atom normally forms four covalent bonds when it combines chemically. Since each carbon atom in acetylene shares a covalent bond with one hydrogen atom, it must share three bonds with the other carbon atom: H—C≡C—H.

22.7. List the following compounds in order from highly ionic through partly ionic and partly covalent to highly covalent: Al_2O_3, CsF, NO_2.

Compounds between elements in Groups I and VII of the periodic table, such as CsF, are always ionic, and those between nonmetals, such as NO_2, are always largely covalent. Aluminum and oxygen are in Group III and VI, hence Al_2O_3 is partly ionic and partly covalent.

22.8. In a perfect covalent bond, the electrons are shared equally by the atoms involved. In what molecules do such bonds occur?

Equal sharing can occur only when the atoms involved have equal attractions for the bonding electrons, which means that the atoms must be the same: for example, H_2, Cl_2, O_2. In all other molecules, the atoms of one of the elements have a greater attraction for the bonding electrons than the others, and the bonds are accordingly polar covalent.

22.9. What kind of bonds would you expect to find in silicon carbide, SiC? How does this accord with the fact that SiC ("Carborundum") is nearly as hard as diamond?

The bonds are covalent. Silicon is just under carbon in Group VI of the periodic table, which suggests that its chemical behavior is similar and therefore that silicon-carbon bonds might well be comparable in character to the carbon-carbon bonds in diamond.

22.10. Oxygen atoms have a stronger attraction for electrons in a covalent bond than either hydrogen or carbon; however, although H_2O molecules are polar, CO_2 molecules are not. Why is there a difference?

The H_2O molecule is "bent," with the O—H bonds nearly perpendicular: H——O. The end of

$$\text{H}$$

the molecule near the oxygen atom is negative and the rest of the molecule is positive because of the greater attraction of the O atom for the shared electrons in each bond. Because the CO_2 molecule is linear, with the C=O bonds in line (O=C=O), the electron distribution is symmetric even though the oxygen atoms have the greater attraction for the shared electrons.

22.11. The two conditions required for metallic bonding to occur in an element are a relatively low ionization energy and the presence of only a few electrons in the outer shells of its atoms. Why should this be?

The low ionization energy means that electrons are easily detached to become part of the electron "sea" that pervades a metal, and the presence of many unoccupied quantum states of relatively low energy in each atom means that the electrons can drift freely from atom to atom.

22.12. Why can metals be deformed with relative ease whereas covalent and ionic solids are quite brittle?

Atoms in a metal can be readily rearranged in position because the bonding occurs by means of a sea of freely moving electrons. In a covalent crystal the bonds are localized between adjacent atoms and must be ruptured to deform the crystal. In an ionic crystal the bonding process requires a configuration of alternate positive and negative ions whose relative positions cannot be altered without breaking the crystal apart.

22.13. The heat of vaporization of calcium is nearly twice that of potassium on a per-atom basis, which suggests that the atoms of calcium are more tightly bound to each other than those of potassium. Can you account for this difference?

Calcium atoms have two electrons in their outer shells to contribute to the electron sea in metallic calcium whereas potassium atoms have only one. Hence twice as many electrons participate in bonding a given number of calcium atoms together than in the case of potassium.

Supplementary Problems

22.14. Which of the fundamental interactions is involved in each of the various bonding mechanisms in molecules and solids?

22.15. How many electrons can occupy a bonding molecular orbital? What must be true of these electrons?

22.16. The *electronegativity* of an element is a measure of the tendency of its atoms to attract shared electrons when they participate in a bond. (*a*) Which group of elements would you expect to have the lowest electronegativities? (*b*) The highest electronegativities?

22.17. Why would you expect the shape of the H_2S molecule to resemble that of the H_2O molecule?

22.18. Distinguish between covalent and ionic bonding and explain why bonds are common that are intermediate in character.

22.19. How many covalent bonds are present between the carbon atom and each oxygen atom in carbon dioxide, CO_2?

22.20. What are the ions present in the crystal structures of the following compounds? CaF_2, KI, K_2S.

22.21. Would you expect the bonds in BF_3 or in NF_3 to be the more ionic in character?

22.22. Would you expect the bonds in CaS or in BeS to be the more ionic in character?

22.23. Would you expect potassium fluoride, KF, or potassium chloride, KCl, to have the higher melting point?

22.24. You are given two solids of almost identical appearance, one of which is held together by ionic bonds and the other by van der Waals bonds. How could you tell them apart?

22.25. Van der Waals forces can hold inert gas atoms together to form solids, but they cannot hold such atoms together to form molecules in the gaseous state. Why not?

22.26. Would you expect the atoms in solid sodium or in solid aluminum to be more tightly bound?

22.27. Mercury and iodine have ionization energies that are nearly the same, yet mercury is a metal and iodine is a nonmetal. Why?

Answers to Supplementary Problems

22.14. The electromagnetic interaction in each case.

22.15. Two electrons of opposite spin.

22.16. the alkali metals; the halogens

22.17. Sulfur is directly under oxygen in Group VI of the periodic table, which suggests that its chemical properties are similar to those of oxygen.

22.18. In a perfect covalent bond, electrons are equally shared between the atoms involved. This can only occur when the atoms are the same, as in H_2 or O_2. In an ionic bond, electrons are completely transferred from one atom to another (or others). In an intermediate bond, the shared electrons favor the atoms of one kind without shifting permanently to them, and the bond is therefore partly covalent and partly ionic.

22.19. Two bonds, so the structure of carbon dioxide is O$=$C$=$O.

22.20. Ca^{++}, F^-; K^+, I^-; K^+, S^{--}

22.21. BF_3

22.22. CaS

22.23. potassium fluoride

22.24. The van der Waals solid will be softer and will melt at a much lower temperature.

22.25. Gas molecules collide frequently, and van der Waals forces are too weak to hold inert gas atoms together when such a collision occurs.

22.26. aluminum

22.27. An iodine atom has only one vacancy in its outer electron shell and forms bonds in which this vacancy is permanently filled. A mercury atom has several vacancies in its outer shell and therefore can participate in metallic bonds with other mercury atoms that involve electrons migrating freely from one atom to another. See also the answer to Problem 22.11.

Chapter 23

Formulas and Equations

CHEMICAL FORMULAS

The *formula* of a compound shows the relative number of atoms of each of the elements in the compound. Thus the formula $BaCl_2$ indicates that there are two chlorine atoms for each barium atom in barium chloride, and the formula Na_2SO_4 indicates that sodium, sulfur, and oxygen atoms are present in the ratios $2:1:4$ in sodium sulfate.

In the case of a compound that exists as individual molecules, it is customary to give in its formula the exact number of atoms of each element in one of the molecules. Thus the formula of hydrogen peroxide is H_2O_2, not just HO, and the formula of benzene is C_6H_6, not just CH.

VALENCE

The *valence* of an element is a way to express the manner in which its atoms combine with atoms of other elements. The valence of an element is most simply defined as the number of electrons each of its atoms gains, loses, or shares in its compounds. It is often possible to predict the formula of an inorganic compound from a knowledge of the valences of its constituent elements.

An atom that donates n electrons to a bond has a valence of $+n$; thus Li has a valence of $+1$ because it forms ionic bonds in which each Li atom gives up one electron to become a Li^+ ion. An atom that acquires m electrons from a bond has a valence of $-m$; thus O has a valence of -2 because it forms ionic bonds in which each O atom picks up two electrons to become an O^{--} ion. In simple compounds the number of positive and negative valences are equal, as the formula for lithium oxide, Li_2O, confirms. Here two Li atoms have a total valence of $+2$ which balances the O valence of -2.

Certain groups of atoms appear together in many compounds. An example is the nitrate group NO_3, which acts as a unit in chemical reactions with the valence -1. Thus sodium nitrate has the formula $NaNO_3$ since the valence of sodium is $+1$, and calcium nitrate has the formula $Ca(NO_3)_2$ since the valence of calcium is $+2$.

Several elements have more than one valence state, corresponding to their ability to exist as ions with different charges. It is customary to refer to the state of lower valence by the suffix *-ous* and to that of higher valence by the suffix *-ic*. For example, iron may form ferrous compounds in which its valence is $+2$ (ferrous chloride has the formula $FeCl_2$) or ferric compounds in which its valence is $+3$ (ferric chloride has the formula $FeCl_3$).

The valences of some common elements and atom groups are given in Table 23-1 together with the symbols of the ions they form. The symbol of the mercurous ion is given as Hg_2^{++} instead of Hg^+ because Hg never appears alone in compounds with a valence of $+1$ but always as Hg_2 with a total valence of $+2$.

Table 23-1

Name	Ion	Valence	Name	Ion	Valence
Hydrogen	H^+	+1	Mercury		
Lithium	Li^+	+1	(mercurous)	Hg_2^{++}	+1
Sodium	Na^+	+1	(mercuric)	Hg^{++}	+2
Cesium	Cs^+	+1	Fluorine	F^-	−1
Potassium	K^+	+1	Chlorine	Cl^-	−1
Silver	Ag^+	+1	Bromine	Br^-	−1
Ammonium group	NH_4^+	+1	Iodine	I^-	−1
Magnesium	Mg^{++}	+2	Oxygen	O^{--}	−2
Calcium	Ca^{++}	+2	Sulfur	S^{--}	−2
Strontium	Sr^{++}	+2	Nitrate group	NO_3^-	−1
Barium	Ba^{++}	+2	Permanganate group	MnO_4^-	−1
Zinc	Zn^{++}	+2	Chlorate group	ClO_3^-	−1
Cadmium	Cd^{++}	+2	Hydroxide group	OH^-	−1
Aluminum	Al^{+++}	+3	Cyanide group	CN^-	−1
Copper			Bicarbonate group	HCO_3^-	−1
(cuprous)	Cu^+	+1	Carbonate group	CO_3^{--}	−2
(cupric)	Cu^{++}	+2	Sulfate group	SO_4^{--}	−2
Iron			Chromate group	CrO_4^{--}	−2
(ferrous)	Fe^{++}	+2	Silicate group	SiO_3^{--}	−2
(ferric)	Fe^{+++}	+3	Phosphate group	PO_4^{---}	−3

COVALENT COMPOUNDS

Electrons are not transferred from one atom to another in a covalent compound but instead are held jointly by the atoms that participate in each bond. The valence of an element in such a compound is considered to be equal to the number of electron pairs which each of its atoms shares with other atoms. Positive and negative valences are not distinguished in covalent compounds, but in each molecule the number of atoms of each kind must be such that the valences all match together. For instance, carbon has a valence of four and hydrogen a valence of one, so the formula of the simplest carbon-hydrogen compound, which is methane, is CH_4. Similarly the formula of the simplest carbon-chlorine compound, which is carbon tetrachloride, is CCl_4. The chemistry of carbon is the subject of Chapter 32. A more detailed method of electronic bookkeeping that employs the notion of oxidation number is discussed in Chapter 28.

CHEMICAL EQUATIONS

A *chemical equation* is a summary of a chemical change of some kind. The symbols and formulas of all the initial substances are placed on the left-hand side of the equation, and those of the final substances are placed on the right-hand side. An arrow indicates the direction of the process. Thus the burning of carbon, which involves the combination of carbon atoms with oxygen molecules to yield carbon dioxide molecules, is represented by the equation

$$C + O_2 \longrightarrow CO_2$$

It is important to keep in mind that being able to write the equation for a chemical process does not mean that the process necessarily can occur; and even if it can occur, the equation by itself does not indicate whatever special conditions may be necessary.

BALANCING AN EQUATION

In every chemical equation the number of atoms of each kind must be the same on both sides. When hydrogen is burned in air, it combines with oxygen to form water. If we simply write down the formulas for each substance, the equation would read

$$H_2 + O_2 \longrightarrow H_2O \qquad \text{[unbalanced equation]}$$

This equation is *unbalanced* because there are two oxygen atoms on the left-hand side but only one on the right-hand side. It is not proper to replace O_2 by O because gaseous oxygen consists of O_2 molecules and not of O atoms. What we do is to note that there are twice as many hydrogen atoms as oxygen atoms in H_2O, hence there must be twice as many hydrogen molecules as oxygen molecules on the left-hand side:

$$2H_2 + O_2 \longrightarrow H_2O \qquad \text{[unbalanced equation]}$$

The equation is still unbalanced since now there are twice as many of both hydrogen and oxygen atoms on the left-hand side as there are on the right-hand side. To obtain a *balanced* equation, all we need do is show two water molecules as the result of the reaction:

$$2H_2 + O_2 \longrightarrow 2H_2O \qquad \text{[balanced equation]}$$

Solved Problems

23.1. What is the formula of potassium iodide?

Potassium has the valence +1 and iodine has the valence −1 (Table 23-1). The formula of potassium iodide is therefore KI, since this provides a match between positive and negative valences.

23.2. What is the formula of ammonium sulfate?

The ammonium group NH_4 has the valence +1 and the sulfate group SO_4 has the valence −2 (Table 23-1). Hence there must be two ammonium groups for each sulfate group, and the formula of ammonium sulfate is $(NH_4)_2SO_4$.

23.3. What are the names and formulas of the two oxides of iron?

As shown in Table 23-1, the two valence states of iron are +2 (ferrous) and +3 (ferric). Since the valence of oxygen is −2, ferrous oxide has the formula FeO, with the +2 valence of each iron atom matching the −2 valence of each oxygen atom. In the case of ferric oxide, the valences of the iron atoms are not the same as those of the oxygen atoms, and in order for the positive and negative valences to match, the formula must be Fe_2O_3. Here the total positive valence is $2 \times (+3) = +6$ and the total negative valence $3 \times (-2) = -6$.

23.4. The formula of lead nitrate is $Pb(NO_3)_2$. What is the valence of lead?

The nitrate group NO_3 has a valence of −1 (Table 23-1). Since two NO_3 groups are associated with each lead atom, the valence of lead must be +2.

23.5. What is the valence of chromium in the compound $K_2Cr_2O_7$?

Potassium has the valence +1 and oxygen has the valence −2 (Table 23-1). The two K atoms have a total valence of +2 and the seven O atoms have a total valence of −14. Thus the total valence of the two Cr atoms must be +12 in order that positive and negative valences match, which means the valence of each Cr atom is +6.

23.6. Which of the following equations is balanced?

(a) $2NH_3 + O_2 \longrightarrow N_2 + 3H_2O$

(b) $3NO_2 + H_2O \longrightarrow 2HNO_3 + NO$

(c) $PCl_5 + H_2O \longrightarrow H_3PO_4 + 5HCl$

(a) This equation is not balanced because there are two O atoms on the left but three on the right. The correct equation is $4NH_3 + 3O_2 \longrightarrow 2N_2 + 6H_2O$.

(b) This equation is balanced since the numbers of the N, O, and H atoms are the same on both sides.

(c) This equation is not balanced because there are only two H atoms and one O atom on the left but eight H atoms and four O atoms on the right. The correct equation is $PCl_5 + 4H_2O \longrightarrow H_3PO_4 + 5HCl$.

23.7. In photosynthesis, plants manufacture carbohydrates from water, which is absorbed from the soil by their roots, and carbon dioxide, which their leaves absorb from the air. The process occurs in a number of steps, and oxygen is a by-product. Find the balanced equation for the overall process by which glucose, $C_6H_{12}O_6$, is produced by photosynthesis.

We begin by writing down a preliminary unbalanced equation that includes all the reactants and products:
$$H_2O + CO_2 \longrightarrow C_6H_{12}O_6 + O_2$$

Since there are 6 carbon atoms on the right, we must have 6 on the left as well:
$$H_2O + 6CO_2 \longrightarrow C_6H_{12}O_2 + O_2$$

There are now 12 hydrogen atoms on the right but only two on the left. Hence we need 6 water molecules:
$$6H_2O + 6CO_2 \longrightarrow C_6H_{12}O_6 + O_2$$

This brings the hydrogen atoms and the carbon atoms into balance, but there are 18 oxygen atoms on the left and only 8 on the right. We cannot change the number of oxygen atoms in the glucose molecule, but we can add the required 10 more oxygen atoms by including 6 oxygen molecules on the right instead of just one:
$$6H_2O + 6CO_2 \longrightarrow C_6H_{12}O_6 + 6O_2$$

The above equation correctly represents the production of glucose by photosynthesis.

23.8. When the gas propane, C_3H_8, is burned in cooking stoves and blowtorches, it combines with oxygen to form carbon dioxide and water vapor. Find the balanced equation for the reaction.

The preliminary unbalanced equation here is
$$C_3H_8 + O_2 \longrightarrow CO_2 + H_2O$$

To balance the 3 carbon atoms on the left we need 3 on the right as well:
$$C_3H_8 + O_2 \longrightarrow 3CO_2 + H_2O$$

In order to match the 8 hydrogen atoms on the left, we need 4 water molecules on the right:
$$C_3H_8 + O_2 \longrightarrow 3CO_2 + 4H_2O$$

Finally we must have 5 oxygen molecules on the left to provide the 10 oxygen atoms on the right:
$$C_3H_8 + 5O_2 \longrightarrow 3CO_2 + 4H_2O$$

This is the required balanced equation.

23.9. The gas butane, C_4H_{10}, is used for the same purposes as propane, and its combination with oxygen also forms carbon dioxide and water. Find the balanced equation for this reaction.

The preliminary unbalanced equation here is

$$C_4H_{10} + O_2 \longrightarrow CO_2 + H_2O$$

To match the 4 carbon atoms on the left we need 4 on the right as well:

$$C_4H_{10} + O_2 \longrightarrow 4CO_2 + H_2O$$

The 10 hydrogen atoms on the left suggest 5 water molecules on the right:

$$C_4H_{10} + O_2 \longrightarrow 4CO_2 + 5H_2O$$

On the right we now have 13 oxygen atoms, which would mean $6\frac{1}{2}$ oxygen molecules on the left. However, we are limited to whole molecules in writing chemical equations. The remedy is to double all the quantities in the above equation, which will lead to a whole number of water molecules on the right and hence to a whole number of oxygen atoms on the left:

$$2C_4H_{10} + 2O_2 \longrightarrow 8CO_2 + 10H_2O$$

Now there are 26 oxygen atoms on the right, which can be provided by 13 oxygen molecules on the left:

$$2C_4H_{10} + 13O_2 \longrightarrow 8CO_2 + 10H_2O$$

This is the required balanced equation.

Supplementary Problems

23.10. What are the formulas of the following aluminum compounds? Aluminum oxide, aluminum nitrate, aluminum hydroxide, aluminum sulfate.

23.11. What are the names and formulas of the two sulfates of iron?

23.12. What is the valence of arsenic, As, in the compound Ag_3AsS_3?

23.13. What is the valence of the element x in the compound Na_3xF_6?

23.14. Most transition elements exhibit several valences in their compounds. What are the valences of manganese in the following compounds? MnO, MnO_2, Mn_2O_3, $KMnO_4$, K_2MnO_4.

23.15. When the gas acetylene, C_2H_2, is burned, it combines with oxygen to form carbon dioxide and water. Find the balanced equation for the reaction.

23.16. When potassium chlorate is heated, it decomposes into potassium chloride and oxygen. Find the formulas for the two potassium compounds and write the balanced equation for the process.

Balance the following equations:

23.17. $Ca + O_2 \longrightarrow CaO$

23.18. $N_2 + H_2 \longrightarrow NH_3$

23.19. $ZnS + O_2 \longrightarrow ZnO + SO_2$

23.20. $FeS_2 + O_2 \longrightarrow Fe_2O_3 + SO_2$

23.21. $H_2S + HNO_3 \longrightarrow S + NO + H_2O$

Answers to Supplementary Problems

23.10. Al_2O_3; $Al(NO_3)_3$; $Al(OH)_3$; $Al_2(SO_4)_3$

23.11. ferrous sulfate, $FeSO_4$; ferric sulfate, $Fe_2(SO_4)_3$

23.12. $+3$

23.13. $+3$

23.14. $+2, +4, +3, +7, +6$

23.15. $2C_2H_2 + 5O_2 \longrightarrow 4CO_2 + 2H_2O$

23.16. $2KClO_3 \longrightarrow 2KCl + 3O_2$

23.17. $2Ca + O_2 \longrightarrow 2CaO$

23.18. $N_2 + 3H_2 \longrightarrow 2NH_3$

23.19. $2ZnS + 3O_2 \longrightarrow 2ZnO + 2SO_2$

23.20. $4FeS_2 + 11O_2 \longrightarrow 2Fe_2O_3 + 8SO_2$

23.21. $3H_2S + 2HNO_3 \longrightarrow 3S + 2NO + 4H_2O$

Chapter 24

Stoichiometry

THE GRAM-ATOM

The samples of chemicals used in both industry and the laboratory involve so many atoms that counting them is out of the question. Instead the mass of a particular sample is used as a measure of its quantity, and it is necessary to relate this mass to the number of atoms of each kind present in the sample. This aspect of chemistry is known as *stoichiometry*.

As mentioned in Chapter 10, atomic masses are expressed in atomic mass units (u), where

$$1 \text{ u} = 1.660 \times 10^{-27} \text{ kg}$$

When the masses of two samples of different elements stand in the same proportion as their atomic masses, they contain the same number of atoms. To make use of this fact, the *gram-atom* is defined as follows: A gram-atom of an element is that amount of it whose mass is equal to its atomic mass expressed in grams. Since the atomic mass of oxygen is 16.00 u, a gram-atom of oxygen has a mass of 16.00 g; since the atomic mass of iron is 55.85 u, a gram-atom of iron has a mass of 55.85 u.

If we know the mass of a sample of a certain element X, we can find the number of gram-atoms present by dividing that mass by the atomic mass of the element with the unit "u" replaced by the equivalent "g/gram-atom":

$$\text{Gram-atoms of } X = \frac{\text{mass of } X}{\text{atomic mass of } X}$$

AVOGADRO'S NUMBER

Because of the way in which the gram-atom is defined, a gram-atom of any element contains the same number of atoms as a gram-atom of any other element. This number is called *Avogadro's number*, N, and its value is

$$N = 6.023 \times 10^{23} \text{ atoms/gram-atom}$$

The number of atoms in a sample of any substance is the number of gram-atoms it contains multiplied by Avogadro's number.

THE MOLE

The *formula mass* of a substance is the sum of the atomic masses of its constituent elements, each multiplied by the number of times it appears in the formula of the substance. Thus the formula mass of molecular oxygen, O_2, is $2 \times 16.00 \text{ u} = 32.00 \text{ u}$, and the formula mass of potassium permanganate, $KMnO_4$, is 158.04 u:

$$\begin{aligned}
\text{Formula mass of K} \quad &= \quad 1 \times 39.10\ u \quad - \quad 39.10\ u \\
\text{Formula mass of Mn} \quad &= \quad 1 \times 54.94\ u \quad = \quad 54.94\ u \\
\text{Formula mass of 4O} \quad &= \quad 4 \times 16.00\ u \quad = \quad \underline{64.00\ u} \\
\text{Formula mass of } KMnO_4 \quad &= \quad 158.04\ u
\end{aligned}$$

Formula mass is a more generally useful concept than molecular mass since most solids and liquids do not consist of individual molecules.

The *mole* is an extension of the idea of the gram-atom: A mole of any substance is that amount of it whose mass is equal to its formula mass expressed in grams. (We use this definition even in the SI system, where the mass unit is the kilogram.) Thus a mole of molecular oxygen has a mass of 32.00 g (0.032 kg) and a mole of potassium permanganate has a mass of 158.04 g (0.15804 kg).

From the definition of the mole, we see that one mole of any substance contains Avogadro's number of formula units of that substance:

$$N = 6.023 \times 10^{23} \text{ formula units/mole}$$

Thus, with the help of the mole, the kinds of calculation that use the gram-atom can be extended to cover all substances whose compositions are known.

Since a gram-atom of an element is the same as a mole of the atoms of that element, the separate term gram-atom is not really necessary and is not used in a number of texts. If the gram-atom is not employed under that name, however, it is important to distinguish between a mole of atoms and a mole of molecules in the case of elements such as oxygen, which exists in nature as O_2 and not as O in the free state.

MASS RELATIONSHIPS IN CHEMICAL PROCESSES

Because of the way in which the mole is defined, a chemical equation can be interpreted in terms of moles as well as in terms of molecules or formula units. Let us consider the burning of acetylene, C_2H_2, in the reaction

$$2C_2H_2 + 5O_2 \longrightarrow 4CO_2 + 2H_2O$$

In terms of molecules, this equation states that two C_2H_2 molecules combine with five O_2 molecules to form four CO_2 molecules and two H_2O molecules. The equation equally correctly states that two moles of C_2H_2 combine with five moles of O_2 to form four moles of CO_2 and two moles of H_2O. Since we can readily determine the mass of a mole of each of the substances involved, we are able to answer questions such as: How much C_2H_2 is required to produce a given mass of CO_2? (See the Solved Problems.)

Solved Problems

24.1. How many gram-atoms of aluminum are present in 5 moles of $MgAl_2O_4$?

Each mole of $MgAl_2O_4$ contains 2 gram-atoms of Al, hence 5 moles contain 10 gram-atoms of Al.

24.2. (a) What is the mass of 75 gram-atoms of uranium? (b) How many atoms of uranium are present?

(a) The atomic mass of uranium is 238.03 u, which means that a gram-atom of U has a mass of 238.03 g. Hence

$$\begin{aligned}
\text{Mass of U} \quad &= \quad \text{gram-atoms of U} \times \text{atomic mass of U} \\
&= \quad 75 \text{ gram-atoms} \times 238.03 \text{ g/gram-atom} \quad = \quad 1.785 \times 10^4 \text{ g}
\end{aligned}$$

(*b*) To find the number of atoms, we multiply the number of gram-atoms by Avogadro's number:

$$\text{Atoms of U} = 75 \text{ gram-atoms} \times 6.023 \times 10^{23} \text{ atoms/gram-atom}$$
$$= 4.52 \times 10^{25} \text{ atoms}$$

24.3. (*a*) How many gram-atoms of Cu are present in 100 g of copper? (*b*) How many atoms of Cu?

(*a*) The atomic mass of Cu is 63.54 u = 63.54 g/gram-atom, hence

$$\text{Gram-atoms of Cu} = \frac{\text{mass of Cu}}{\text{atomic mass of Cu}} = \frac{100 \text{ g}}{63.54 \text{ g/gram-atom}}$$
$$= 1.574 \text{ gram-atoms}$$

(*b*) There are two ways to find the number of Cu atoms in 100 g of Cu. The first way is to multiply the number of gram-atoms in the sample by Avogadro's number:

$$\text{Atoms of Cu} = 1.574 \text{ gram-atoms} \times 6.023 \times 10^{23} \text{ atoms/gram-atom}$$
$$= 9.48 \times 10^{23} \text{ atoms}$$

The second way is to find first the actual mass of a Cu atom and then divide the total mass of Cu present by this mass:

$$\text{Mass of Cu atom} = 63.54 \text{ u} \times 1.660 \times 10^{-27} \text{ kg/u} = 1.055 \times 10^{-25} \text{ kg}$$

$$\text{Atoms of Cu} = \frac{\text{mass of Cu}}{\text{mass of Cu atom}} = \frac{0.1 \text{ kg}}{1.055 \times 10^{-25} \text{ kg}} = 9.48 \times 10^{23} \text{ atoms}$$

If the number of gram-atoms in a sample is already known, the first procedure is obviously simpler.

24.4. How many gram-atoms are present in a sample that contains 10^{24} atoms of any element?

Since 1 gram-atom of any element contains Avogadro's number of atoms, we have

$$\text{Gram-atoms of } X = \frac{\text{number of atoms}}{\text{atoms/gram-atom}} = \frac{10^{24} \text{ atoms}}{6.023 \times 10^{23} \text{ atoms/gram-atom}}$$
$$= 1.66 \text{ gram-atoms}$$

24.5. (*a*) Find the mass of 9.4 moles of ethylene, C_2H_4. (*b*) How many carbon atoms are present?

(*a*) The formula mass of ethylene is

$$2C = 2 \times 12.01 \text{ u} = 24.02 \text{ u}$$
$$4H = 4 \times 1.008 \text{ u} = \underline{4.03 \text{ u}}$$
$$28.05 \text{ u} = 28.05 \text{ g/mole}$$

Hence the required mass is

$$\text{Mass of } C_2H_4 = \text{moles of } C_2H_4 \times \text{formula mass of } C_2H_4$$
$$= 9.4 \text{ moles} \times 28.05 \text{ g/mole} = 264 \text{ g}$$

(*b*) Two gram-atoms of carbon are present in each mole of C_2H_4, hence there are $2 \times 9.4 = 18.8$ gram-atoms of carbon present in 9.4 moles of C_2H_4. The number of carbon atoms is

$$\text{Atoms of C} = \text{gram-atoms of C} \times \text{Avogadro's number}$$
$$= 18.8 \text{ gram-atoms} \times 6.023 \times 10^{23} \text{ atoms/gram-atom}$$
$$= 1.13 \times 10^{25} \text{ atoms}$$

24.6. How many moles are present in 500 lb of glucose, $C_6H_{12}O_6$?

The formula mass of glucose is

$$6C = 6 \times 12.01 \text{ u} = 72.06 \text{ u}$$
$$12H = 12 \times 1.008 \text{ u} = 12.10 \text{ u}$$
$$6O = 6 \times 16.00 \text{ u} - \underline{96.00 \text{ u}}$$
$$180.16 \text{ u} = 180.16 \text{ g/mole}$$

Since the mass of 1 lb is 454 g, the mass of glucose here is $500 \text{ lb} \times 454 \text{ g/lb} = 2.27 \times 10^5 \text{ g}$ and

$$\text{Moles of } C_6H_{12}O_6 = \frac{\text{mass of } C_6H_{12}O_6}{\text{formula mass of } C_6H_{12}O_6} = \frac{2.27 \times 10^5 \text{ g}}{180.16 \text{ g/mole}} = 1260 \text{ moles}$$

24.7. What is the percentage by mass of nitrogen in lead nitrate, $Pb(NO_3)_2$?

We consider for convenience 1 mole of lead nitrate. To find its formula mass, we note that there are two NO_3 groups, hence two N atoms and six O atoms per formula unit. The formula mass is therefore

$$Pb = 1 \times 207.19 \text{ u} = 207.19 \text{ u}$$
$$2N = 2 \times 14.01 \text{ u} = 28.02 \text{ u}$$
$$6O = 6 \times 16.00 \text{ u} = \underline{96.00 \text{ u}}$$
$$331.21 \text{ u} = 331.21 \text{ g/mole}$$

The number of grams of N per mole of $Pb(NO_3)_2$ is 28.02 g, hence in one mole the proportion of nitrogen is

$$\frac{\text{Mass of N}}{\text{Mass of } Pb(NO_3)_2} = \frac{28.02 \text{ g}}{331.21 \text{ g}} = 0.0846 = 8.46\%$$

Of course, the proportion of nitrogen is the same no matter what mass of lead nitrate we choose.

24.8. (a) How much chlorine is needed to react with 50 g of sodium to form sodium chloride, NaCl? (b) How much sodium chloride is produced? (c) What is the percentage composition of sodium chloride by mass?

(a) In NaCl there are equal numbers of Na and Cl atoms, hence there must be equal numbers of gram-atoms of Na and of Cl. Since the atomic mass of Na is 22.99 u = 22.99 g/gram-atom, the number of gram-atoms in 50 g of sodium is

$$\text{Gram-atoms of Na} = \frac{\text{mass of Na}}{\text{atomic mass of Na}} = \frac{50 \text{ g}}{22.99 \text{ g/gram-atom}}$$
$$= 2.17 \text{ gram-atoms}$$

The atomic mass of chlorine is 35.46 u = 35.46 g/gram-atom and so the mass of 2.17 gram-atoms of chlorine is

$$\text{Mass of Cl} = \text{gram-atoms of Cl} \times \text{atomic mass of Cl}$$
$$= 2.17 \text{ gram-atoms} \times 35.46 \text{ g/gram-atom} = 77 \text{ g}$$

(b) The reaction of 50 g of sodium and 77 g of chlorine produces $50 \text{ g} + 77 \text{ g} = 127 \text{ g}$ of sodium chloride.

(c) The proportion of sodium in NaCl is

$$\frac{\text{Mass of Na}}{\text{Mass of NaCl}} - \frac{50 \text{ g}}{127 \text{ g}} - 0.39 - 39\%$$

The proportion of chlorine is

$$\frac{\text{Mass of Cl}}{\text{Mass of NaCl}} = \frac{77 \text{ g}}{127 \text{ g}} = 0.61 = 61\%$$

24.9. (a) How many gram-atoms of hydrogen are needed to react with 1 gram-atom of nitrogen to form ammonia, NH_3? (b) How many grams of hydrogen are needed to react with 70 g of nitrogen to form ammonia? (c) What is the percentage composition by mass of ammonia?

(a) Since there are three atoms of hydrogen for each atom of nitrogen in NH_3, there must be three gram-atoms of hydrogen for each gram-atom of nitrogen.

(b) The number of gram-atoms in 70 g of nitrogen is

$$\text{Gram-atoms of N} = \frac{\text{mass of N}}{\text{atomic mass of N}} = \frac{70\text{ g}}{14.01\text{ g/gram-atom}}$$
$$= 5.0\text{ gram-atoms}$$

To react with 5.0 gram-atoms of nitrogen there must be three times as many gram-atoms of hydrogen, or 15.0 gram-atoms. The corresponding mass of hydrogen is

$$\text{Mass of H} = \text{gram-atoms of H} \times \text{atomic mass of H}$$
$$= 15.0\text{ gram-atoms} \times 1.008\text{ g/gram-atom} = 15\text{ g}$$

(c) The mass of ammonia formed in (b) is $15\text{ g} + 70\text{ g} = 85\text{ g}$. The proportion of hydrogen is therefore

$$\frac{\text{Mass of H}}{\text{Mass of NH}_3} = \frac{15\text{ g}}{85\text{ g}} = 0.18 = 18\%$$

and the proportion of nitrogen is

$$\frac{\text{Mass of N}}{\text{Mass of NH}_3} = \frac{70\text{ g}}{85\text{ g}} = 0.82 = 82\%$$

24.10. (a) How many gram-atoms of sulfur are needed to react with one gram-atom of oxygen to form sulfur dioxide, SO_2? (b) How many grams of sulfur are needed to react with 200 g of oxygen to form sulfur dioxide?

(a) There are two atoms of oxygen for each atom of sulfur in SO_2, which means two gram-atoms of oxygen for each gram-atom of sulfur. Hence half a gram-atom of sulfur is needed for each gram-atom of oxygen.

(b) The number of gram-atoms in 200 g of oxygen is

$$\text{Gram-atoms of O} = \frac{\text{mass of O}}{\text{atomic mass of O}} = \frac{200\text{ g}}{16.00\text{ g/gram-atom}}$$
$$= 12.5\text{ gram-atoms}$$

The number of gram-atoms of S needed is half this, or 6.25 gram-atoms, and the corresponding mass of sulfur is

$$\text{Mass of S} = \text{gram-atoms of S} \times \text{atomic mass of S}$$
$$= 6.25\text{ gram-atoms} \times 32.06\text{ g/gram-atom} = 200\text{ g}$$

24.11. When the gas acetylene, C_2H_2, is burned, it combines with oxygen to form carbon dioxide and water in the reaction

$$2C_2H_2 + 5O_2 \longrightarrow 4CO_2 + 2H_2O$$

(a) How many grams of oxygen are needed for the complete combustion of 200 g of acetylene? (b) How many grams of carbon dioxide are produced?

(a) The formula masses of acetylene and oxygen are as follows:

$$C_2H_2: \quad 2C = 2 \times 12.01\text{ u} = 24.02\text{ u}$$
$$2H = 2 \times 1.008\text{ u} = \underline{2.02\text{ u}}$$
$$26.04\text{ u} = 26.04\text{ g/mole}$$
$$O_2: \quad 2O = 2 \times 16.00\text{ u} = 32.00\text{ u} = 32.00\text{ g/mole}$$

The number of moles in 200 g of acetylene is

$$\text{Moles of } C_2H_2 \;=\; \frac{\text{mass of } C_2H_2}{\text{formula mass of } C_2H_2} \;=\; \frac{200\text{ g}}{26.04\text{ g/mole}} \;=\; 7.68\text{ moles}$$

According to the equation of the process, 5 moles of oxygen are needed for every 2 moles of acetylene, or 2.5 moles of oxygen per mole of acetylene. The number of moles of oxygen needed here is therefore 2.5×7.68 moles $= 19.20$ moles, and their mass is

$$\text{Mass of } O_2 \;=\; \text{moles of } O_2 \times \text{formula mass of } O_2$$
$$=\; 19.20\text{ moles} \times 32.00\text{ g/mole} \;=\; 614\text{ g}$$

Thus 614 g of oxygen is needed for the complete combustion of 200 g of acetylene.

(b) Since 4 moles of CO_2 are produced when 2 moles of C_2H_2 are burned, 2 moles of CO_2 are produced per mole of C_2H_2. Hence 2×7.68 moles $= 15.36$ moles of CO_2 are produced when 7.68 moles of C_2H_2 are burned. The formula mass of CO_2 is

$$C \;=\; 1 \times 12.01\text{ u} \;=\; 12.01\text{ u}$$
$$2O \;=\; 2 \times 16.00\text{ u} \;=\; \underline{32.00\text{ u}}$$
$$44.01\text{ u} \;=\; 44.01\text{ g/mole}$$

and so the mass of carbon dioxide produced here is

$$\text{Mass of } CO_2 \;=\; \text{moles of } CO_2 \times \text{formula mass of } CO_2$$
$$=\; 15.36\text{ moles} \times 44.01\text{ g/mole} \;=\; 676\text{ g}$$

24.12. When hydrogen is burned in oxygen, water is formed according to the reaction

$$2H_2 + O_2 \;\longrightarrow\; 2H_2O$$

(a) If a mixture of 1000 g of hydrogen and 1000 g of oxygen is ignited, how much water will be produced? (b) Will any hydrogen or oxygen be left over, and if so, how much of which of them?

(a) The formula masses of H_2, O_2, and H_2O are respectively 2.02 g/mole, 32.00 g/mole, and 18.02 g/mole. The number of moles of hydrogen and oxygen present are as follows:

$$\text{Moles of } H_2 \;=\; \frac{\text{mass of } H_2}{\text{formula mass of } H_2} \;=\; \frac{1000\text{ g}}{2.02\text{ g/mole}} \;=\; 495\text{ moles}$$

$$\text{Moles of } O_2 \;=\; \frac{\text{mass of } O_2}{\text{formula mass of } O_2} \;=\; \frac{1000\text{ g}}{32.00\text{ g/mole}} \;=\; 31.25\text{ moles}$$

According to the formula for the reaction, 2 moles of hydrogen are required for each mole of oxygen, and two moles of water are produced. Since we have 31.25 moles of O_2 to start with, 2×31.25 moles $= 62.5$ moles of H_2 can be burned to produce 62.5 moles of H_2O. The mass of H_2O produced is therefore

$$\text{Mass of } H_2O \;=\; \text{moles of } H_2O \times \text{formula mass of } H_2O$$
$$=\; 62.5\text{ moles} \times 18.02\text{ g/mole} \;=\; 1126\text{ g}$$

(b) All of the oxygen is consumed, but only 62.5 moles of hydrogen out of the 495 moles we started with. Thus 495 moles $-$ 62.5 moles $= 432.5$ moles of hydrogen are left over, and the mass of hydrogen left over is

$$\text{Mass of } H_2 \;=\; \text{moles of } H_2 \times \text{formula mass of } H_2$$
$$=\; 432.5\text{ moles} \times 2.02\text{ g/mole} \;=\; 874\text{ g}$$

The same result is obtained simply by noting that, since 1126 g of water were formed from 2000 g of initial reactants, 2000 g $-$ 1126 g $= 874$ g must be left over.

24.13. When potassium chlorate is heated, it decomposes into potassium chloride and oxygen in the reaction

$$2KClO_3 \longrightarrow 2KCl + 3O_2$$

How much potassium chlorate is needed to produce 1 g of oxygen?

The formula masses of oxygen and potassium chlorate are as follows:

$$O_2: \qquad 2O = 2 \times 16.00 \text{ u} = \quad 32.00 \text{ u} = 32.00 \text{ g/mole}$$

$$KClO_3: \qquad K = 1 \times 39.10 \text{ u} = \quad 39.10 \text{ u}$$
$$Cl = 1 \times 35.46 \text{ u} = \quad 35.46 \text{ u}$$
$$3O = 3 \times 16.00 \text{ u} = \quad \underline{48.00 \text{ u}}$$
$$122.56 \text{ u} = 122.56 \text{ g/mole}$$

The number of moles of O_2 in 1 g of oxygen is

$$\text{Moles of } O_2 = \frac{\text{mass of } O_2}{\text{formula mass of } O_2} = \frac{1 \text{ g}}{32.00 \text{ g/mole}} = 0.0313 \text{ mole}$$

Since 2 moles of $KClO_3$ produces 3 moles of O_2, $\frac{2}{3}$ mole of $KClO_3$ produces 1 mole of O_2, and $\frac{2}{3} \times 0.0313$ mole $= 0.0208$ mole of $KClO_3$ produces 0.0313 mole of O_2. Another way to obtain this result is to use the proportion

$$\frac{\text{Moles of } O_2}{\text{Moles of } KClO_3} = \frac{\text{moles of } O_2}{\text{moles of } KClO_3}$$

$$\frac{3}{2} = \frac{0.0313 \text{ mole}}{\text{moles of } KClO_3}$$

$$\text{Moles of } KClO_3 = \frac{2}{3} \times 0.0313 \text{ mole} = 0.0208 \text{ mole}$$

Now we find the required mass of $KClO_3$ in the usual way:

$$\text{Mass of } KClO_3 = \text{moles of } KClO_3 \times \text{formula mass of } KClO_3$$
$$= 0.0208 \text{ moles} \times 122.56 \text{ g/mole} = 2.55 \text{ g}$$

24.14. (*a*) How much sulfur is needed to manufacture 100 g of sodium sulfate, Na_2SO_4? (*b*) How much sodium?

(*a*) The formula mass of sodium sulfate is

$$2Na = 2 \times 22.99 \text{ u} = \quad 45.98 \text{ u}$$
$$S = 1 \times 32.06 \text{ u} = \quad 32.06 \text{ u}$$
$$4O = 4 \times 16.00 \text{ u} = \quad \underline{64.00 \text{ u}}$$
$$142.04 \text{ u} = 142.04 \text{ g/mole}$$

The number of moles in 100 g of this compound is

$$\text{Moles of } Na_2SO_4 = \frac{\text{mass of } Na_2SO_4}{\text{formula mass of } Na_2SO_4} = \frac{100 \text{ g}}{142.04 \text{ g/mole}} = 0.704 \text{ mole}$$

From its formula, each mole of sodium sulfate contains 1 gram-atom of sulfur, so that 0.704 mole of sodium sulfate contains 0.704 gram-atom of sulfur. The corresponding mass of sulfur is

$$\text{Mass of S} = \text{gram-atoms of S} \times \text{atomic mass of S}$$
$$= 0.704 \text{ gram-atom} \times 32.06 \text{ g/gram-atom} = 22.6 \text{ g}$$

(*b*) Each mole of Na_2SO_4 contains two gram-atoms of sodium, hence 0.704 mole of Na_2SO_4 contains 2×0.704 gram-atom $= 1.408$ gram-atoms of sodium. The corresponding mass of sodium is

$$\text{Mass of Na} = \text{gram-atoms of Na} \times \text{atomic mass of Na}$$
$$= 1.408 \text{ gram-atoms} \times 22.99 \text{ g/gram-atom} = 32.4 \text{ g}$$

24.15. The *empirical formula* of a compound expresses the relative numbers of atoms of the various elements present using the smallest whole numbers possible. Sometimes the empirical formula of a compound is the same as its molecular formula if the compound exists as molecules (H_2O, NH_3, CO_2), sometimes it is not (H_2O_2, N_2H_2, C_6H_{14}). If the analysis of a sample of an unknown compound shows it to consist of **128.2 g** of sulfur and 8.06 g of hydrogen, find the empirical formula of the compound.

The atomic masses of sulfur and hydrogen are respectively 32.06 u and 1.008 u, so the sample consists of

$$\text{Gram-atoms of S} = \frac{\text{mass of S}}{\text{atomic mass of S}} = \frac{128.2 \text{ g}}{32.06 \text{ g/gram-atom}} = 4.00 \text{ gram-atoms}$$

$$\text{Gram-atoms of H} = \frac{\text{mass of H}}{\text{atomic mass of H}} = \frac{8.06 \text{ g}}{1.008 \text{ g/gram-atom}} = 8.00 \text{ gram-atoms}$$

There are twice as many gram-atoms of H as there are of S, hence twice as many atoms of H as there are of S. The empirical formula of the compound is accordingly H_2S, which is hydrogen sulfide.

24.16. A sample of an unknown compound is analyzed and found to contain 57.54 g of bromine, 17.29 g of carbon, and 3.63 g of hydrogen. Find its empirical formula.

The atomic masses of bromine, carbon, and hydrogen are respectively 79.91 u, 12.01 u, and 1.008 u, so the sample consists of

$$\text{Gram-atoms of Br} = \frac{\text{mass of Br}}{\text{atomic mass of Br}} = \frac{57.54 \text{ g}}{79.91 \text{ g/gram-atom}} = 0.72 \text{ gram-atom}$$

$$\text{Gram-atoms of C} = \frac{\text{mass of C}}{\text{atomic mass of C}} = \frac{17.29 \text{ g}}{12.01 \text{ g/gram-atom}} = 1.44 \text{ gram-atoms}$$

$$\text{Gram-atoms of H} = \frac{\text{mass of H}}{\text{atomic mass of H}} = \frac{3.63 \text{ g}}{1.008 \text{ g/gram-atom}} = 3.60 \text{ gram-atoms}$$

To find the smallest set of whole numbers that can be used in the formula of this compound, we first divide through by 0.72, the smallest number of gram-atoms, to see if such a set then results. We obtain

$$\text{Br:} \quad \frac{0.72 \text{ gram-atom}}{0.72} = 1 \text{ gram-atom}$$

$$\text{C:} \quad \frac{1.44 \text{ gram-atoms}}{0.72} = 2 \text{ gram-atoms}$$

$$\text{H:} \quad \frac{3.60 \text{ gram-atoms}}{0.72} = 5 \text{ gram-atoms}$$

Evidently the empirical formula here is C_2H_5Br, which is ethyl bromide.

24.17. A sample of an unknown compound is analyzed and found to contain 100.9 g of carbon and 22.6 g of hydrogen. Find its empirical formula.

The atomic masses of carbon and hydrogen are respectively 12.01 u and 1.008 u, so the sample consists of

$$\text{Gram-atoms of C} = \frac{\text{mass of C}}{\text{atomic mass of C}} = \frac{100.9 \text{ g}}{12.01 \text{ g/gram-atom}} = 8.4 \text{ gram-atoms}$$

$$\text{Gram-atoms of H} = \frac{\text{mass of H}}{\text{atomic mass of H}} = \frac{22.6 \text{ g}}{1.008 \text{ g/gram-atom}} = 22.4 \text{ gram-atoms}$$

When we divide through by 8.4, we find

$$\text{C:} \quad \frac{8.4 \text{ gram-atoms}}{8.4} = 1 \text{ gram-atom}$$

$$\text{H:} \quad \frac{22.4 \text{ gram-atoms}}{8.4} = 2.67 \text{ gram-atoms}$$

Although the formula $CH_{2.67}$ correctly represents the composition of this compound in terms of gram-atoms, it makes no sense in terms of atoms. But all is not lost. Further study shows that the carbon and hydrogen gram-atom contents are in the ratio

$$\frac{C}{H} = \frac{1}{2.67} = \frac{1}{2\frac{2}{3}} = \frac{1}{8/3} = \frac{3}{8}$$

From this it is clear that the required formula here is C_3H_8, which is propane.

Supplementary Problems

24.18. How many gram-atoms of carbon are present in 4 moles of C_3H_8?

24.19. How many gram-atoms of sulfur are present in 10 moles of $Al_2(SO_4)_3$?

24.20. (a) What is the mass of 1 gram-atom of aluminum? (b) How many atoms does it contain?

24.21. (a) What is the mass of 200 gram-atoms of lithium? (b) How many Li atoms are present?

24.22. (a) How many gram-atoms of sulfur are present in 5 kg of sulfur? (b) How many sulfur atoms?

24.23. A pound of zinc contains 4.18×10^{24} atoms. How many gram-atoms is this?

24.24. Ammonia is produced by the reaction $N_2 + 3H_2 \longrightarrow 2NH_3$. How many moles of nitrogen and how many moles of hydrogen are needed to produce 1 mole of ammonia?

24.25. (a) Find the mass of 80 moles of sulfuric acid, H_2SO_4. (b) How many atoms of oxygen are present?

24.26. (a) How many moles of O_2 are present in 1 kg of oxygen? (b) How many gram-atoms of O? (c) How many molecules of O_2? (d) How many atoms of O?

24.27. (a) How much calcium is present in 15 g of calcium oxide, CaO? (b) What percentage by mass is this?

24.28. (a) How much carbon is present in 20 kg of ethyl alcohol, C_2H_5OH? (b) What percentage by mass is this?

24.29. How many moles are present in 120 kg of boric acid, H_3BO_3?

24.30. (a) How many gram-atoms of sulfur are needed to react with 1 gram-atom of hydrogen to form hydrogen sulfide, H_2S? (b) How many grams of sulfur are needed to react with 40 g of hydrogen to form hydrogen sulfide? (c) What is the percentage composition by mass of hydrogen sulfide?

24.31. The combustion of propane proceeds according to the equation $C_3H_8 + 5O_2 \longrightarrow 3CO_2 + 4H_2O$. (a) How many grams of oxygen are needed for the complete combustion of 40 g of propane? (b) How many grams of water are produced?

24.32. (a) How much oxygen is needed to react with 500 g of carbon to produce carbon monoxide, CO? (b) How much carbon monoxide is produced? (c) What is the percentage by mass of oxygen in carbon monoxide?

24.33. (a) How much oxygen is needed to react with 500 g of mercury to produce mercuric oxide, HgO? (b) How much mercuric oxide is produced? (c) What is the percentage by mass of oxygen in mercuric oxide?

24.34. The neutralization of hydrochloric acid, HCl, by sodium hydroxide, NaOH, proceeds according to the equation HCl + NaOH $\longrightarrow$ NaCl + H$_2$O. (a) If 50 g of HCl is mixed with 100 g of NaOH, how much NaCl is produced? (b) How much of HCl or NaOH will be left over?

24.35. When ferric oxide, Fe$_2$O$_3$, reacts with hydrogen at a high temperature, iron and water vapor are produced: Fe$_2$O$_3$ + 3H$_2$ $\longrightarrow$ 2Fe + 3H$_2$O. How much ferric oxide and how much hydrogen are required to produce 1 ton (2000 lb) of iron?

Samples of a set of unknown compounds are analyzed and the following compositions are determined. Find the empirical formulas of these compounds.

24.36. 3.410 g of Hg and 0.272 g of O.

24.37. 10.58 g of Na and 3.68 g of O.

24.38. 4.13 g of O, 3.95 g of Na, and 2.42 g of Si.

24.39. 33.46 g of Al and 29.76 g of O.

24.40. 10.56 g of O, 7.93 g of C, and 0.89 g of H.

Answers to Supplementary Problems

24.18. 12 gram-atoms

24.19. 30 gram-atoms

24.20. 26.98 g; 6.023×10^{23} atoms

24.21. 1.39 kg; 1.20×10^{26} atoms

24.22. 156 gram-atoms; 9.39×10^{25} atoms

24.23. 6.94 gram-atoms

24.24. 0.5 moles of N$_2$ and 1.5 moles of H$_2$

24.25. 7.85 kg; 1.93×10^{26} atoms

24.26. 31.25 moles; 62.50 gram-atoms;
        1.88×10^{25} molecules; 3.76×10^{25} atoms

24.27. 10.7 g; 71.3%

24.28. 10.4 kg; 52%

24.29. 1940 moles

24.30. 0.5 gram-atom; 636 g; 5.9% H, 94.1% S

24.31. 145 g; 65 g

24.32. 666 g; 1166 g; 57%

24.33. 40 g; 540 g; 7.4%

24.34. 80 g; 45 g of NaOH

24.35. 2859 lb of Fe$_2$O$_3$; 108 lb of H$_2$

24.36. HgO

24.37. Na$_2$O

24.38. Na$_2$SiO$_3$

24.39. Al$_2$O$_3$

24.40. C$_3$H$_4$O$_3$

Chapter 25

Gas Stoichiometry

GAS VOLUMES

Equal volumes of all gases, under the same conditions of temperature and pressure, contain the same number of molecules and therefore the same number of moles. This observation is most useful stated in reverse: under given conditions of temperature and pressure, the volume of a gas is proportional to the number of moles present.

As an illustration, let us consider the reaction

$$N_2 + 3H_2 \longrightarrow 2NH_3$$

This equation states that one mole of N_2 combines with three moles of H_2 to yield two moles of NH_3 (ammonia). In terms of the respective gas volumes, under the same conditions of temperature and pressure the volume of H_2 required to combine with a certain volume of N_2 is three times as great, and the volume of NH_3 produced is twice the volume of N_2.

MOLAR VOLUME

For convenience, a temperature of 0 °C (273 K) and a pressure of 1 atm (1.013×10^5 N/m² = 14.7 lb/in²) are taken as the standard temperature and pressure (STP); Charles's and Boyle's laws permit measurements made at other temperatures and pressures to be reduced to their equivalents at STP. Experimentally it is found that one mole of any gas at STP occupies a volume of 22.4 liters. Thus the *molar volume* of a gas is 22.4 liters at STP. This observation makes it possible to deal with gas volumes in chemical reactions. If a certain reaction is known to produce 2.5 moles of a gas, for instance, we know that at STP the volume of the gas will be

$$V = \text{moles of gas} \times \text{molar volume} = 2.5 \text{ moles} \times 22.4 \text{ liters/mole} = 56 \text{ liters}$$

UNIVERSAL GAS CONSTANT

According to the ideal gas law (Chapter 10), the pressure, volume, and temperature of a gas sample obey the relationship $pV/T =$ constant. We can find the value of the constant in terms of the number of moles n of gas in the sample by making use of the fact that the molar volume at STP is 22.4 liters. At STP we have $T = 0$ °C $= 273$ K, $p = 1$ atm, and $V = n \times 22.4$ liters/mole, so that

$$\frac{pV}{T} = \frac{n \times 1 \text{ atm} \times 22.4 \text{ liters/mole}}{273 \text{ K}} = nR$$

where R, the *universal gas constant*, has the value

$$R = 0.0821 \text{ atm-liter/mole-K}$$

In SI units, in which the unit of p is the N/m² and the unit of V is the m³,

$$R = 8.31 \text{ J/mole-K}$$

The complete ideal gas law is usually written in the form

$$pV = nRT$$

Solved Problems

25.1. How many liters of oxygen at STP are needed to combine with 100 liters of hydrogen sulfide at STP in the reaction $2H_2S + 3O_2 \longrightarrow 2H_2O + 2SO_2$?

Since the O_2 and the H_2S are at the same temperature and pressure, the ratio of volumes of the gases is equal to the ratio of the number of moles of each:

$$\frac{\text{Volume of } O_2}{\text{Volume of } H_2S} = \frac{\text{moles of } O_2}{\text{moles of } H_2S}$$

$$\frac{\text{Volume of } O_2}{100 \text{ liters}} = \frac{3 \text{ moles}}{2 \text{ moles}}$$

$$\text{Volume of } O_2 = \frac{3}{2} \times 100 \text{ liters} = 150 \text{ liters}$$

25.2. How many liters of oxygen at STP are evolved when 200 g of potassium chlorate is heated? The equation of the process is $2KClO_3 \longrightarrow 2KCl + 3O_2$ and the formula mass of $KClO_3$ is 122.56 g/mole.

The number of moles in 200 g of $KClO_3$ is

$$\text{Moles of } KClO_3 = \frac{\text{mass of } KClO_3}{\text{formula mass of } KClO_3} = \frac{200 \text{ g}}{122.56 \text{ g/mole}} = 1.63 \text{ moles}$$

From the equation of the process, 3 moles of O_2 are evolved when 2 moles of $KClO_3$ are heated, or 1.5 moles of O_2 for each mole of $KClO_3$. Hence the number of moles of O_2 evolved here is 1.5×1.63 moles = 2.45 moles, and

$$\text{Volume of } O_2 = \text{moles of } O_2 \times \text{molar volume}$$

$$= 2.45 \text{ moles} \times 22.4 \text{ liters/mole} = 54.9 \text{ liters}$$

25.3. When sodium bicarbonate (baking soda), $NaHCO_3$, is heated, it decomposes into sodium carbonate, water, and carbon dioxide in the reaction

$$2NaHCO_3 \longrightarrow Na_2CO_3 + H_2O + CO_2$$

How much sodium bicarbonate must be decomposed to yield 2 liters of CO_2 at STP? The formula mass of $NaHCO_3$ is 84 g/mole.

The number of moles contained in 2 liters of CO_2 at STP is

$$\text{Moles of } CO_2 = \frac{\text{volume of } CO_2}{\text{molar volume}} = \frac{2 \text{ liters}}{22.4 \text{ liters/mole}} = 0.0893 \text{ mole}$$

From the equation of the process, two moles of $NaHCO_3$ must be decomposed to produce one mole of CO_2, hence 2×0.0893 mole = 0.179 mole of $NaHCO_3$ must be decomposed to produce 0.0893 mole of CO_2. The corresponding mass of $NaHCO_3$ is

$$\text{Mass of } NaHCO_3 = \text{moles of } NaHCO_3 \times \text{formula mass of } NaHCO_3$$

$$= 0.179 \text{ moles} \times 84 \text{ g/mole} = 15 \text{ g}$$

25.4. The reaction of ferric oxide with hydrogen proceeds according to the equation

$$Fe_2O_3 + 3H_2 \longrightarrow 2Fe + 3H_2O$$

(a) How much ferric oxide can be reduced to iron by 1000 liters of hydrogen at 400 °C and atmospheric pressure? (b) How much iron is produced? The formula mass of Fe_2O_3 is 159.7 g/mole.

(*a*) The number of moles of hydrogen must first be calculated from the ideal gas law $pV = nRT$. Since $p = 1$ atm, $V = 1000$ liters, and $T = 400\ °C = 673$ K,

$$n = \frac{pV}{RT} = \frac{1 \text{ atm} \times 1000 \text{ liters}}{0.0821 \text{ atm-liter/mole-K} \times 673 \text{ K}} = 18.1 \text{ moles}$$

From the equation of the process, $\frac{1}{3}$ mole of Fe_2O_3 reacts with each mole of H_2, hence the number of moles of Fe_2O_3 that can be reduced here is $\frac{1}{3} \times 18.1$ moles = 6.03 moles. The mass of 6.03 moles of Fe_2O_3 is

$$\text{Mass of } Fe_2O_3 = \text{moles of } Fe_2O_3 \times \text{formula mass of } Fe_2O_3$$
$$= 6.03 \text{ moles} \times 159.7 \text{ g/mole} = 963 \text{ g}$$

(*b*) Since 3 moles of H_2 produces 2 gram-atoms of Fe, each mole of H_2 produces $\frac{2}{3}$ gram-atom of Fe, and 18.1 moles of H_2 produce $\frac{2}{3} \times 18.1 = 12.1$ gram-atoms of Fe. The atomic mass of Fe is 55.85 g/gram-mole, hence

$$\text{Mass of Fe} = \text{gram-atoms of Fe} \times \text{atomic mass of Fe}$$
$$= 12.1 \text{ gram-atoms} \times 55.85 \text{ g/gram-atom} = 676 \text{ g}$$

25.5. (*a*) What volume does 1 g of ammonia, NH_3, occupy at STP? (*b*) What volume does it occupy at 100 °C and a pressure of 1.2 atm?

(*a*) The molecular mass of NH_3 is

$$1N = 1 \times 14.01 \text{ u} = 14.01 \text{ u}$$
$$3H = 3 \times 1.008 \text{ u} = \underline{3.02 \text{ u}}$$
$$17.03 \text{ u} = 17.03 \text{ g/mole}$$

so the number of moles in 1 g of NH_3 is

$$\text{Moles of } NH_3 = \frac{\text{mass of } NH_3}{\text{molecular mass of } NH_3} = \frac{1 \text{ g}}{17.03 \text{ g/mole}} = 0.0587 \text{ mole}$$

The volume at STP is therefore

$$\text{Volume of } NH_3 = \text{moles of } NH_3 \times \text{molar volume}$$
$$= 0.0587 \text{ mole} \times 22.4 \text{ liters/mole} = 1.32 \text{ liters}$$

(*b*) From the ideal gas law,

$$\frac{p_1 V_1}{T_1} = \frac{p_2 V_2}{T_2} \quad \text{or} \quad V_2 = \frac{p_1 V_1 T_2}{p_2 T_1}$$

Here $p_1 = 1$ atm, $V_1 = 1.32$ liters, $T_1 = 0\ °C = 273$ K and $p_2 = 1.2$ atm, $V_2 = ?$, $T_2 = 100\ °C = 373$ K.   Hence

$$V_2 = \frac{1 \text{ atm} \times 1.32 \text{ liters} \times 373 \text{ K}}{1.2 \text{ atm} \times 273 \text{ K}} = 1.50 \text{ liters}$$

25.6. What is the mass of 40 liters of uranium hexafluoride, UF_6, at 500 °C and 4 atm pressure?

The most direct way to solve this problem is to use the ideal gas law to find the number of moles of UF_6 present in the sample. Since $pV = nRT$ and $T = 500\ °C = 773$ K we have

$$n = \frac{pV}{RT} = \frac{4 \text{ atm} \times 40 \text{ liters}}{0.0821 \text{ atm-liter/mole-K} \times 773 \text{ K}} = 2.52 \text{ moles}$$

The molecular mass of UF_6 is

$$1U = 1 \times 238.03 \text{ u} = 238.03 \text{ u}$$
$$6F = 6 \times 19.00 \text{ u} = \underline{114.00 \text{ u}}$$
$$352.03 \text{ u} = 352.03 \text{ g/mole}$$

so the mass of UF_6 is

$$\text{Mass of } UF_6 = \text{moles of } UF_6 \times \text{molecular mass of } UF_6$$
$$= 2.52 \text{ moles} \times 352.03 \text{ g/mole} = 887 \text{ g}$$

25.7. At an altitude of 10,000 m, the atmosphere normally has a temperature of about −50 °C and a pressure of about 0.26 atm. Find the mass of helium needed to fill a balloon whose volume at this altitude is to be 120 m³.

We first use the ideal gas law to find the number of moles of He needed. Since 1 atm = 1.013×10^5 N/m², $p = 0.26 \times 1.013 \times 10^5$ N/m² $= 0.263 \times 10^5$ N/m²; $V = 120$ m³; and $T = -50$ °C $= 223$ K. From $pV = nRT$ we have

$$n = \frac{pV}{RT} = \frac{0.263 \times 10^5 \text{ N/m}^2 \times 120 \text{ m}^3}{8.31 \text{ J/mole-K} \times 223 \text{ K}} = 1.70 \times 10^3 \text{ moles}$$

Helium is a monatomic gas and its molecular mass is equal to its atomic mass of 4.00 u = 4.00 g/gram-atom. Since the number of moles of He is the same as the number of gram-atoms of He,

Mass of He = gram-atoms of He × atomic mass of He

= 1.70×10^3 moles × 4.00 g/mole = 6.8×10^3 g = 6.8 kg

25.8. Find the density in g/liter of ethylene, C_2H_4, at STP.

At STP one mole of any gas occupies 22.4 liters. The molecular mass of C_2H_4 is

2C = 2 × 12.01 u = 24.02 u

4H = 4 × 1.008 u = 4.03 u

28.05 u = 28.05 g/mole

One mole of C_2H_4 therefore has a density at STP of

$$d = \frac{m}{V} = \frac{28.05 \text{ g}}{22.4 \text{ liters}} = 1.25 \text{ g/liter}$$

25.9. What is the density of oxygen at 20 °C and 5 atm pressure?

It is simplest here to use the ideal gas law to find the mass of 1 liter of O_2 under the specified conditions. The number of moles in 1 liter of O_2 at $T = 20$ °C $= 293$ K and $p = 5$ atm is, from $pV = nRT$,

$$n = \frac{pV}{RT} = \frac{5 \text{ atm} \times 1 \text{ liter}}{0.0821 \text{ atm-liter/mole-K} \times 293 \text{ K}} = 0.208 \text{ moles}$$

Since the molecular mass of O_2 is 2 × 16.00 u = 32.00 u = 32.00 g/mole, the mass here is

Mass of O_2 = moles of O_2 × molecular mass of O_2

= 0.208 moles × 32.00 g/mole = 6.66 g

Hence the density of the gas is

$$d = \frac{m}{V} = \frac{6.66 \text{ g}}{1 \text{ liter}} = 6.66 \text{ g/liter}$$

There are 10^3 g per kg and 10^3 liters per m³, which means that the density in SI units is 6.66 kg/m³.

25.10. A sample of an unknown gas has a mass of 28.1 g and occupies 4.8 liters at STP. What is its molecular mass?

Since 1 mole of any gas occupies 22.4 liters at STP, this sample must consist of

4.8 liters/(22.4 liters/mole) = 0.214 moles

Hence   Molecular mass = $\frac{\text{mass of sample}}{\text{moles of sample}}$ = $\frac{28.1 \text{ g}}{0.214 \text{ mole}}$ = 131 g/mole = 131 u

Supplementary Problems

25.11. (a) How many liters of oxygen at STP are needed to combine with 6 liters of carbon monoxide at STP to form carbon dioxide in the reaction $2CO + O_2 \longrightarrow 2CO_2$? (b) How many liters of carbon dioxide are produced as a result?

25.12. How many cubic feet of oxygen at 20 °C and 4 atm pressure are needed for the complete combustion of 5 cubic feet of propane at the same temperature and pressure in the reaction $C_3H_8 + 5O_2 \longrightarrow 3CO_2 + 4H_2O$?

25.13. Hydrogen peroxide decomposes into water and oxygen in the reaction $2H_2O_2 \longrightarrow 2H_2O + O_2$. What mass of hydrogen peroxide is needed to provide 10 liters of oxygen at STP?

25.14. Lithium oxide, Li_2O, combines with carbon dioxide to form lithium carbonate in the reaction $Li_2O + CO_2 \longrightarrow Li_2CO_3$ and accordingly can be used to remove exhaled CO_2 in closed living spaces such as those in submarines and space craft. How many liters of CO_2 at STP can 1 kg of Li_2O absorb?

25.15. How many liters of oxygen at 27 °C and 1 atm pressure are needed to combine with 40 g of sulfur to form sulfur dioxide, SO_2?

25.16. (a) What volume does 8.2 moles of fluorine, F_2, occupy at STP? (b) What volume does it occupy at 40 °C and a pressure of 2.5 atm?

25.17. (a) What volume does 5 g of methane, CH_4, occupy at STP? (b) What volume does it occupy at 0 °C and a pressure of 0.5 atm? (c) What volume does it occupy at 80 °C and a pressure of 2 atm?

25.18. (a) What volume does 20 g of CO_2 occupy at STP? (b) What volume does it occupy at −20 °C and a pressure of 4 atm?

25.19. What is the mass of 4 liters of ammonia, NH_3, at STP?

25.20. (a) Find the mass of 12 liters of chlorine, Cl_2, at 40 °C and 0.8 atm pressure. (b) What is its density under those conditions?

25.21. A balloon is filled with 50 m³ (5×10^4 liters) of hydrogen at STP. What is the mass of the hydrogen?

25.22. Find the density of sulfur dioxide, SO_2, at STP.

25.23. Find the density of nitrogen, N_2, at 120 °C and a pressure of 66,600 N/m².

25.24. One liter of an unknown gas has a mass of 2.9 g at STP. Find its molecular mass.

Answers to Supplementary Problems

25.11. 3 liters; 6 liters

25.12. 25 ft³

25.13. 30.4 g

25.14. 750 liters

25.15. 30.8 liters

25.16. 184 liters; 84.4 liters

25.17. 6.98 liters; 13.96 liters; 4.51 liters

25.18. 10.2 liters; 2.36 liters

25.19. 3.04 g

25.20. 26.5 g; 2.21 g/liter = 2.21 kg/m³

25.21. 4.5 kg

25.22. 2.86 g/liter = 2.86 kg/m³

25.23. 0.571 kg/m³

25.24. 65 u

Chapter 26

Solutions

SOLVENT AND SOLUTE

When a solid or gas is dissolved in a liquid, the solid or gas is called the *solute* and the liquid is called the *solvent*. When one liquid is dissolved in another, the liquid whose quantity is greater is usually considered the solvent; if the two liquids are present in equal amounts, the designations of solvent and solute are arbitrary.

SOLUBILITY

The *solubility* of a substance is the maximum amount of it that will dissolve in a specified quantity of solvent at a certain temperature. Solubilities are commonly expressed in grams of solute per 100 grams of solvent, in moles of solute per kilogram of solvent, or in moles of solute per liter of solution. A *saturated* solution is one that contains the greatest concentration of solute possible. The solubilities of most solids increase with temperature; the solubilities of gases in water usually decrease with temperature, and those of gases in nonpolar liquids may increase or decrease depending upon the nature of the substances.

The extent to which a certain solute dissolves in a certain solvent depends a great deal upon their electrical characters. A *polar molecule* has an asymmetrical charge distribution, with one end positively charged and the other end negatively charged; a *polar liquid*, such as water, consists of polar molecules clustered together with oppositely charged ends adjacent. A substance composed of polar molecules, such as alcohol or sugar, dissolves readily in a polar liquid since the solute molecules can readily join the aggregates of solvent molecules.

A substance composed of nonpolar molecules, such as fat, does not dissolve in a polar liquid because its molecules cannot join the aggregates of polar molecules in the liquid nor can any of the polar molecules in the liquid break loose and mix with the nonpolar molecules of the solute. On the other hand, a nonpolar substance can dissolve in a nonpolar liquid such as gasoline or carbon tetrachloride. As a general rule, "like dissolves like."

IONS IN SOLUTION

An ionic compound in the solid state consists of an array of separate positive and negative ions. Such a compound only dissolves in a highly polar liquid, since only polar molecules can exert the electrical forces needed to break the ions loose from each other. The resulting solution contains ions, not molecules. When NaCl dissolves in water, for instance, the result is a solution of separate Na^+ and Cl^- ions. Substances that *dissociate* (separate into ions) in solution are called *electrolytes* because their solutions conduct electric current. The extent of dissociation varies with the substance concerned: some, such as NaCl, are completely dissociated when dissolved in water; others, such as $HC_2H_3O_2$ (acetic acid), are only slightly dissociated.

When an electrolyte is dissolved, each ion then behaves independently of the others. For instance, let us consider what happens when NaCl and $AgNO_3$ are added to water. First, the two compounds dissociate into ions:

$$NaCl \longrightarrow Na^+ + Cl^-$$

$$AgNO_3 \longrightarrow Ag^+ + NO_3^-$$

Then the Ag^+ ions and the Cl^- ions react to form the insoluble compound AgCl, leaving Na^+ and NO_3^- ions in solutions:

$$Na^+ + Cl^- + Ag^+ + NO_3^- \longrightarrow AgCl + Na^+ + NO_3^-$$

CONCENTRATION

There are various ways to express the concentration of solute in a solution. A particularly useful one is *molarity*. The molarity of a solution is the number of moles of solute per liter of solution. A solution that contains 2 moles of NaCl per liter is designated $2M$ NaCl, and one that contains 0.4 moles of H_2SO_4 (sulfuric acid) per liter is designated $0.4M$ H_2SO_4. In equation form,

$$\text{Molarity} = \frac{\text{moles of solute}}{\text{liters of solution}}$$

When a certain number of moles of a compound is required for a laboratory procedure, it is simple to just pour out the corresponding volume of a solution of known molarity.

Another common measure of concentration is *molality*: The molality of a solution is the number of moles of solute per kilogram of solvent. A solution that contains 6 moles of HNO_3 (nitric acid) per kg of water is designated $6m$ HNO_3. In equation form,

$$\text{Molality} = \frac{\text{moles of solute}}{\text{kg of solvent}}$$

COLLIGATIVE PROPERTIES OF SOLUTIONS

Certain properties of solutions depend only upon the concentration of particles (molecules or ions) of solute present and not upon their chemical nature. Two examples of such *colligative properties* are the depression of the freezing point of the solvent and the elevation of its boiling point. In the case of water the freezing point is lowered by 1.86 °C for each mole of solute particles per kg of water, and the boiling point is raised by 0.52 °C for each mole of solute particles per kg of water. (The boiling point is lowered, however, if the solute is a volatile liquid such as alcohol.) Thus a $1m$ solution of $C_6H_{12}O_6$ (glucose) in water lowers its freezing point to −1.86 °C since the glucose molecules remain intact, but a $1m$ solution of NaCl in water lowers its freezing point more than this since 1 mole of NaCl dissociates into 1 mole of Na^+ and 1 mole of Cl^- ions. The reduction of freezing point in a dilute NaCl solution corresponds to 2×1.86 °C per mole of NaCl per kg of water; in a concentrated solution, the electrical interaction of the ions somewhat decreases the amount of freezing-point reduction.

Solved Problems

26.1. How can you tell whether a solution of KCl is saturated or not?

There are two simple methods. One is to add more KCl to a sample of the solution and see whether it dissolves; if it does not, the solution is saturated. The other method is to cool the solution and see if any KCl crystallizes out; if it does, the solution is saturated, since the solubility of KCl decreases with decreasing temperature.

26.2. DDT is found to concentrate in the fat of animals, birds, and fish. Also, DDT tends to remain in the soil despite heavy rain that washes away other contaminants. What do these observations indicate about the nature of the DDT molecule?

Since DDT is not very soluble in water, which is a highly polar liquid, but is soluble in fat, whose molecules are nonpolar, the conclusion is that DDT molecules are nonpolar.

26.3. What is the molarity of a solution that contains 75 g of barium chloride, $BaCl_2$, per liter?

The formula mass of $BaCl_2$ is

$$1Ba = 1 \times 137.34 \text{ u} = 137.34 \text{ u}$$
$$2Cl = 2 \times 35.46 \text{ u} = \underline{70.92 \text{ u}}$$
$$208.26 \text{ u} = 208.26 \text{ g/mole}$$

so the number of moles of $BaCl_2$ in 75 g is

$$\text{Moles of } BaCl_2 = \frac{\text{mass of } BaCl_2}{\text{formula mass of } BaCl_2} = \frac{75 \text{ g}}{208.26 \text{ g/mole}} = 0.36 \text{ mole}$$

Hence the concentration of the solution is 0.36 mole/liter = 0.36 M.

26.4. How many moles of sulfuric acid, H_2SO_4, are present in 120 ml of a 2M solution? (1000 ml = 1 liter)

The volume of the solution is 0.12 liter and its molarity is 2M = 2 moles/liter. Since

$$\text{Molarity} = \frac{\text{moles}}{\text{liters}}$$

we find that the number of moles present is

$$\text{Moles} = \text{molarity} \times \text{liters} = 2 \text{ moles/liter} \times 0.12 \text{ liter} = 0.24 \text{ moles}$$

26.5. How many ml of 2M sulfuric acid are needed to provide 0.082 mole of this acid?

Since Molarity = moles ÷ liters, we have

$$\text{Liters} = \frac{\text{moles}}{\text{molarity}} = \frac{0.082 \text{ mole}}{2 \text{ moles/liter}} = 0.041 \text{ liter} = 41 \text{ ml}$$

26.6. How much silver nitrate, $AgNO_3$, should be dissolved to make 1 liter of a 2M solution? The formula mass of $AgNO_3$ is 170 g/mole.

A 2M solution contains 2 moles of solute per liter, which in this case means

$$\text{Mass of } AgNO_3 = \text{moles of } AgNO_3 \times \text{formula mass of } AgNO_3$$
$$= 2 \text{ moles} \times 170 \text{ g/mole} = 340 \text{ g}$$

26.7. How much calcium chloride, $CaCl_2$, is needed to prepare 50 ml of a 1.2 M solution? The formula mass of $CaCl_2$ is 111 g/mole.

The volume of the solution is 0.05 liter and its molarity is 1.5 M = 1.5 moles/liter. The number of moles of $CaCl_2$ needed is therefore

$$\text{Moles} = \text{molarity} \times \text{liters} = 1.5 \text{ moles/liter} \times 0.05 \text{ liter} = 0.075 \text{ moles}$$

Hence $\text{Mass of } CaCl_2 = \text{moles of } CaCl_2 \times \text{formula mass of } CaCl_2$
$$= 0.075 \text{ mole} \times 111 \text{ g/mole} = 8.33 \text{ g}$$

26.8. How many ml of an $0.8\,M$ solution of potassium iodide, KI, are needed to provide 4 g of KI? The formula mass of KI is 166 g/mole.

The number of moles of KI required is

$$\text{Moles of KI} = \frac{\text{mass of KI}}{\text{formula mass of KI}} = \frac{4\text{ g}}{166\text{ g/mole}} = 0.024\text{ mole}$$

Hence

$$\text{Liters} = \frac{\text{moles}}{\text{molarity}} = \frac{0.024\text{ mole}}{0.8\text{ mole/liter}} = 0.030\text{ liter} = 30\text{ ml}$$

26.9. The solubility of CO_2 in water at STP is 0.33 g per 100 g of water. What is the molal concentration of a saturated solution of CO_2 at STP?

A concentration of 0.33 g per 100 g of H_2O is the same as 3.3 g per kg. The molecular mass of CO_2 is

$$1C = 1 \times 12.01\text{ u} = 12.01\text{ u}$$
$$2O = 2 \times 16.00\text{ u} = 32.00\text{ u}$$
$$\overline{\phantom{2O = 2 \times 16.00\text{ u} = {}}44.01\text{ u}} = 44.01\text{ g/mole}$$

so the number of moles in 3.3 g of CO_2 is

$$\text{Moles of }CO_2 = \frac{\text{mass of }CO_2}{\text{molecular mass of }CO_2} = \frac{3.3\text{ g}}{44.01\text{ g/mole}} = 0.075\text{ mole}$$

The molal concentration of CO_2 is therefore 0.075 mole/kg of water $= 0.075\,m$

26.10. Which would freeze at a lower temperature, a $1m$ solution of KCl or a $1m$ solution of $CaCl_2$?

Since $KCl \longrightarrow K^+ + Cl^-$ when it dissolves, the KCl solution has a $2m$ concentration of ions. Since $CaCl_2 \longrightarrow Ca^{++} + 2Cl^-$ when it dissolves, the $CaCl_2$ solution has a $3m$ concentration of ions. Hence the $CaCl_2$ solution freezes at a lower temperature than the KCl solution.

26.11. Twenty grams of glucose, $C_6H_{12}O_6$, are dissolved in 50 g of water. Glucose does not dissociate into ions, and its molecular mass is 180 u. (a) What is the freezing point of the solution? (b) What is its boiling point?

(a) The number of moles in 20 g of glucose is

$$\text{Moles of }C_6H_{12}O_6 = \frac{\text{mass of }C_6H_{12}O_6}{\text{molecular mass of }C_2H_{12}O_6} = \frac{20\text{ g}}{180\text{ g/mole}} = 0.11\text{ mole}$$

Since $50\text{ g} = 0.05$ kg, the molality of the solution is

$$\text{Molality} = \frac{\text{moles of solute}}{\text{kg of solvent}} = \frac{0.11\text{ moles of }C_6H_{12}O_6}{0.05\text{ kg of water}} = 2.2\,m$$

A $1m$ solution reduces the freezing point of water by 1.86 °C, hence a $2.2\,m$ solution reduces it by 2.2×1.86 °C $= 4.1$ °C to -4.1 °C.

(b) A $1m$ solution raises the boiling point of water by 0.52 °C, hence a $2.2\,m$ solution raises it by 2.2×0.52 °C $= 1.1$ °C to 101.1 °C.

26.12. Ethylene glycol, $C_2H_4(OH)_2$, is widely used as an antifreeze in car radiators. What concentration, in grams per kg of water, is required to protect a radiator from freezing at temperatures down to -25 °C? (The molecule does not dissociate into ions.)

Each mole of solute per kg of water reduces its freezing point by 1.86 °C, hence

$$\frac{25\text{ °C}}{1.86\text{ °C/mole}} = 13.4\text{ moles}$$

of ethylene glycol per kg of water are needed to reduce its freezing point by 25 °C. The molecular mass of ethylene glycol is

$$2C = 2 \times 12.01 \text{ u} = 24.02 \text{ u}$$
$$6H = 6 \times 1.008 \text{ u} = 6.05 \text{ u}$$
$$2O = 2 \times 16.00 \text{ u} = \underline{32.00 \text{ u}}$$
$$62.07 \text{ u} = 62.07 \text{ g/mole}$$

The corresponding mass of ethylene glycol is

$$\text{Mass of } C_2H_4(OH)_2 = \text{moles of } C_2H_4(OH)_2 \times \text{molecular mass of } C_2H_4(OH)_2$$
$$= 13.4 \text{ moles} \times 62.07 \text{ g/mole} = 832 \text{ g}$$

The required concentration of ethylene glycol is 832 g/kg of water.

26.13. The freezing point of a $0.1 \, m$ solution of acetic acid, $HC_2H_3O_2$, in water is -0.188 °C. Is acetic acid a strong or a weak electrolyte?

A solution with a $1m$ concentration of particles has a freezing point of -1.86 °C and so a solution with a $0.1 \, m$ concentration has a freezing point of -0.186 °C. If acetic acid dissociated completely into H^+ and $C_2H_3O_2^-$ ions, the solution would have a $0.2 \, m$ concentration of particles and a freezing point of -0.372 °C. Since the freezing point of $0.1 \, m$ acetic acid is -0.188 °C, it is only slightly dissociated, and therefore is a weak electrolyte.

26.14. Three hundred grams of sucrose (cane sugar) is dissolved in 500 g of water. The resulting solution boils at 100.91 °C. What is the approximate molecular mass of sucrose?

The boiling point of a $1m$ solution is 0.52 °C higher than that of pure water, hence the molality of the present solution is

$$\text{Molality} = \frac{0.91 \text{ °C}}{0.52 \text{ °C}/m} = 1.75 \, m$$

Hence the number of moles of sucrose in the solution is

$$\text{Moles of solute} = \text{molality} \times \text{kg of solvent}$$
$$= 1.75 \text{ moles/kg} \times 0.5 \text{ kg} = 0.875 \text{ moles}$$

and
$$\text{Molecular mass of sucrose} = \frac{\text{mass of sucrose}}{\text{moles of sucrose}} = \frac{300 \text{ g}}{0.875 \text{ moles}}$$
$$= 343 \text{ g/mole} = 343 \text{ u}$$

Supplementary Problems

26.15. What is the difference between a molecular ion and a polar molecule?

26.16. How can you tell whether a substance is an electrolyte or not?

26.17. Find the molarity of a solution that contains 80 g of hydrochloric acid, HCl, per liter.

26.18. Find the molarity of a solution that contains 30 g of ethyl alcohol, C_2H_5OH, per 100 ml.

26.19. How many ml of $6M$ hydrochloric acid are needed to provide 0.5 mole of this acid?

26.20. (a) How many moles of potassium permanganate, $KMnO_4$, are present in 250 ml of a $1.6 \, M$ solution? (b) How many grams?

26.21. How many grams of sodium thiosulfate, $Na_2S_2O_3$, are needed to prepare 600 ml of a $2.5 \, M$ solution?

26.22. (a) How many ml of a $2.5 \, M$ solution of sodium hydroxide, NaOH, are needed to provide 0.25 mole of NaOH? (b) To provide 20 g of NaOH?

26.23. The solubility of lithium carbonate, Li_2CO_3, at 20 °C is 1.33 g/100 g of water. Find the molal concentration of a saturated solution of Li_2CO_3 at 20 °C.

26.24. Would the addition of 1 kg of methyl alcohol, CH_3OH, or 1 kg of ethyl alcohol, C_2H_5OH, be more effective in reducing the freezing point of the water in a car radiator? Why?

26.25. The chief solute in sea water is NaCl, which is present at a concentration of about 3.5% by mass. Calculate the freezing and boiling points of sea water.

26.26. Ten grams of urea, $CO(NH_2)_2$, are dissolved in 40 grams of water. Urea does not dissociate into ions in solution. (*a*) What is the freezing point of the solution? (*b*) What is its boiling point?

26.27. An antifreeze solution is prepared by adding 4 kg of ethylene glycol, $C_2H_4(OH)_2$, to 10 kg of water. What is the freezing point of this solution?

26.28. How much methyl alcohol, CH_3OH, should be added to 12 kg of water to prepare an antifreeze solution that can protect a car radiator down to −20 °C?

26.29. Aluminum sulfate, $Al_2(SO_4)_3$, dissociates completely into Al^{+++} and SO_4^{--} ions when dissolved in water. (*a*) What is the molal concentration of ions in a 0.1 m solution of aluminum sulfate in water? (*b*) What is the approximate freezing point of the solution? (*c*) What is its approximate boiling point?

26.30. A dilute solution of HF is prepared by adding 0.2 g of it to 100 g of water. The freezing point of the solution is −0.25 °C. Does HF dissociate into H^+ and F^- ions completely, partially, or not at all at this concentration?

26.31. A solution consisting of equal parts by mass of glycerin and water is observed to freeze at −20.2 °C. Find the molecular mass of glycerin.

26.32. Ten grams of urea are dissolved in 40 grams of water and the resulting solution is observed to boil at 102.16 °C. Find the molecular mass of urea.

Answers to Supplementary Problems

26.15. A molecular ion is a molecule that has gained or lost one or more electrons and so has a net charge. A polar molecule is electrically neutral but has an asymmetrical distribution of charge and so one end is negative and the other positive.

26.16. A solution of an electrolyte conducts electric current, a solution of a nonelectrolyte does not.

26.17. 2.2 *M*

26.18. 6.5 *M*

26.19. 83 ml

26.20. 0.4 mole; 63.2 g

26.21. 237 g

26.22. 100 ml; 200 ml

26.23. 0.18 *m*

26.24. The methyl alcohol would be more effective since its molecular mass is smaller than that of ethyl alcohol and the molality of the resulting solution is therefore greater than that of the ethyl alcohol solution.

26.25. −1.15 °C; 100.32 °C

26.26. −7.74 °C; 102.16 °C

26.27. −12 °C

26.28. 4.13 kg

26.29. 0.5 *m*; −0.93 °C; 100.26 °C

26.30. partially

26.31. 92 u

26.32. 60 u

Chapter 27

Acids and Bases

DISSOCIATION OF WATER

Water dissociates into H^+ and OH^- ions to a very small extent:

$$H_2O \longrightarrow H^+ + OH^-$$

In pure water the concentration of H^+ ions is $1.0 \times 10^{-7} M$, that is, 1.0×10^{-7} mole/liter, and the concentration of OH^- ions is the same.

The H^+ ion is a bare proton, and it is considerably smaller than all other ions and neutral atoms since the latter have electron clouds around their nuclei. Because of its small size, the electric field near a proton is very strong, and it therefore tends to become firmly attached to the negative end of a water molecule. Thus it is more realistic to write H_3O^+ rather than H^+; the H_3O^+ ion is called the *hydronium* ion. Actually, more than one H_2O molecule is attached to each H^+ ion, and indeed all other ions in water usually are *hydrated* by having clusters of H_2O molecules around them as well. In most reactions that involve ions in water solution the hydration of the ions is not significant to the final result, and for simplicity the equations of these reactions include only the ions themselves. It is therefore customary to write H^+ for the hydrogen ion even though in reality the H^+ ion never exists as such in solution.

ACIDS AND BASES

An *acid solution* is one in which the concentration of H^+ ions exceeds that of OH^- ions; a *basic solution* is one in which the concentration of OH^- ions exceeds that of H^+ ions. An *acid* is a substance which, when added to water, releases H^+ ions; a *base* is a substance which, when added to water, releases OH^- ions. (There are other, more general definitions of acid and base, but the above ones are sufficient for many purposes.)

The "strength" of an acid or base refers to its degree of dissociation. Thus hydrochloric acid, HCl, is a strong acid because it dissociates completely in water:

$$HCl \longrightarrow H^+ + Cl^-$$

Acetic acid, $HC_2H_3O_2$, is a weak acid because (see Problem 26.13) relatively few of its molecules dissociate in water:

$$HC_2H_3O_2 \longrightarrow H^+ + C_2H_3O_2^-$$

Similarly sodium hydroxide, NaOH, is a strong base because it dissociates completely in water:

$$NaOH \longrightarrow Na^+ + OH^-$$

Ammonia, NH_3, is a weak base even though its formula does not include an OH group. When it is dissolved in water, some of the NH_3 molecules react with H_2O molecules to produce OH^- ions:

$$NH_3 + H_2O \longrightarrow NH_4^+ + OH^-$$

THE pH SCALE

In a neutral aqueous solution the H^+ and OH^- concentrations are both $10^{-7} M$; in an acid solution, the H^+ concentration is greater than $10^{-7} M$ and the OH^- concentration is less; in a basic solution, the H^+ concentration is less than $10^{-7} M$ and the OH^- concentration is more. The H^+ concentration by itself is thus a measure of the degree of acidity or basicity of a solution.

If the H^+ concentration in moles/liter is written $[H^+]$, the quantity pH is defined as follows:

$$pH = -\log [H^+]$$

If the H^+ concentration is 10^{-x}, the corresponding pH is x. In a neutral solution, $[H^+] = 10^{-7}$, hence its pH is 7. In an acid solution, $[H^+] > 10^{-7}$ and the pH is less than 7; in a basic solution, $[H^+] < 10^{-7}$ and the pH is more than 7. The pH scale is indicated in Fig. 27-1.

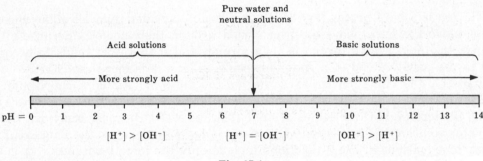

Fig. 27-1

NEUTRALIZATION

The slight degree of dissociation of H_2O signifies that H^+ and OH^- ions have a strong tendency to join together to form H_2O molecules. When an acid and a base are placed in the same solution, the H^+ ions from the acid combine with the OH^- ions from the base to form H_2O, leaving the negative ion of acid and the positive ion of the base in solution. This process is called *neutralization*. For example, when HCl and NaOH solutions are mixed, the result is

$$HCl + NaOH \longrightarrow H^+ + Cl^- + Na^+ + OH^- \longrightarrow H_2O + Na^+ + Cl^-$$

The essential neutralization reaction is

$$H^+ + OH^- \longrightarrow H_2O$$

If equal numbers of moles of HCl and NaOH are mixed, the resulting solution contains only Na^+ and Cl^- ions. Evaporating the solution will yield crystals of NaCl, which is the *salt* sodium chloride. Every salt consists of the positive ion of a base and the negative ion of an acid, and can be formed by mixing the corresponding acid and base together and evaporating the resulting solution.

GRAM-EQUIVALENTS AND NORMALITY

A *gram-equivalent* of an acid is that amount of the acid which can provide 1 mole of H^+ ions; a gram-equivalent of a base is that amount of the base which can provide 1 mole of OH^- ions. Thus a gram-equivalent of an acid or base provides Avogadro's number of H^+ or OH^- ions to a solution. Since the dissociation of HCl proceeds as $HCl \longrightarrow H^+ + Cl^-$, 1 gram-equivalent of HCl is equal to 1 mole of HCl. Since the dissociation of sulfuric acid, H_2SO_4, proceeds as $H_2SO_4 \longrightarrow 2H^+ + SO_4^{--}$, 1 mole of H_2SO_4 can provide 2 moles of H^+, and so 1 gram-equivalent of H_2SO_4 is equal to $\frac{1}{2}$ mole of H_2SO_4.

The complete neutralization of an acid (or base) occurs when it is mixed with the same number of gram-equivalents of a base (or acid). Thus 1 mole of HCl or $\frac{1}{2}$ mole of H_2SO_4 will neutralize 1 mole of NaOH, and 1 mole of NaOH or $\frac{1}{2}$ mole of $Ca(OH)_2$ will neutralize 1 mole of HCl.

The *normality* of an acid or basic solution is the number of gram-equivalents it contains per liter. A $1N$ solution of any acid contains 1 mole/liter of H^+ ions, a $1N$ solution of any base contains 1 mole/liter of OH^- ions.

Solved Problems

27.1. What proportion of the H_2O molecules in a sample of pure water is dissociated into H^+ and OH^- ions at any time?

The molecular mass of H_2O is 18 u and there are 1000 g in a liter of water, hence the number of moles of H_2O in a liter of water is

$$\text{Moles of } H_2O \;=\; \frac{\text{mass of } H_2O}{\text{molecular mass of } H_2O} \;=\; \frac{1000\text{ g}}{18\text{ g/mole}} \;=\; 55\text{ moles}$$

The concentration of H_2O is therefore $55M$. Since the concentrations of H^+ and OH^- ions are both $1.0 \times 10^{-7}\,M$, the fraction of H_2O that dissociates is

$$\frac{\text{Dissociated molecules}}{\text{Total } H_2O \text{ molecules}} \;=\; \frac{1.0 \times 10^{-7}\,M}{55M} \;=\; 2 \times 10^{-9}$$

Thus two H_2O molecules per billion are dissociated.

27.2. The H^+ concentration in vinegar is 10^{-3} moles/liter. (*a*) Is vinegar an acid or a base? (*b*) What is its pH?

(*a*) An H^+ concentration of $10^{-3}\,M$ is greater than $10^{-7}\,M$, hence vinegar is an acid.

(*b*) The pH of a solution in which the H^+ concentration is $10^{-x}\,M$ is x, hence the pH of vinegar is 3.

27.3. The pH of cow's milk is 6.6. Is it acidic or basic?

A pH less than 7 signifies an acidic solution.

27.4. (*a*) The hydrochloric acid in 100 ml of a $0.1\,M$ HCl solution is completely dissociated. What is the pH of the solution? (*b*) The solution is diluted by the addition of 900 ml of pure water. What is its new pH? (*c*) How much $0.1\,M$ NaOH must be added to the original HCl solution to neutralize it? (*d*) How much $0.1\,M$ NaOH must be added to the diluted HCl solution to neutralize it?

(*a*) The H^+ concentration is $0.1\,M = 10^{-1}\,M$. Since a solution whose H^+ concentration is $10^{-x}\,M$ has a pH of x, the pH here is 1.

(*b*) The original solution contained 0.1 mole/liter $\times$ 0.1 liter $=$ 0.01 mole of HCl. The new solution contains the same 0.01 mole of HCl but now in a total volume of 1 liter, hence its concentration is 0.01 mole/liter $= 0.01\,M = 10^{-2}\,M$ and the pH is 2.

(*c*) Since 0.01 mole of HCl is present, 0.01 mole of NaOH is needed to neutralize it. The volume of NaOH solution needed is therefore

$$\text{Liters of solution} \;=\; \frac{\text{moles of NaOH}}{\text{molarity}} \;=\; \frac{0.01\text{ mole}}{0.1\,M} \;=\; 0.1\text{ liter} \;=\; 100\text{ ml}$$

(*d*) The diluted HCl solution contains the same 0.01 mole of HCl as the original solution did, hence the same 100 ml of $0.1\,M$ NaOH will also neutralize it.

27.5. What salts are formed by the neutralization of calcium hydroxide, $Ca(OH)_2$, by (a) HCl, (b) H_2SO_4, and (c) H_3PO_4? Give the equation for the overall process in each case.

(a) Calcium chloride, $CaCl_2$: $2HCl + Ca(OH)_2 \longrightarrow CaCl_2 + 2H_2O$

(b) Calcium sulfate, $CaSO_4$: $H_2SO_4 + Ca(OH)_2 \longrightarrow CaSO_4 + 2H_2O$

(c) Calcium phosphate, $Ca_3(PO_4)_2$: $2H_3PO_4 + 3Ca(OH)_2 \longrightarrow Ca_3(PO_4)_2 + 6H_2O$

27.6. One liter of $1M$ HCl is mixed with 1 liter of $0.5M$ NaOH. (a) What are the molar concentrations of the various ions in the resulting solution? (b) Is the solution acidic or basic?

(a) The entire 0.5 mole of OH^- from the dissociation of the NaOH is neutralized by half of the 1 mole of H^+ from the dissociation of the HCl, so what is left in solution is 0.5 mole of H^+, 0.5 mole of Na^+, and 1 mole of Cl^-. Since there are 2 liters of solution and Molarity = moles/liter, the solution contains $0.25M$ H^+, $0.25M$ Na^+, and $0.5M$ Cl^-.

(b) The solution is acidic because the H^+ concentration exceeds the OH^- concentration.

27.7. It is observed that, when a salt containing the negative ion of a weak acid is dissolved in water, the solution is basic. For example, a solution of sodium acetate, $NaC_2H_3O_2$, is basic. Explain.

Sodium acetate dissociates in water to form Na^+ and $C_2H_3O_2^-$ ions:

$$NaC_2H_3O_2 \longrightarrow Na^+ + C_2H_3O_2^-$$

Some of the acetate ions then react with water to form undissociated acetic acid, thus liberating OH^- ions:

$$C_2H_3O_2^- + H_2O \longrightarrow HC_2H_3O_2 + OH^-$$

This reaction occurs because acetic acid is weak and therefore can exist undissociated in solution. The sodium acetate solution therefore contains Na^+, $C_2H_3O_2^-$, and OH^- ions, plus $HC_2H_3O_2$ molecules, and is basic.

27.8. The Al^{+++} ion tends to form $AlOH^{++}$ ions in water solution, whereas Na^+ and Cl^- ions remain free. (a) Is a solution of $AlCl_3$ acidic, basic, or neutral? (b) Is a solution of NaCl acidic, basic, or neutral?

(a) When $AlCl_3$ is dissolved in water, it first dissociates into Al^{+++} and Cl^- ions:

$$AlCl_3 \longrightarrow Al^{+++} + 3Cl^-$$

Some of the Al^{+++} ions then react with water to form $AlOH^{++}$ and H^+ ions:

$$Al^{+++} + H_2O \longrightarrow H^+ + AlOH^{++}$$

Hence the solution contains Al^{+++}, $AlOH^{++}$, H^+, and Cl^- ions, and is acidic.

(b) When NaCl is dissolved, it dissociates into Na^+ and Cl^- ions which do not react further. Hence the solution is neutral.

27.9. What is the mass of a gram-equivalent of phosphoric acid, H_3PO_4?

The formula mass of H_3PO_4 is

$$3H = 3 \times 1.008\,u = 3.02\,u$$
$$1P = 1 \times 30.98\,u = 30.98\,u$$
$$4O = 4 \times 16.00\,u = \underline{64.00\,u}$$
$$98.00\,u = 98.00\text{ g/mole}$$

Since 1 formula unit of H_3PO_4 can yield 3 H^+ ions, 1 gram-equivalent of this acid is $\frac{1}{3}$ mole or 32.67 g.

27.10. How much potassium hydroxide, KOH, is needed to prepare 400 ml of a $2N$ solution?

Since the dissociation of KOH proceeds as $KOH \longrightarrow K^+ + OH^-$, 1 gram-equivalent of KOH = 1 mole of KOH and the concentration of a $2N$ solution is 2 gram-equivalents/liter = 2 moles/liter. Because

$$Molarity = \frac{moles\ of\ solute}{liters\ of\ solution}$$

we have for the number of moles of KOH required here

$$Moles\ of\ KOH = molarity \times liters\ of\ solution$$
$$= 2\ moles/liter \times 0.4\ liter = 0.8\ moles$$

The formula mass of KOH is $(39.10 + 16.00 + 1.01)$ u = 56.11 u = 56.11 g/mole, hence the required mass is

$$Mass\ of\ KOH = moles\ of\ KOH \times formula\ mass\ of\ KOH$$
$$= 0.8\ moles \times 56.11\ g/mole = 44.9\ g$$

27.11. How much sulfuric acid, H_2SO_4, is needed to prepare 5 liters of a $6N$ solution?

Since the dissociation of H_2SO_4 proceeds as $H_2SO_4 \longrightarrow 2H^+ + SO_4^{--}$, 1 gram-equivalent of $H_2SO_4 = \frac{1}{2}$ mole of H_2SO_4 and the concentration of a $6N$ solution is 6 gram-equivalents/liter = 3 moles/liter. The number of moles of H_2SO_4 required here is

$$Moles\ of\ H_2SO_4 = molarity \times liters\ of\ solution$$
$$= 3\ moles/liter \times 5\ liters = 15\ moles$$

The formula mass of H_2SO_4 is

$$2H = 2 \times 1.008\ u = 2.02\ u$$
$$1S = 1 \times 32.06\ u = 32.06\ u$$
$$4O = 4 \times 16.00\ u = \underline{64.00\ u}$$
$$98.08\ u = 98.08\ g/mole$$

and so the required mass is

$$Mass\ of\ H_2SO_4 = moles\ of\ H_2SO_4 \times formula\ mass\ of\ H_2SO_4$$
$$= 15\ moles \times 98.08\ g/mole = 1471\ g$$

27.12. (a) How many ml of $10N$ HCl are needed to neutralize 50 ml of a $2N$ solution of NaOH? (b) How many ml of $10N$ HCl are needed to neutralize 50 ml of a $2N$ solution of $Ca(OH)_2$?

(a) Since the normality of a solution equals the number of gram-equivalents of solute it contains per liter, the number of gram-equivalents in two different solutions A and B will be the same if

$$Normality\ of\ A \times volume\ of\ A = normality\ of\ B \times volume\ of\ B$$

Hence we have

$$Volume\ of\ HCl = \frac{normality\ of\ NaOH \times volume\ of\ NaOH}{normality\ of\ HCl} = \frac{2N \times 50\ ml}{10N} = 10\ ml$$

(b) Two solutions with the same normality and the same volume contain the same number of gram-equivalents, so the amount of HCl is again 10 ml.

Supplementary Problems

27.13. The pH of lemon juice is about 2 and that of tomato juice is about 4. Which is more strongly acidic?

27.14. H_2SO_4, HCl, and NaOH dissociate completely in solution, $HC_2H_3O_2$ dissociates partially, and a fraction of the NH_3 in an ammonia solution reacts with water to form NH_4^+ and OH^- ions. (a) Arrange 0.1 M solutions of H_2SO_4, HCl, NaOH, $HC_2H_3O_2$, and NH_3 in order of increasing pH. (b) Where would pure water fit into the sequence?

27.15. The H^+ concentration of a borax solution is $10^{-9} M$. (a) Is borax an acid or a base? (b) What is the pH of this solution?

27.16. What salt is formed when HCl is neutralized by a solution of NH_3?

27.17. What salts are formed by the following neutralizations? Give the equation for the overall process in each case. (a) H_2SO_4 and NaOH; (b) H_3BO_3 and $Ca(OH)_2$; (c) $HC_2H_3O_2$ and KOH.

27.18. Forty ml of $5M$ H_2SO_4 are added to 60 ml of $2M$ KOH. What are the molar concentrations of the various ions in the resulting solution? Is the solution acidic or basic?

27.19. Why is a solution of ammonium chloride, NH_4Cl, acidic?

27.20. Nitric acid is a strong acid. Is a solution of sodium nitrate acidic, basic, or neutral?

27.21. Benzoic acid, $HC_7H_5O_2$, is a weak acid. Is a solution of sodium benzoate acidic, basic, or neutral?

27.22. What is the mass of a gram-equivalent of nitric acid, HNO_3?

27.23. What is the mass of a gram-equivalent of barium hydroxide, $Ba(OH)_2$?

27.24. How much ammonia, NH_3, is needed to prepare 1.5 liters of a $0.1 N$ solution?

27.25. How much $Ca(OH)_2$ is needed to prepare 1 liter of a $0.5 N$ solution?

27.26. How many liters of $0.5 N$ H_2SO_4 are needed to neutralize 1.5 liters of $2N$ KOH?

27.27. (a) What is the pH of a $1N$ solution of HNO_3? (b) What is the pH of a $1N$ solution of H_2SO_4? Both acids are completely dissociated in solution.

Answers to Supplementary Problems

27.13. Lemon juice is more strongly acid because it has a lower pH.

27.14. (a) H_2SO_4, HCl, $HC_2H_3O_2$, NH_3, NaOH. (b) Between $HC_2H_3O_2$ and NH_3.

27.15. base; 9

27.16. Ammonium chloride, NH_4Cl.

27.17. (a) Sodium sulfate, Na_2SO_4: $H_2SO_4 + 2NaOH \longrightarrow Na_2SO_4 + 2H_2O$

(b) Calcium borate, $Ca_3(BO_3)_2$: $2H_3BO_3 + 3Ca(OH)_2 \longrightarrow Ca_3(BO_3)_2 + 6H_2O$

(c) Potassium acetate, $KC_2H_3O_2$: $HC_2H_3O_2 + KOH \longrightarrow KC_2H_3O_2 + H_2O$

27.18. $2.8 M$ H^+, $1.2 M$ K^+, $2M$ SO_4^{--}; acid

27.19. Ammonium chloride dissociates in water to form NH_4^+ and Cl^- ions. Some of the ammonium ions then dissociate to form ammonia molecules and hydrogen ions: $NH_4^+ \longrightarrow NH_3 + H^+$. The solution thus contains NH_4^+, H^+, and Cl^- ions, plus NH_3 molecules, and is acidic.

27.20. neutral   **27.24.** 2.56 g

27.21. basic   **27.25.** 18.5 g

27.22. 63.02 g   **27.26.** 6 liters

27.23. 85.68 g   **27.27.** 0; 0

Chapter 28

Oxidation and Reduction

OXIDATION-REDUCTION REACTIONS

A chemical reaction in which electrons are transferred from one atom or group of atoms to another is called an *oxidation-reduction reaction*. *Oxidation* refers to the loss of electrons; *reduction* refers to the gain of electrons. Thus the combination of Na atoms and Cl_2 molecules to form NaCl, which consists of Na^+ and Cl^- ions, involves the oxidation of the Na atoms and the reduction of the Cl atoms:

$$2Na + Cl_2 \longrightarrow 2Na^+ + 2Cl^- \longrightarrow 2NaCl$$

An *oxidizing agent* is a substance that brings about the oxidation of another one; thus an oxidizing agent gains electrons from the substance being oxidized, and is itself reduced. A *reducing agent* is a substance that brings about the reduction of another one; thus a reducing agent furnishes electrons to the substance being reduced, and is itself oxidized. For example, chlorine is a powerful oxidizing agent because the Cl atoms in a Cl_2 molecule have a strong tendency to pick up electrons and become Cl^- ions. Similarly sodium is a powerful reducing agent because Na atoms have a strong tendency to lose electrons and become Na^+ ions.

OXIDATION NUMBER

A quantity called *oxidation number* (or *oxidation state*) is useful in keeping track of the electron shifts in an oxidation-reduction reaction. During oxidation, the oxidation number of a substance increases; during reduction, the oxidation number decreases (Fig. 28-1).

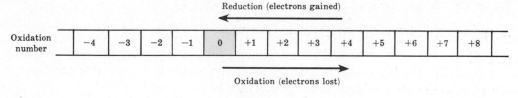

Fig. 28-1

The oxidation number of an atom or group of atoms is assigned on the basis of the following rules:

1. The oxidation number of an atom in its elemental state is 0. Thus the oxidation numbers of each hydrogen atom in H_2, of each copper atom in metallic Cu, and of each sulfur atom in S_8 are all 0.

2. The oxidation number of a monatomic ion is equal to its actual charge. Thus the oxidation number of H^+ is +1, that of Al^{+++} is +3, and that of O^{--} is −2.

3. When hydrogen is combined with a nonmetal, its oxidation number is +1; when hydrogen is combined with a metal, its oxidation number is −1.

4. In nearly all compounds the oxidation number of oxygen is -2; the only exceptions are in the various peroxides (such as hydrogen peroxide, H_2O_2) where its oxidation number is -1, and in F_2O where its oxidation number is $+2$.

5. The elements in Group I of the periodic table (the alkali metals) always have the oxidation number $+1$.

6. The elements in Group II of the periodic table always have the oxidation number $+2$.

7. Fluorine always has the oxidation number -1; the other halogens (Group VII of the periodic table) have the oxidation number -1 except in certain compounds with other halogens or with oxygen.

8. The sum of the oxidation numbers in a compound is 0, and their sum in an atomic group equals the charge on the group.

Many elements exhibit two or more different oxidation numbers in the various compounds of which they are part. Iron, for instance, has two oxidation numbers, $+2$ in ferrous compounds and $+3$ in ferric compounds. Thus ferrous chloride has the formula $FeCl_2$ and ferric chloride has the formula $FeCl_3$. Nitrogen has the oxidation numbers -3, $+1$, $+2$, $+3$, $+4$, and $+5$ in its compounds, although only the -3 and $+5$ states are common.

GRAM-EQUIVALENTS

The notion of gram-equivalent can be applied to oxidation-reduction reactions as well as to acid-base reactions. A gram-equivalent of an oxidizing agent is that amount of the agent which can absorb 1 mole of electrons in a specified reaction; a gram-equivalent of a reducing agent is that amount of the agent which can donate 1 mole of electrons in a specified reaction. Thus Avogadro's number of electrons is transferred when 1 gram-equivalent of a substance is oxidized or reduced. In every oxidation-reduction reaction, the number of gram-equivalents of the oxidized and reduced substances is the same.

In the reaction

$$Zn + 2H^+ \longrightarrow Zn^{++} + H_2$$

the H^+ ion is the oxidizing agent $(2H^+ + 2e^- \longrightarrow H_2)$ and the Zn atom is the reducing agent $(Zn \longrightarrow Zn^{++} + 2e^-)$. Since each H^+ ion gains one electron, a gram-equivalent of H^+ is equal to 1 mole of H^+; since each Zn atom loses two electrons, a gram-equivalent of Zn is equal to $\frac{1}{2}$ mole of Zn (or $\frac{1}{2}$ gram-atom of Zn, which is the same thing here).

Solved Problems

28.1. Find the oxidation number of sulfur in the compounds H_2S, SO_2, H_2SO_4, and $Na_2S_4O_6$.

(a) The oxidation number of H when in combination with nonmetals is $+1$, hence the oxidation number sum of H_2 is $+2$ and the oxidation number of S is -2.

(b) The oxidation number of O is -2, hence the oxidation number sum of O_2 is -4 and the oxidation number of S is $+4$.

(c) The oxidation number of H is $+1$ and that of O is -2, hence the oxidation number sum of H_2O_4 is $2 \times (+1) + 4 \times (-2) = +2 - 8 = -6$ and the oxidation number of S is $+6$.

(d) The oxidation number of Na is $+1$ and that of O is -2, hence the oxidation number sum of Na_2O_6 is -10. The oxidation number sum of S_4 is therefore $+10$ and the oxidation number of each S is $+10/4$.

28.2. Find the oxidation number of each atom in the ions NH_4^+ (ammonium), NO_3^- (nitrate), and PO_4^{---} (phosphate).

(a) The oxidation number of H is +1, hence the oxidation number sum of H_4 is +4. Since the ion as a whole has an oxidation number of +1, the oxidation number of N is −3.

(b) The oxidation number of O is −2, hence the oxidation number sum of O_3 is −6. Since the ion as a whole has an oxidation number of −1, the oxidation number of N is +5.

(c) The oxidation number of O is −2, hence the oxidation number sum of O_4 is −8. Since the ion as a whole has an oxidation number of −3, the oxidation number of P is +5.

28.3. Is neutralization an oxidation-reduction reaction?

The essential reaction in neutralization is the formation of H_2O molecules from H^+ and OH^- ions:

$$H^+ + OH^- \longrightarrow H_2O$$

The oxidation number of H^+ is +1. In OH^-, the oxidation number of O is −2 and that of H is +1. In H_2O, the oxidation number of each H atom is +1 and that of the O atom is −2. Since there is no change in the oxidation number of the hydrogen or oxygen atoms, neutralization is not an oxidation-reduction reaction.

28.4. The gases nitrogen and hydrogen react under the proper circumstances to yield ammonia in the reaction

$$N_2 + 3H_2 \longrightarrow 2NH_3$$

Is this an oxidation-reduction reaction? If so, which elements are oxidized and which are reduced?

The oxidation numbers of N in N_2 and of H in H_2 are both 0. The oxidation number of H in NH_3 is +1 and that of N is −3. Since the oxidation numbers of N and H have changed, this is an oxidation-reduction reaction. The oxidation number of H has gone from 0 to +1, so it is oxidized; the oxidation number of N has gone from 0 to −3, so it is reduced.

28.5. In the compound FeS_2 (iron pyrites, or "fool's gold") iron has the oxidation number +2. Iron pyrites reacts with oxygen at a high temperature to form ferric oxide and sulfur dioxide in the reaction

$$4FeS_2 + 11O_2 \longrightarrow 2Fe_2O_3 + 8SO_2$$

Which elements are oxidized and which are reduced in this reaction?

The oxidation number of Fe goes from +2 in FeS_2 to +3 in Fe_2O_3, so it is oxidized. The oxidation number of S goes from −1 in FeS_2 to +4 in SO_2, so it is also oxidized. The oxidation number of O goes from 0 in O_2 to −2 in both Fe_2O_3 and SO_2, so it is reduced.

28.6. A *displacement reaction* is an oxidation-reduction reaction in which one element displaces another from solution. In each of the following displacement reactions identify (a) the element that is oxidized, (b) the element that is reduced, (c) the oxidizing agent, and (d) the reducing agent:

1. $Zn + Cu^{++} \longrightarrow Zn^{++} + Cu$

2. $Fe + 2H^+ \longrightarrow Fe^{++} + H_2$

3. $Cl_2 + 2Br^- \longrightarrow 2Cl^- + Br_2$

An element which loses electrons in a reaction acts as a reducing agent and is itself oxidized; an element which gains electrons in a reaction acts as an oxidizing agent and is itself reduced. Hence we have:

1. (a) and (d) Zn,   (b) and (c) Cu^{++}
2. (a) and (d) Fe,   (b) and (c) H^+
3. (a) and (d) Br^-,   (b) and (c) Cl_2

28.7. On the basis of the following displacement reactions, arrange zinc, iron, and copper in the order of their strength as reducing agents:

$$Zn + Cu^{++} \rightarrow Zn^{++} + Cu$$
$$Fe + Cu^{++} \rightarrow Fe^{++} + Cu$$
$$Zn + Fe^{++} \rightarrow Zn^{++} + Fe$$

Since zinc reduces both copper and iron ions, zinc is the strongest reducing agent. Since copper ions are reduced by both zinc and iron, copper is the weakest reducing agent. In order of strength as reducing agents, then, we have Zn, Fe, Cu.

28.8. Zinc and iron dissolve in acid solutions with the evolution of H_2, but copper in general does not. Where does hydrogen belong in the sequence Zn, Fe, Cu of reducing agents?

Both zinc and iron reduce H^+ to H_2, but copper does not. Hence hydrogen is a weaker reducing agent than zinc and iron, but is stronger than copper. The sequence is therefore Zn, Fe, H, Cu.

28.9. In the reaction $2HNO_3 + 3H_2S \rightarrow 2NO + 4H_2O + 3S$

H_2S acts as a reducing agent. What is the mass of a gram-equivalent of H_2S in this reaction? The molecular mass of H_2S is 34.08 u.

The oxidation numbers of the H atoms in H_2S do not change in the reaction, but the oxidation number of the S atom increases from -2 to 0. Each molecule of H_2S therefore gives up 2 electrons, so a gram-equivalent is equal to $\frac{1}{2}$ mole. The mass of a gram-equivalent of H_2S is therefore

$$\text{Mass of } H_2S = \text{moles of } H_2S \times \text{molecular mass of } H_2S$$
$$= \frac{1}{2} \text{ mole} \times 34.08 \text{ g/mole} = 17.04 \text{ g}$$

28.10. Ferric oxide reacts with hydrogen at high temperatures to yield iron and water:

$$Fe_2O_3 + 3H_2 \rightarrow 2Fe + 3H_2O$$

(a) In this reaction, which elements are oxidized and which are reduced? (b) What is the mass of a gram-equivalent of ferric oxide and that of a gram-equivalent of hydrogen? The formula mass of Fe_2O_3 is 159.70 u and the molecular mass of H_2 is 2.016 u. (c) How many grams of hydrogen are needed to reduce 100 g of Fe_2O_3?

(a) On the left-hand side of the equation, the oxidation number of O is -2, so that of Fe must be $+3$. In H_2 the oxidation number of H is 0. On the right-hand side of the equation, the oxidation number of Fe is 0 and in H_2O the oxidation numbers of H and O are respectively $+1$ and -2. The oxidation number of Fe has gone from $+3$ to 0, so it is reduced; the oxidation number of H has gone from 0 to $+1$, so it is oxidized. The oxidation number of O is not changed in this reaction, so it is neither oxidized nor reduced.

(b) Since the oxidation number of each Fe atom goes from $+3$ to 0 in the reaction while the oxidation numbers of the O atoms do not change, each formula unit of Fe_2O_3 experiences a decrease in oxidation number of 6, which means a gain of 6 electrons. A gram-equivalent of Fe_2O_3 is therefore equal to $\frac{1}{6}$ mole and its mass is

$$\text{Mass of } Fe_2O_3 \ = \ \text{moles of } Fe_2O_3 \ \times \ \text{formula mass of } Fe_2O_3$$

$$= \ \frac{1}{6} \text{ mole} \ \times \ 159.70 \text{ g/mole} \ = \ 26.62 \text{ g}$$

The oxidation number of each H atom goes from 0 to +1, and so each molecule of H_2 experiences an increase in oxidation number of 2, which means a loss of 2 electrons. A gram-equivalent of H_2 is therefore equal to $\frac{1}{2}$ mole and its mass is

$$\text{Mass of } H_2 \ = \ \text{moles of } H_2 \ \times \ \text{molecular mass of } H_2$$

$$= \ \frac{1}{2} \text{ mole} \ \times \ 2.016 \text{ g/mole} \ = \ 1.008 \text{ g}$$

(c) The number of gram-equivalents in 100 g of Fe_2O_3 is

$$\text{Gram-equivalents of } Fe_2O_3 \ = \ \frac{\text{mass of } Fe_2O_3}{\text{mass per gram-equivalent}} \ = \ \frac{100 \text{ g}}{26.62 \text{ g}} \ = \ 3.76$$

The number of gram-equivalents of H_2 must be the same, so

$$\text{Mass of } H_2 \ = \ \text{gram-equivalents of } H_2 \ \times \ \text{mass per gram-equivalent}$$

$$= \ 3.76 \times 1.008 \text{ g} \ = \ 3.79 \text{ g}$$

An advantage of proceeding in this way is that it is not necessary to know the balanced equation of the reaction. In this particular case, since the balanced equation is given, we could equally well have proceeded on the basis that 3 moles of H_2 are required for each mole of Fe_2O_3.

28.11. The gas propane, C_3H_8, is widely used as a fuel. When propane burns, it combines with oxygen to form carbon dioxide and water. (a) Which elements are oxidized and which are reduced? (b) What is the mass of a gram-equivalent of propane and that of a gram-equivalent of oxygen in this reaction? (c) How many kg of oxygen are needed for the complete combustion of 1.5 kg of propane?

(a) In C_3H_8, the C and H atoms have the respective oxidation numbers $-\frac{8}{3}$ and +1; in O_2, each O atom has the oxidation number 0; in CO_2, the C and O atoms have the respective oxidation numbers +4 and -2; and in H_2O, the H and O atoms have the respective oxidation numbers +1 and -2. Since the oxidation number of C increases from $-\frac{8}{3}$ to +4, it is oxidized; since the oxidation number of O decreases from 0 to -2, it is reduced; and since the oxidation number of H does not change, it is neither oxidized nor reduced.

(b) The oxidation number of each C atom goes from $-\frac{8}{3}$ to +4 in the reaction, which is an increase of

$$\frac{8}{3} + 4 \ = \ \frac{8}{3} + \frac{12}{3} \ = \ \frac{20}{3}$$

Since the oxidation numbers of the H atoms do not change, each molecule of C_3H_8 experiences an increase in oxidation number of $(20/3) \times 3 = 20$, which means a gain of 20 electrons. A gram-equivalent of C_3H_8 is therefore equal to 1/20 mole. The molecular mass of C_3H_8 is

$$3C \ = \ 3 \times 12.01 \text{ u} \ = \ 36.03 \text{ u}$$
$$8H \ = \ 8 \times 1.008 \text{ u} \ = \ \underline{8.06 \text{ u}}$$
$$44.09 \text{ u} \ = \ 44.09 \text{ g/mole}$$

so that the mass of a gram-equivalent of C_3H_8 is

$$\text{Mass of } C_3H_8 \ = \ \text{moles of } C_3H_8 \ \times \ \text{molecular mass of } C_3H_8$$

$$= \ \frac{1}{20} \text{ mole} \ \times \ 44.09 \text{ g/mole} \ = \ 2.205 \text{ g}$$

The oxidation number of each O atom goes from 0 to -2, and so each molecule of O_2 experiences a decrease in oxidation number of 4, which means a gain of 4 electrons. A gram-equivalent of O_2 is therefore equal to $\frac{1}{4}$ mole. The molecular mass of O_2 is 2×16.00 u $= 32.00$ u $= 32.00$ g/mole, so that the mass of a gram-equivalent of O_2 is

$$\text{Mass of } O_2 \ = \ \text{moles of } O_2 \ \times \ \text{molecular mass of } O_2$$

$$= \ \frac{1}{4} \text{ mole} \ \times \ 32.00 \text{ g/mole} \ = \ 8.00 \text{ g}$$

(c) The number of gram-equivalents in 1.5 kg of propane is

$$\text{Gram-equivalents of } C_3H_8 \;=\; \frac{\text{mass of } C_3H_8}{\text{mass per gram-equivalent}} \;=\; \frac{1500 \text{ g}}{22.05 \text{ g}} \;=\; 68.03$$

The number of gram-equivalents of O_2 must be the same, so

$$\text{Mass of } O_2 \;=\; \text{gram-equivalents of } O_2 \times \text{mass per gram-equivalent}$$
$$=\; 68.03 \times 8.00 \text{ g} \;=\; 544 \text{ g} \;=\; 0.544 \text{ kg}$$

It was not necessary to find the balanced equation of the reaction.

Supplementary Problems

28.12. Find the oxidation numbers of nitrogen in the compounds N_2H_4, N_2O, N_2O_4, and HNO_3.

28.13. Find the oxidation numbers of each atom in the ions HCO_3^- (bicarbonate), SiO_4^{-4} (silicate), and $Cr_2O_7^{--}$ (dichromate).

28.14. In mercurous compounds mercury has the oxidation number +1 and in mercuric compounds it has the oxidation number +2. Give the formulas of mercurous oxide and mercuric oxide.

28.15. The gases ammonia and hydrogen chloride react to form ammonium chloride in the reaction $NH_3 + HCl \longrightarrow NH_4Cl$. Is this an oxidation-reduction reaction? If so, which elements are oxidized and which are reduced?

28.16. Certain rocket motors use the reaction between hydrazine, N_2H_4, and hydrogen peroxide, H_2O_2, both of which are liquids, for propulsion. The reaction yields N_2 and H_2O as products, $N_2H_4 + 2H_2O_2 \longrightarrow N_2 + 4H_2O$, and evolves a great deal of energy. Is this an oxidation-reduction reaction? If so, which elements are oxidized and which are reduced?

In each of the following oxidation-reduction reactions identify (a) the element that is oxidized, (b) the element that is reduced, (c) the oxidizing agent, and (d) the reducing agent.

28.17. $Cl_2 + 2Fe^{++} \longrightarrow 2Cl^- + 2Fe^{+++}$

28.18. $MnO_2 + 4H^+ + 2Cl^- \longrightarrow Mn^{++} + 2H_2O + Cl_2$

28.19. $O_2 + 4H^+ + 4I^- \longrightarrow 2H_2O + 2I_2$

28.20. $Fe_2O_3 + 3CO \longrightarrow 2Fe + 3CO_2$

28.21. On the basis of the following displacement reactions, arrange chlorine, fluorine, and bromine in the order of their strength as oxidizing agents:

$$Cl_2 + 2Br^- \longrightarrow 2Cl^- + Br_2$$
$$F_2 + 2Br^- \longrightarrow 2F^- + Br_2$$
$$F_2 + 2Cl^- \longrightarrow 2F^- + Cl_2$$

28.22. Aluminum and oxygen react to form aluminum oxide, Al_2O_3. (a) What is the change in the oxidation number of aluminum in this reaction? (b) What is the mass of a gram-equivalent of aluminum in this reaction?

28.23. Potassium permanganate, $KMnO_4$, reacts with HCl to form $MnCl_2$, KCl, Cl_2, and H_2O. (a) What is the change in the oxidation number of manganese in this reaction? (b) What is the mass of a gram-equivalent of potassium permanganate in this reaction?

28.24. In an oxyacetylene torch acetylene, C_2H_2, combines with oxygen to form carbon dioxide and water. (a) Which elements are oxidized and which are reduced? (b) How many moles are there in a gram-equivalent of acetylene in this reaction and what is its mass? (c) How many moles are there in a gram-equivalent of oxygen in this reaction and what is its mass? (d) How many grams of oxygen are needed for the combustion of 50 g of acetylene?

Answers to Supplementary Problems

28.12. $-2, +1, +4, +5$

28.13. (a) H, $+1$; C, $+4$; O, -2

      (b) Si, $+4$; O, -2

      (c) Cr, $+6$; O, -2

28.14. Hg_2O, HgO

28.15. This is not an oxidation-reduction reaction since the oxidation numbers of N, H, and Cl do not change.

28.16. This is an oxidation-reduction reaction in which N is oxidized and O is reduced.

28.17. (a) Fe, (b) Cl, (c) Cl_2, (d) Fe^{++}

28.18. (u) Cl, (b) Mn, (c) MnO_2, (d) Cl^-

28.19. (a) I, (b) O, (c) O_2, (d) I^-

28.20. (a) C, (b) Fe, (c) Fe_2O_3, (d) CO

28.21. F, Cl, Br

28.22. (a) The oxidation number increases from 0 to $+3$.   (b) 8.99 g.

28.23. (a) The oxidation number decreases from $+7$ to $+2$.   (b) 31.61 g.

28.24. (a) Carbon is oxidized and oxygen is reduced. (b) $\frac{1}{10}$ mole $= 2.60$ g. (c) $\frac{1}{4}$ mole $= 8.00$ g. (d) 154 g.

Chapter 29

Electrochemistry

ELECTROLYSIS

An oxidation-reduction reaction that will not occur spontaneously can be brought about by electrical means in the process of *electrolysis*. In electrolysis the transfer of electrons from the substance being oxidized to the substance being reduced takes place by virtue of a potential difference supplied by an outside source of electrical energy. In an electrolytic cell the potential difference is applied across two electrodes immersed in a bath containing positive and negative ions, either a molten salt or oxide or a water solution of an electrolyte. The oxidation process occurs at the positive electrode, which is called the *anode*; the reduction process occurs at the negative electrode, which is called the *cathode*.

AN EXAMPLE OF ELECTROLYSIS

The electrolysis of molten sodium chloride is depicted in Fig. 29-1. The cathode attracts Na^+ ions and reduces them to Na atoms by the reaction

$$Na^+ + e^- \longrightarrow Na$$

The anode attracts Cl^- ions and oxidizes them to Cl atoms by the reaction

$$Cl^- \longrightarrow Cl + e^-$$

The electrons removed from the Cl^- ions at the anode pass through the external electrical circuit to the cathode where they are transferred to the Na^+ ions. The overall reaction is

$$2NaCl \longrightarrow 2Na + Cl_2$$

The sodium, a liquid at the temperature of molten NaCl, collects around the cathode and the chlorine bubbles up from the anode.

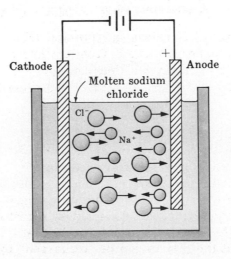

Fig. 29-1

If a solution of NaCl in water rather than molten NaCl is used, chlorine again is liberated at the anode but hydrogen and not sodium is liberated at the cathode. The reason is that H$^+$ ions have a greater affinity for electrons than Na$^+$ ions. Thus the electrolysis of an NaCl solution yields H$_2$ and Cl$_2$ with Na$^+$ and OH$^-$ ions remaining behind. Solid NaOH can be recovered by evaporation. This process is widely used industrially to prepare H$_2$, Cl$_2$, and NaOH.

ELECTROPLATING

Electrolysis can be used to deposit a layer of metal on any electrically conducting surface, such as another metal or a graphite coating on a nonmetallic object. In *electroplating* the object to be plated is used as the cathode and a bar of the plating metal is used as the anode. A salt of the plating metal is dissolved in water or other suitable solvent to make the conducting bath in which the process takes place. At the anode, atoms of the plating metal lose electrons and go into solution as positive ions; at the cathode, positive ions of the plating metal arrive and become atoms by absorbing electrons. The net result is the transfer of metal atoms from the anode to the cathode.

FARADAY'S LAWS

Faraday's laws of electrolysis state that the mass m of an element set free at either electrode is proportional to two quantities:

1. The total amount of charge Q passed through the cell, which is equal to the product It of the current I and the time t;

2. The gram-atomic mass A of the element divided by the number v of electron charges (+ or −) on its ions, which is equal to the gram-equivalent mass of the element.

Faraday's laws can be combined in the formula

$$m = \frac{QA}{Fv} = \frac{ItA}{Fv}$$

where the constant F represents the total charge of 1 mole of electrons, that is, of Avogadro's number of electrons:

$$F = 9.65 \times 10^4 \text{ C/mole} = 96,500 \text{ C/mole}$$

The amount of charge 96,500 C is called the *faraday* (F). Thus the passage of 1 F of charge will liberate 1 gram-atom of element X from a solution that contains X^+ or X^- ions; $\frac{1}{2}$ gram-atom of element Y from a solution that contains Y^{++} or Y^{--} ions; $\frac{1}{3}$ gram-atom of element Z from a solution that contains Z^{+++} or Z^{---} ions, and so forth.

GALVANIC CELLS

A *galvanic* (or *voltaic*) cell is the opposite of an electrolytic cell: instead of a nonspontaneous oxidation-reduction reaction being forced to occur by an outside source of electrical energy, a spontaneous oxidation-reduction reaction provides electrical energy for use outside the cell. What is done is to physically separate the oxidation and reduction parts of the reaction so that the electrons involved do not pass directly between the reacting substances but instead travel through an external circuit first. In a *storage battery* the reactions are reversible, so that the battery can be "recharged" by electrolysis when the initial reactants are exhausted. (All batteries can be recharged in this way in principle, but the construction of certain types of battery prevents this.) In a *fuel cell* the reactants are fed in and the products removed continuously, so the cell is never exhausted and does not need recharging.

Solved Problems

29.1. What happens when a charging current is passed through a fully-charged storage battery?

    The water content of the electrolyte undergoes electrolysis with the production of H_2 and O_2 gases.

29.2. In what fundamental way is a fuel cell different from a galvanic cell?

    A galvanic cell cannot function when its initial supply of reactants is exhausted and must be recharged or replaced. A fuel cell can be supplied continuously with reactants.

29.3. Express the faraday in terms of ampere-hours.

    Since $1 A = 1 C/s$, the charge equivalent to 1 A-hr is

$$1 \text{ A-hr} = 1 A \times 1 \text{ hr} = 1 C/s \times 3600 s = 3600 C$$

and

$$1 F = \frac{96{,}500 \text{ C}}{3600 \text{ C/A-hr}} = 26.8 \text{ A-hr}$$

29.4. A 12-A current is passed through an electrolytic cell in which the anode is a zinc bar and the electrolyte contains Zn^{++} ions. How much zinc is plated on the cathode after 2 hr?

    The gram-atomic mass of zinc is 65.37 g and $v = 2$ since the Zn^{++} ion carries a charge of $+2e$. Two hours $= 2 \times 60$ s/min $\times 60$ min/hr $= 7200$ s, and so the mass of zinc deposited is

$$m = \frac{ItA}{Fv} = \frac{12 A \times 7200 s \times 65.37 g}{96{,}500 C \times 2} = 29.3 \text{ g}$$

29.5. A 20-A current is passed through a solution that contains Cd^{++} ions. How long should the current flow to deposit 50 g of cadmium?

    The gram-atomic mass of cadmium is 112.4 g and $v = 2$ since the Cd^{++} ion carries a charge of $+2e$. From $m = ItA/Fv$ we have

$$t = \frac{mFv}{IA} = \frac{50 g \times 96{,}500 C \times 2}{20 A \times 112.4 g} = 4293 s = 71 \text{ min } 33 \text{ s}$$

29.6. Aluminum is refined commercially by the electrolysis of a solution of aluminum oxide, Al_2O_3, in molten cryolite, a mineral whose formula is Na_3AlF_6. The container, a carbon-lined iron box, is the cathode, and the anodes are carbon. The passage of current through the cell provides enough heat to keep the electrolyte at about 1000 °C, at which temperature both the cryolite and the deposited aluminum are molten. Oxygen is liberated at the anode where it reacts to form carbon monoxide and carbon dioxide. If 5.6 g of aluminum are produced when a current of 100 A is applied for 10 min, what is the charge on an aluminum ion in the electrolyte?

    The gram-atomic mass of aluminum is 26.98 g, hence from $m = ItA/Fm$ we have

$$v = \frac{ItA}{Fm} = \frac{100 A \times 600 s \times 26.98 g}{96{,}500 C \times 5.6 g} = 3$$

    The charge on each aluminum ion is therefore $+3e$, and the solution contains Al^{+++} ions.

29.7. A steel sheet 1 m long and 30 cm wide is to be plated with 0.02 mm of copper. The density of copper is 8.9 g/cm³. How long should the sheet remain in an electroplating bath that contains Cu^{++} ions in which the current is 100 A?

The area of each side of the steel sheet is 100 cm × 30 cm = 3000 cm². Since it has two sides, the total area is 6000 cm². The volume of copper to be plated is the area multiplied by the thickness of 0.02 mm = 0.002 cm, so V = 6000 cm² × 0.002 cm = 12 cm³. The mass of copper needed is

$$m = dV = 8.9 \text{ g/cm}^3 \times 12 \text{ cm}^3 = 106.8 \text{ g}$$

The gram-atomic mass of copper is 63.54 g and $v = 2$ since Cu^{++} ions are involved. From $m = ItA/Fv$ we have

$$t = \frac{mFv}{IA} = \frac{106.8 \text{ g} \times 96,500 \text{ C} \times 2}{100 \text{ A} \times 63.54 \text{ g}} = 1622 \text{ s} = 27 \text{ min } 2 \text{ s}$$

29.8. A current of 50 A is passed through a bath of molten NaCl. (a) How many grams of metallic sodium are produced per hour? (b) How many liters of chlorine gas at STP are evolved per hour?

(a) The gram-atomic mass of Na is $A = 22.99$ g; molten NaCl contains Na^+ ions, so $v = 1$; $I = 50$ A; $t = 1$ hr = 3600 s; and $F = 96,500$ C. Hence

$$m = \frac{ItA}{Fv} = \frac{50 \text{ A} \times 3600 \text{ s} \times 22.99 \text{ g}}{96,500 \text{ C} \times 1} = 43 \text{ g}$$

(b) The number of gram-atoms of Cl evolved per hour is

$$\text{Gram-atoms} = \frac{It}{Fv} = \frac{50 \text{ A} \times 3600 \text{ s}}{96,500 \text{ C} \times 1} = 1.86$$

There are two gram-atoms of Cl per mole of Cl_2, hence $\frac{1}{2} \times 1.86$ moles = 0.93 moles of Cl_2 are evolved per hour. Since 1 mole of any gas occupies 22.4 liters at STP, the volume of Cl_2 evolved per hour is

$$V = 0.93 \text{ moles} \times 22.4 \text{ liters/mole} = 20.8 \text{ liters}$$

Note that it was not necessary to know the atomic mass of chlorine to find this result.

29.9. The electrolysis of water produces gaseous H_2 and O_2. (a) What volume of H_2 at STP is liberated by a current of 10 A in an hour? (b) What volume of O_2?

(a) Since each H^+ ion requires one electron to be converted to an H atom, the number of gram-atoms of H liberated is

$$\text{Gram-atoms} = \frac{It}{Fv} = \frac{10 \text{ A} \times 3600 \text{ s}}{96,500 \text{ C} \times 1} = 0.373$$

There are two gram-atoms of H per mole of H_2, hence $\frac{1}{2} \times 0.373$ mole = 0.187 mole of H_2 is liberated in an hour. Since 1 mole of any gas occupies 22.4 liters at STP, the volume of H_2 liberated is

$$V = 0.187 \text{ mole} \times 22.4 \text{ liters/mole} = 4.2 \text{ liters}$$

(b) Because the electrolysis of water proceeds as $2H_2O \longrightarrow 2H_2 + O_2$, half as many moles of O_2 are produced as are moles of H_2, and hence the volume of O_2 is half the volume of H_2. The volume of O_2 is therefore 2.1 liters.

29.10. The *electrochemical equivalent* of an element is the mass of it that is deposited from a suitable electrolytic cell per second by a current of 1 A; that is, the mass deposited per coulomb of charge. If z is the electrochemical equivalent of a certain element, then $m = zIt$. (a) Find the electrochemical equivalent of silver in mg/C when the electrolyte contains Ag^+ ions. (b) Find the electrochemical equivalent of oxygen when the electrolyte contains O^{--} ions.

(a) The general formula for the mass of an element liberated by electrolysis is $m = ItA/Fv$, and in the case where the electrochemical equivalent is used, $m = zIt$. Both formulas give the same quantity m, so

$$zIt = \frac{ItA}{Fv} \quad \text{and} \quad z = \frac{A}{Fv}$$

For silver, $A = 107.87$ g, $v = 1$, and 1 mg $= 0.001$ g, hence

$$z = \frac{A}{Fv} = \frac{107.87 \text{ g}}{96{,}500 \text{ C} \times 1} = 0.001117 \text{ g/C} = 1.117 \text{ mg/C}$$

(b) Here $A = 16.00$ g and $v = 2$, hence

$$z = \frac{16.00 \text{ g}}{96{,}500 \text{ C} \times 2} = 0.0000829 \text{ g/C} = 0.0829 \text{ mg/C}$$

29.11. The electrochemical equivalent of potassium is 0.405 mg/C. (a) How much potassium is liberated when a current of 25 A is passed through a suitable electrolyte for 20 min? (b) How long will it take for 1 kg of potassium to be liberated?

(a) Since $t = 20$ min $\times 60$ s/min $= 1200$ s,

$$m = zIt = 0.405 \text{ mg/C} \times 25 \text{ A} \times 1200 \text{ s} = 12{,}150 \text{ mg} = 12.15 \text{ g}$$

(b) 1 kg $= 10^3$ g $= 10^6$ mg, hence

$$t = \frac{m}{zI} = \frac{10^6 \text{ mg}}{0.405 \text{ mg/C} \times 25 \text{ A}} = 9.88 \times 10^4 \text{ s} = 27.4 \text{ hr}$$

29.12. A Daniell cell consists of a zinc electrode immersed in a zinc sulfate solution that is separated by a porous barrier from a copper electrode immersed in a copper sulfate solution. The barrier permits ions to diffuse through it when the cell is operating, but it prevents the two solutions from mixing together. The overall reaction in a Daniell cell is

$$\text{Zn} + \text{Cu}^{++} \longrightarrow \text{Zn}^{++} + \text{Cu}$$

(a) What reactions occur at the electrodes? (b) Which electrode is positive and which is negative?

(a) The overall reaction takes place in two steps called *half-reactions*:

$$\text{Zn} \longrightarrow \text{Zn}^{++} + 2e^-$$
$$\text{Cu}^{++} + 2e^- \longrightarrow \text{Cu}$$

The first of these half-reactions occurs at the zinc electrode and the second at the copper electrode. The electrons set free by the solution of zinc atoms as Zn^{++} ions pass through the external circuit to the copper electrode where they combine with Cu^{++} ions to form copper atoms. (See Fig. 29-2.)

(b) Electrons are liberated from Zn atoms at the zinc electrode, so it is negative. Electrons are absorbed by Cu^{++} ions at the copper electrode, so it is positive.

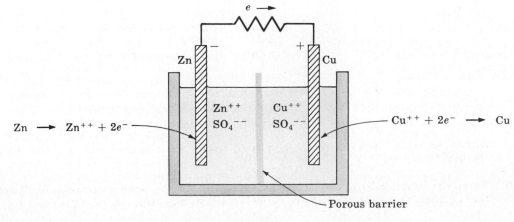

Fig. 29-2

29.13. How much zinc goes into solution in a Daniell cell when it delivers 200 C of charge?

The calculation proceeds in exactly the same way as if the zinc were being plated out instead of going into solution. The gram-atomic mass of zinc is 65.37 g and $v = 2$ since the zinc goes into solution as Zn^{++} ions, and so

$$m = \frac{QA}{Fv} = \frac{200 \text{ C} \times 65.37 \text{ g}}{96,500 \text{ C} \times 2} = 0.068 \text{ g}$$

29.14. In a fuel cell, the gaseous reactants are fed in through hollow electrodes made of an inert, conducting material with microscopic pores that enable the gases to enter the electrolyte gradually. In a typical hydrogen-oxygen fuel cell the electrolyte is a solution of KOH. At one electrode H_2 molecules combine with OH^- ions to form water and liberate electrons, and at the other electrode O_2 molecules combine with H_2O molecules and electrons to form OH^- ions. (*a*) What half-reactions occur at the electrodes? (*b*) What is the overall reaction in the cell? (*c*) Which electrode is positive and which is negative?

(*a*) At the hydrogen electrode the half-reaction is

$$H_2 + 2OH^- \longrightarrow 2H_2O + 2e^-$$

At the oxygen electrode the half-reaction is

$$O_2 + 2H_2O + 4e^- \longrightarrow 4OH^-$$

(*b*) The first half-reaction must occur twice for each time the second occurs in order that the numbers of electrons and of OH^- ions balance. The overall reaction is therefore

$$2H_2 + O_2 \longrightarrow 2H_2O$$

(*c*) The hydrogen electrode is negative and the oxygen electrode is positive.

Supplementary Problems

29.15. Which of the following metals will be deposited in the greatest mass when one coulomb of charge is passed through appropriate electrolytic cells? Aluminum, chromium, nickel, silver. The ions of these metals in solution are respectively Al^{+++}, Cr^{+++}, Ni^{++}, Ag^+.

29.16. A 20-A current is passed through an electrolytic cell in which the anode is an iron bar and the electrolyte contains Fe^{+++} ions. How much iron is plated on the cathode in 40 min?

29.17. An 8-A current is passed through a solution that contains Cu^{++} ions. How long should the current continue in order to deposit 10 g of copper?

29.18. A bowl whose total surface area is 0.15 m^2 is used as the cathode in a chromium plating bath in which the electrolyte contains Cr^{+++} ions. The current is 5 A and the bowl is plated for 3 hr. How thick is the layer of chromium on the bowl? The density of chromium is 7.2 g/cm^3.

29.19. A spoon whose total surface area is 140 cm^2 is to be plated with a layer of silver 0.01 mm thick. The plating current is 1.2 A, silver goes into solution as Ag^+ ions, and the density of silver is 10.5 g/cm^3. How long should the spoon stay in the electroplating bath?

29.20. When platinum is being electroplated, it is found that 0.506 mg of platinum is deposited for each coulomb of charge passed through the plating bath. What is the charge on each platinum ion in the bath?

29.21. Three electrolytic cells are connected in series, so that the same current passes through each of them. Aluminum, lead, and silver are plated out in the cells, whose electrolytes contain respectively Al^{+++}, Pb^{++}, and Ag^+ ions. If 100 g of silver is deposited after a period of time, how much aluminum and lead are deposited?

29.22. A total of 10^6 C of charge is passed through a bath of molten NaCl. (a) How many grams of sodium metal are produced? (b) How many liters of chlorine gas at STP are liberated?

29.23. The electrolysis of an HCl (hydrochloric acid) solution liberates H_2 at the cathode and Cl_2 at the anode. (a) What volume of H_2 at STP is liberated by the passage of 2 faradays of charge? (b) What volume of Cl_2?

29.24. The electrochemical equivalent of lead is 1.074 mg/C. How much lead is deposited from a plating bath by the passage of 50 A for 1 hr?

29.25. Gold has an atomic mass of 197 u and its most common oxidation number is +3. What is the electrochemical equivalent of gold in mg/C?

29.26. The electrochemical equivalent of magnesium is 0.126 mg/C. For how many hours should a current of 100 A be passed through an appropriate electrolytic cell to plate out 1 kg of magnesium?

29.27. Each cell of a lead-acid storage battery consists of electrodes of pure lead interleaved with electrodes of lead dioxide, PbO_2, both immersed in a bath of sulfuric acid, H_2SO_4. The half-reaction at the lead electrode is

$$Pb + SO_4^{--} \longrightarrow PbSO_4 + 2e^-$$

and the half-reaction at the lead dioxide electrode is

$$PbO_2 + SO_4^{--} + 4H^+ + 2e^- \longrightarrow PbSO_4 + 2H_2O$$

The effect of drawing current from the cell is thus to deposit lead sulfate on both sets of electrodes. (a) Which electrode is positive and which is negative? (b) What is the overall reaction in the cell? (c) What half-reactions occur at each electrode when the battery is being recharged?

29.28. How many grams of lead are converted to lead sulfate in a lead-acid galvanic cell for each ampere-hour of charge delivered by the cell?

29.29. The reactants in a certain fuel cell are methane, CH_4, and oxygen, and the electrolyte is molten KOH. At one electrode CH_4 is oxidized to CO_3^{--} ions and at the other O_2 is reduced to OH^- ions. (a) What half-reactions occur at the electrodes? (b) What is the overall reaction in the cell? (c) Which electrode is positive and which is negative?

Answers to Supplementary Problems

29.15. silver

29.16. 9.26 g

29.17. 63.3 min

29.18. 0.0090 mm

29.19. 18 min 16 s

29.20. $+4e$

29.21. 8.3 g; 96 g

29.22. 238 g; 116 liters

29.23. 22.4 liters; 22.4 liters

29.24. 193 g

29.25. 0.680 mg/C

29.26. 22 hr

29.27. (a) The Pb electrode is negative and the PbO_2 electrode is positive.

(b) $Pb + PbO_2 + 4H^+ + 2SO_4^{--} \longrightarrow 2PbSO_4 + 2H_2O$

(c) At the negative electrode, $PbSO_4 + 2e^- \longrightarrow Pb + SO_4^{--}$
At the positive electrode, $PbSO_4 + 2H_2O \longrightarrow PbO_2 + SO_4^{--} + 4H^+ + 2e^-$

29.28. 3.86 g

29.29. (a) $CH_4 + 10OH^- \longrightarrow CO_3^{--} + 7H_2O + 8e^-$; $O_2 + 2H_2O + 4e^- \longrightarrow 4OH^-$

(b) $CH_4 + 2OH^- + 2O_2 \longrightarrow CO_3^{--} + 3H_2O$

(c) The methane electrode is positive and the oxygen electrode is negative.

Chapter 30

Chemical Energy

HEAT OF REACTION

A chemical change in which energy is liberated is called *exothermic*; a chemical change in which energy is absorbed is called *endothermic*. A reaction is exothermic when the bonds between the atoms in the product substances are stronger than those in the reactant substances; a reaction is endothermic when the bonds between the atoms in the product substances are weaker than those in the reactant substances.

The amount of energy liberated or absorbed in a chemical process that takes place at constant pressure is the *heat of reaction*, ΔH, and its value for a given process represents the energy change that occurs when the number of moles of each participating substance equals the coefficient of its symbol in the balanced equation for the process. For instance, the burning of methane to produce carbon dioxide and water is an exothermic process in which 213 kcal of heat is given off per mole of CH_4:

$$CH_4 + 2O_2 \longrightarrow CO_2 + 2H_2O + 213 \text{ kcal}$$

The decomposition of calcium carbonate into calcium oxide and carbon dioxide when it is heated is an example of an endothermic reaction:

$$CaCO_3 + 42.5 \text{ kcal} \longrightarrow CaO + CO_2$$

By convention, the heat of reaction of an endothermic process is considered positive and that of an exothermic process is considered negative. Thus the above reactions could alternatively be written

$$CH_4 + 2O_2 \longrightarrow CO_2 + 2H_2O \qquad \Delta H = -213 \text{ kcal}$$
$$CaCO_3 \longrightarrow CaO + CO_2 \qquad \Delta H = 42.5 \text{ kcal}$$

HEAT OF FORMATION

The *heat of formation* of a substance is the energy change involved in the formation of one mole of the substance from the elements it contains. Since

$$C + O_2 \longrightarrow CO_2 + 94.1 \text{ kcal}$$

the heat of formation of carbon dioxide is $\Delta H_f = -94.1$ kcal.

Heats of formation depend upon the physical states of the reacting elements and upon that of the final compound. Unless otherwise specified, elements and compound are assumed to be in their normal stable states at 25 °C and 1 atm pressure. In the case of the formation of water from gaseous H_2 and O_2, for instance, more heat is liberated when the water is a liquid at the end of the process than when it is a vapor:

$$H_2 + \tfrac{1}{2}O_2 \longrightarrow H_2O \text{ (liquid)} \qquad H_f = -68.3 \text{ kcal}$$
$$H_2 + \tfrac{1}{2}O_2 \longrightarrow H_2O \text{ (gas)} \qquad H_f = -57.8 \text{ kcal}$$

The difference is due to the heat of vaporization of water. The above equations are written with the coefficient $\tfrac{1}{2}$ in front of O_2 because heats of formation are always given per mole of the product and we cannot write just $H_2 + O \longrightarrow H_2O$ since gaseous oxygen consists of O_2 molecules and not of O atoms.

HESS'S LAW

It is possible to predict a particular heat of reaction from the heats of formation of the various reactants and products by using *Hess's law*. This law states that the heat of reaction ΔH is equal to the sum of the heats of formation of the products minus the sum of the heats of formation of the reactants, with elements considered to have no heats of formation even if they occur as molecules. Using the capital Greek letter sigma, Σ, as the symbol for "sum of", the above law is given in equation form as

$$\Delta H \;=\; \Sigma\, \Delta H_f(\text{products}) - \Sigma\, \Delta H_f(\text{reactants})$$

Solved Problems

30.1. The heat of formation of CO_2 from C and O_2 is -94.5 kcal when the carbon is in the form of diamond and -94.1 kcal when it is in the form of graphite. Which form of carbon is more stable at 25 °C and 1 atm?

> More energy is given off in the combustion of diamond than in the combustion of graphite, hence diamond is a less stable form of carbon than graphite under the specified conditions.

30.2. Heats of reaction represent energy changes that occur in processes that take place at constant pressure. In the decomposition of nitrogen dioxide,

$$NO_2 \;\longrightarrow\; \tfrac{1}{2}N_2 + O_2 + 8.1 \text{ kcal}$$

the products consist of $1\frac{1}{2}$ moles of gas for each mole of NO_2 gas, so at constant pressure (and temperature) the volume of the products is $1\frac{1}{2}$ times the initial volume of NO_2. If instead the reaction occurs in a closed container at constant volume, how will the heat liberated be affected?

> At constant pressure, work is done by the expansion of the N_2 and O_2 products, so the heat evolved, 8.1 kcal/mole, is less than the actual amount of energy ΔE liberated in the process. At constant volume, the pressure rises but no work is done, so the entire amount ΔE is liberated as heat.

30.3. Find the heat required to decompose 400 g of calcium carbonate in the reaction $CaCO_3 \longrightarrow CaO + CO_2$ in which $\Delta H = 42.5$ kcal.

> The heat of reaction holds for 1 mole of $CaCO_3$. The formula mass of $CaCO_3$ is
>
> $$1Ca = 1 \times 40.08 \text{ u} = 40.08 \text{ u}$$
> $$1C = 1 \times 12.01 \text{ u} = 12.01 \text{ u}$$
> $$3O = 3 \times 16.00 \text{ u} = \underline{48.00 \text{ u}}$$
> $$100.09 \text{ u} = 100.09 \text{ g/mole}$$
>
> so the number of moles in 400 g of $CaCO_3$ is
>
> $$\text{Moles of } CaCO_3 = \frac{\text{mass of } CaCO_3}{\text{formula mass of } CaCO_3} = \frac{400 \text{ g}}{100.09 \text{ g/mole}} = 4.00 \text{ moles}$$
>
> and the heat required is
>
> $$\text{Total heat} = \text{moles} \times \text{heat of reaction} = 4.00 \text{ moles} \times 42.5 \text{ kcal/mole} = 170 \text{ kcal}$$

30.4. Octane, C_8H_{18}, is the chief constituent of gasoline. The molecular mass of octane is 114 g/mole and its heat of combustion is about 11,300 kcal/kg. Find the heat of combustion of octane per mole.

The number of moles in 1000 g of octane is

$$\text{Moles of } C_8H_{18} = \frac{\text{mass of } C_8H_{18}}{\text{molecular mass of } C_8H_{18}} = \frac{1000 \text{ g}}{114 \text{ g/mole}} = 8.77 \text{ moles}$$

and so

$$\Delta H = \frac{-11,300 \text{ kcal/kg}}{8.77 \text{ moles/kg}} = -1290 \text{ kcal/mole}$$

30.5. Neutralization is an exothermic process in which $\Delta H = -13.7$ kcal:

$$H^+ + OH^- \longrightarrow H_2O + 13.7 \text{ kcal}$$

How much heat is given off when 0.5 liter of $1M$ HCl is added to 1 liter of $1M$ NaOH?

Since a $1M$ concentration corresponds to 1 mole/liter, 0.5 liter of $1M$ HCl contains 0.5 mole of HCl. When this amount of HCl is added to 1 liter of $1M$ NaOH, the result is the formation of 0.5 mole of H_2O, with 0.5 mole of OH^- ions remaining in solution with 0.5 mole of Cl^- ions and 1 mole of Na^+ ions. The heat given off is

$$\text{Total heat} = \text{moles} \times \text{heat of reaction} = 0.5 \text{ mole} \times 13.7 \text{ kcal/mole} = 6.85 \text{ kcal}$$

30.6. The heats of formation of CO and CO_2 are respectively -26.4 kcal and -94.1 kcal. (a) Find the heat of reaction for

$$2CO + O_2 \longrightarrow 2CO_2$$

(b) What is the heat of reaction per mole of CO? (c) Is the reaction endothermic or exothermic?

(a) The heat of formation of O_2 is 0 by definition. To find the heat of reaction we must multiply the heats of formation of CO and CO_2 by 2 since both compounds have the coefficient 2 in the equation. The heat of reaction is

$$
\begin{aligned}
\Delta H &= \Sigma \, \Delta H_f(\text{products}) - \Sigma \, \Delta H_f(\text{reactants}) \\
&= 2 \times \Delta H_f(CO_2) - 2 \times \Delta H_f(CO) - \Delta H_f(O_2) \\
&= 2 \times (-94.1 \text{ kcal}) - 2 \times (-26.4 \text{ kcal}) - 0 \\
&= -188.2 \text{ kcal} + 52.8 \text{ kcal} = -135.4 \text{ kcal}
\end{aligned}
$$

(b) The above heat of reaction corresponds to the oxidation of two moles of CO. In the case of 1 mole of CO, ΔH is $\frac{1}{2} \times (-135.4 \text{ kcal}) = -67.7$ kcal, and the equation for the process is

$$CO + \tfrac{1}{2}O_2 \longrightarrow CO_2 \qquad \Delta H = -67.7 \text{ kcal}$$

(c) Since ΔH is negative, the reaction is exothermic.

30.7. When steam is passed over hot coal, the reaction

$$H_2O + C \longrightarrow CO + H_2$$

takes place. The resulting mixture of carbon monoxide and hydrogen, called "water gas," was once widely used as a fuel. (a) Given that the heat of formation of H_2O as a gas is -57.8 kcal and the heat of formation of CO is -26.4 kcal, find the heat of reaction for manufacturing water gas. (b) How much water is needed to convert 1000 kg of coal to water gas? (c) How much heat? Assume that the coal is 95% carbon.

(a) The heat of reaction is

$$
\begin{aligned}
\Delta H &= \Sigma \, \Delta H_f(\text{products}) - \Sigma \, \Delta H_f(\text{reactants}) \\
&= \Delta H_f(CO) + \Delta H_f(H_2) - \Delta H_f(H_2O) - \Delta H_f(C) \\
&= -26.4 \text{ kcal} + 0 - (-57.8 \text{ kcal}) - 0 \\
&= -26.4 \text{ kcal} + 57.8 \text{ kcal} = 31.4 \text{ kcal}
\end{aligned}
$$

The reaction is evidently endothermic.

(b) The gram-atomic mass of carbon is 12.01 g, so the number of gram-atoms in $950 \text{ kg} = 9.50 \times 10^5 \text{ g}$ of carbon is

$$\text{Gram-atoms of C} = \frac{\text{mass of C}}{\text{atomic mass of C}} = \frac{9.50 \times 10^5 \text{ g}}{12.01 \text{ g/gram-atom}}$$

$$= 7.91 \times 10^4 \text{ gram-atoms}$$

Each atom of carbon combines with a molecule of water, so the number of moles of H_2O required is also 7.91×10^4. Since the molecular mass of H_2O is 18.02 g/mole, the mass of H_2O required is

$$\text{Mass of } H_2O = \text{moles of } H_2O \times \text{molecular mass of } H_2O$$

$$= 7.91 \times 10^4 \text{ moles} \times 18.02 \text{ g/mole}$$

$$= 1.425 \times 10^6 \text{ g} = 1425 \text{ kg}$$

(c) Each gram-atom of carbon requires 31.4 kcal of heat to react. Hence the heat needed is

$$\text{Total heat} = \text{gram-atoms} \times \text{heat of reaction}$$

$$= 7.91 \times 10^4 \text{ gram-atoms} \times 31.4 \text{ kcal/gram-atom}$$

$$= 2.48 \times 10^6 \text{ kcal}$$

30.8. The combustion of propane, C_3H_8, proceeds according to the equation

$$C_3H_8 + 5O_2 \longrightarrow 3CO_2 + 4H_2O \qquad \Delta H = -530.7 \text{ kcal}$$

where the H_2O is assumed in liquid form. The heats of formation of CO_2 and H_2O are respectively -94.1 kcal and -68.3 kcal. Find the heat of formation of propane.

We start from the equation for the heat of reaction of the burning of propane, which is

$$\Delta H = \Sigma \Delta H_f(\text{products}) - \Sigma \Delta H_f(\text{reactants})$$

$$= 3 \times \Delta H_f(CO_2) + 4 \times \Delta H_f(H_2O) - \Delta H_f(C_3H_8) - 5 \times \Delta H_f(O_2)$$

Now we solve for $\Delta H_f(C_3H_8)$ and find that, since $\Delta H_f(O_2) = 0$,

$$\Delta H_f(C_3H_8) = 3 \times \Delta H_f(CO_2) + 4 \times \Delta H_f(H_2O) - \Delta H$$

$$= 3 \times (-94.1 \text{ kcal}) + 4 \times (-68.3 \text{ kcal}) - (-530.7 \text{ kcal})$$

$$= -282.3 \text{ kcal} - 273.2 \text{ kcal} + 530.7 \text{ kcal} = -24.8 \text{ kcal}$$

30.9. To break up H_2 molecules into H atoms at 25 °C and 1 atm requires 104 kcal per mole of H_2; to break up F_2 molecules into F atoms requires 37 kcal per mole of F_2, and to break up HF molecules into H and F atoms requires 135 kcal per mole of HF. (a) Of the molecules H_2, F_2, and HF, which is the most stable and which is the least stable? (b) What is the heat of formation of HF from H_2 and F_2?

(a) The greater the energy required to break apart a molecule, the more stable it is. Hence the most stable molecule is HF and the least stable is F_2.

(b) The formation of HF may be considered to take place in three steps:

$$H_2 + 104 \text{ kcal} \longrightarrow H + H$$

$$F_2 + 37 \text{ kcal} \longrightarrow F + F$$

$$H + H + F + F \longrightarrow 2HF + 2 \times 135 \text{ kcal}$$

Adding the three equations together yields

$$H_2 + F_2 + 104 \text{ kcal} + 37 \text{ kcal} + 2H + 2F \longrightarrow 2HF + 2 \times 135 \text{ kcal} + 2H + 2F$$

$$H_2 + F_2 + 141 \text{ kcal} \longrightarrow 2HF + 270 \text{ kcal}$$

$$H_2 + F_2 \longrightarrow 2HF + 129 \text{ kcal}$$

Hence the heat of formation of HF is -129 kcal for 2 moles or -64.5 kcal per mole.

Supplementary Problems

30.10. At constant pressure, the reaction of H_2 and O_2 to form water vapor, $2H_2 + O_2 \longrightarrow 2H_2O$, liberates 57.8 kcal per mole H_2O. If the reaction takes place at constant volume, will the heat liberated be greater or less than 57.8 kcal? Why?

30.11. The heat of formation of ammonia, NH_3, is -11.0 kcal. How much energy is evolved in the manufacture of 800 g of ammonia from gaseous nitrogen and hydrogen?

30.12. Nitroglycerine, $C_3H_5(NO_3)_3$, decomposes explosively to form N_2, O_2, CO_2, and H_2O with a heat of reaction of -431 kcal. (a) Find the energy released when 1 kg of nitroglycerine explodes. (b) Why is nitroglycerine a powerful explosive even though its heat of reaction is not extraordinarily large?

30.13. Natural gas consists largely of methane, CH_4, whose heat of combustion is 9.5 kcal per liter at STP. Find the heat of combustion of methane in (a) kcal/mole and (b) kcal/kg.

30.14. The heats of combustion of acetylene (C_2H_2), ethylene (C_2H_4), and ethane (C_2H_6) are respectively 312 kcal, 32 kcal, and 368 kcal. (a) Which gas has the highest heat of combustion per liter at STP? (b) Per kilogram?

30.15. Ferric oxide and powdered aluminum react to form aluminum oxide and molten iron in what is known as the *thermite reaction*. The heats of formation of Fe_2O_3 and of Al_2O_3 are respectively -196.5 kcal and -399.1 kcal. Find the heat evolved per mole of Fe_2O_3.

30.16. The reduction of ferric oxide to metallic iron by carbon follows the equation $3C + 2Fe_2O_3 \longrightarrow 4Fe + 3CO_2$. (a) Given that the heats of formation of Fe_2O_3 and of CO_2 are respectively -196.5 kcal and -94.1 kcal, find the heat of reaction for the above process. (b) Is the process exothermic or endothermic?

30.17. Hydrogen peroxide, H_2O_2, is unstable and gradually decomposes into water and oxygen: $2H_2O_2 \longrightarrow 2H_2O + O_2 + 47.0$ kcal. The heat of formation of liquid water is -68.3 kcal. (a) Find the heat of formation of H_2O_2. (b) How much heat is given off when 10 g of H_2O_2 decomposes?

30.18. The reaction between hydrazine and hydrogen peroxide is sometimes used in rocket propulsion: $N_2H_4 + 2H_2O_2 \longrightarrow N_2 + 4H_2O$, $\Delta H = -170$ kcal. The heats of formation of H_2O_2 and H_2O are respectively -45 kcal and -68 kcal. Find the heat of formation of N_2H_4.

30.19. Given that $H_2 + 104$ kcal $\longrightarrow$ H + H, $Cl_2 + 58$ kcal $\longrightarrow$ Cl + Cl, and HCl + 103 kcal $\longrightarrow$ H + Cl, find the heat of formation of HCl from H_2 and Cl_2.

Answers to Supplementary Problems

30.10. The heat liberated at constant volume will be less than 57.8 kcal because at constant pressure the volume of H_2O is less than the combined volumes of the H_2 and O_2 reactants. This means that at constant pressure the volume decreases, hence work is done by the atmosphere *on* the system, which adds to the energy given off. At constant volume no work is done by or on the system.

30.11. 517 kcal

30.12. (a) 1900 kcal. (b) The gaseous products of the explosion occupy much more volume at 25 °C and 1 atm pressure than the original liquid, but the considerable work done during their expansion is not included in the heat of reaction. The total energy released is actually quite considerable.

30.13. 213 kcal/mole; 1.33×10^4 kcal/kg

30.14. ethane; ethane

30.15. 202.6 kcal

30.16. 110.7 kcal; endothermic

30.17. 44.8 kcal; 6.9 kcal

30.18. -12 kcal

30.19. -22 kcal

Chapter 31

Reaction Rates and Equilibrium

REACTION RATES

The rate at which a chemical reaction takes place depends upon a number of factors. An important one is the nature of the reactants. A reaction in which bonds between atoms in the reactants must initially be broken will be slower than a reaction in which no bonds need be broken. An example of the latter is acid-base neutralization, which is very rapid.

The concentration of the reactants also influences the speed of a reaction. The greater the concentration of reactants in a gas or liquid solution, the more frequent the collisions between the reacting particles (atoms, molecules, or ions), and, in general, the faster the reaction. Thus increasing the pressure increases the rate of a gas-phase reaction. In a process that involves reactants in different phases, the reaction rate depends upon the area of contact between the phases. Thus powdered iron rusts more rapidly than solid iron.

ACTIVATION ENERGY

Exothermic reactions give off energy, yet few of them take place spontaneously when the appropriate reactants are brought together. A certain initial *activation energy* must be provided in order to start the reaction going, after which the evolved heat can continue it. The existence of activation energy follows from the need to disrupt the bonds holding the atoms of the reactants together in order that they can then recombine in a different way to form the products. A small activation energy means a rapid reaction, other things being equal, and a large activation energy means a slow reaction.

Reaction rates increase with temperature because the number of particles which have sufficient kinetic energy to react depends upon the temperature. If molecule A strikes molecule B hard enough, the *activated complex AB* results which may break up into the new molecules C and D or back into the original molecules A and B. The higher the temperature, the greater the number of reactant particles with enough energy to form activated complexes after colliding with each other, and the faster the reaction. As a rule of thumb, a 10 °C rise in temperature approximately doubles the speed of a chemical reaction.

A *catalyst* is a substance that increases the rate of a chemical reaction without itself being permanently altered. Catalysts affect reaction rates by lowering the activation energy required.

EQUILIBRIUM

Many chemical reactions do not result in the complete conversion of the reactants to the products. As the concentration of products builds up, they undergo reverse reactions in which the original reactants are produced, so that an *equilibrium* is eventually established in which the forward and reverse reactions occur at the same rate. An equilibrium is represented by a double arrow:

$$A + B \rightleftarrows C + D$$

At equilibrium in a closed system at a fixed temperature, the concentrations of the various reactants and products remain constant as the effect of the reverse reaction balances the effect of the forward reaction.

The relative concentrations of reactants and products at equilibrium depends upon the nature of the reaction and the conditions under which it takes place. At equilibrium the forward and reverse reactions must occur at the same rate. Suppose the forward reaction is faster than the reverse reaction when the concentrations of the various substances are the same. The forward reaction will therefore proceed more than halfway to completion so that the concentration of the products is great enough to compensate for the slower intrinsic rate of the reverse reaction. If the reverse reaction is the faster one, all else equal, then the forward reaction will proceed less than halfway to completion. (Of course, the designations of "forward" and "reverse" are arbitrary since any equilibrium can be approached from either side: $A + B \rightleftarrows C + D$ can just as well be written $C + D \rightleftarrows A + B$.)

LE CHÂTELIER'S PRINCIPLE

According to *Le Châtelier's principle,* when the conditions under which a system is in equilibrium are changed, the equilibrium will shift in such a way as to tend to counteract the change. In other words, an equilibrium system will respond to a stress of some kind by shifting so as to relieve the stress. There are three chief ways in which an equilibrium system can be altered.

1. **Change the concentration** of one or more of the components. Increasing the concentration of one of the reactants, for example, increases the speed of the forward reaction and hence shifts the equilibrium closer to completion. Removing one of the products has the same effect because it reduces the speed of the reverse reaction.

2. **Change the temperature.** If one of the reactions is endothermic and the other is exothermic, raising the temperature favors the endothermic reaction since it absorbs energy and hence tends to restore the original conditions of the equilibrium. Lowering the temperature has the opposite effect.

3. **Change the pressure.** In the case of an equilibrium involving gases in which the number of reactant molecules is different from the number of product molecules, an increase in pressure favors the reaction that yields the smaller number of molecules since this tends to counteract the effect of the rise in pressure. Reducing the pressure has the opposite consequence.

The use of a catalyst can increase the rate at which equilibrium is reached, but because it affects both forward and reverse reactions equally it cannot change the location of the equilibrium, that is, the relative concentrations of reactants and products.

Solved Problems

31.1. What is the general condition for a reaction to go to completion instead of an equilibrium being established? Give examples of liquid-phase reactions that go to completion.

> When one of the products of a reaction leaves the system, the reaction must go to completion since the reverse reaction cannot then occur. A reaction in a liquid will go to completion if one of the products is (*a*) a gas which escapes; (*b*) an insoluble precipitate; or (*c*) composed of molecules that do not dissociate when the reaction involves ions.

31.2. One mole each of CO and NO_2 are placed in a closed container, and the reaction $CO + NO_2 \longrightarrow CO_2 + NO$ occurs. Will the reaction be faster in a small or a large container?

> In a small container, the reactants will collide with each other more frequently, and so the reaction will proceed faster.

31.3. Why does adding an acid to water reduce the concentration of OH^- ions as well as increase the concentration of H^+ ions?

The dissociation of water proceeds as $H_2O \rightleftarrows H^+ + OH^-$. Increasing the concentration of H^+ ions increases the rate of the reverse reaction and hence shifts the equilibrium in the direction of a greater number of H_2O molecules at the expense of the OH^- concentration.

31.4. (*a*) What effect will an increase in pressure have on the yield of water vapor in the reaction $H_2 + CO_2 \rightleftarrows CO + H_2O$? (*b*) What effect will it have on the rate at which equilibrium is established?

(*a*) Since the numbers of molecules of reactants and products are the same, changing the pressure will not affect the concentration of any of the substances involved in this equilibrium system.

(*b*) Increasing the pressure will increase the rate at which collisions between the various molecules occur and hence will increase the rate at which equilibrium is established.

31.5. The reaction of carbon and steam to form a mixture of carbon monoxide and hydrogen ("water gas") proceeds as $C + H_2O \rightleftarrows CO + H_2$. Will a change in pressure affect the yield of water gas? What should the change be to increase the yield?

Since the volume of water gas is double the volume of the steam that went into its production (the volume of carbon is not significant compared with the gas volumes), a decrease in pressure will increase the yield of water gas.

31.6. When most solids are dissolved in water, the solution becomes cold, indicating that the process is endothermic. How is this observation related to the increase in the solubility of such solids with increasing temperature?

Consider a solid added to some water in excess of the amount that will dissolve at a certain temperature. This is an equilibrium system with some of the solid continuing to dissolve while simultaneously some of the dissolved matter precipitates out at the same rate. Increasing the temperature changes the conditions of the equilibrium, and the system reacts by favoring the dissolving of more solid since this absorbs heat and thus tends to restore the temperature to its original value.

31.7. The synthesis of ammonia from nitrogen and hydrogen in the Haber process proceeds as $N_2 + 3H_2 \rightleftarrows 2NH_3 + 22$ kcal. (*a*) What is the effect of an increase in temperature on the rate at which equilibrium is attained? (*b*) On the yield of NH_3?

(*a*) Increasing the temperature increases the rate at which equilibrium is reached by increasing the concentration of activitated complexes.

(*b*) The reaction is exothermic, so increasing the temperature favors the reverse reaction and so decreases the yield.

31.8. Acetic acid dissociates only partially in solution, so that the equilibrium $HC_2H_3O_2 \rightleftarrows H^+ + C_2H_3O_2^-$ exists. (*a*) What happens to the concentration of acetate ions when some concentrated HCl, a strong acid, is added to an acetic acid solution? (*b*) What happens when some NaOH, a strong base, is added?

(*a*) The HCl increases the H^+ concentration and so favors the reverse reaction. The concentration of acetate ions therefore decreases.

(*b*) The OH^- ions neutralize some of the H^+ ions, which reduces the H^+ concentration and so favors the forward reaction. The concentration of acetate ions therefore increases.

Supplementary Problems

31.9. An increase in temperature increases the rate of exothermic as well as endothermic reactions. Why?

31.10. Why does food stored in a refrigerator spoil less rapidly than food stored at room temperature?

31.11. Heat is given off when a gas dissolves in a liquid. How do the solubilities of gases vary with temperature?

31.12. The solubility of lithium carbonate, Li_2CO_3, in water is observed to decrease with temperature. Is the process by which Li_2CO_3 dissolves endothermic or exothermic?

31.13. The solubility of KCl in water approximately doubles between 0 °C and 100 °C whereas that of NaCl increases only slightly. Which salt absorbs the greater amount of heat per mole when it dissolves?

31.14. The oxidation of the gas sulfur dioxide proceeds as $2SO_2 + O_2 \longrightarrow 2SO_3$. (a) How is the reaction rate affected by an increase in the O_2 concentration? (b) By an increase in temperature? (c) By an increase in pressure? (d) By the presence of an appropriate catalyst?

31.15. (a) What effect will a decrease in pressure have on the yield of methyl alcohol, CH_3OH, in the reaction $CO + 2H_2 \rightleftharpoons CH_3OH$? (b) What effect will it have on the rate at which equilibrium is established? All three substances are gases in the equilibrium system.

31.16. Nitric oxide is produced in the reaction $N_2 + O_2 + 43$ kcal $\rightleftharpoons 2NO$. (a) What is the effect on the yield of NO of increasing the pressure? (b) Increasing the temperature? (c) Increasing the concentration of O_2? (d) Adding a catalyst?

31.17. Silver nitrate, $AgNO_3$, is very soluble in water, silver acetate, $AgC_2H_3O_2$, is slightly soluble, and silver chloride, AgCl, is virtually insoluble. When an excess of silver acetate is added to water, a saturated solution results in which the equilibrium $AgC_2H_3O_2 \rightleftharpoons Ag^+ + C_2H_3O_2^-$ exists. (a) What is the effect on the concentration of acetate ions of adding $AgNO_3$ to the solution? (b) What is the effect of adding NaCl to the solution?

Answers to Supplementary Problems

31.9. Both kinds of reaction have activation energies, so increasing the temperature increases the number of activated complexes and hence increases the reaction rate in both cases.

31.10. Food spoilage involves chemical reactions, and reaction rates increase with increasing temperature.

31.11. They decrease with increasing temperature.

31.12. exothermic

31.13. KCl

31.14. Each of these changes increases the reaction rate.

31.15. The yield and the rate of attainment of equilibrium will both decrease.

31.16. (a) No effect. (b) The yield is increased. (c) The yield is increased. (d) No effect.

31.17. (a) The acetate ion concentration decreases. (b) The acetate ion concentration increases.

Chapter 32

Organic Chemistry

CARBON BONDS

The ability of carbon atoms to form covalent bonds with one another as well as with other atoms is responsible for the number and diversity of carbon compounds. A carbon atom can form one, two, or three bonds with another carbon atom. Using a dash to represent a covalent bond, the structures of ethane, ethylene (properly *ethene*), and acetylene (properly *ethyne*) are

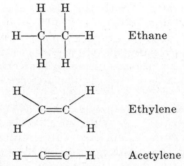

In all compounds except CO, carbon atoms participate in four bonds each.

An organic compound whose molecules have only single carbon—carbon bonds is said to be *saturated*; *unsaturated* compounds have molecules with one or more double or triple bonds. The latter molecules are highly reactive since the second and third bonding orbitals extend well outside the region between the carbon nuclei and the electrons they contain can shift with little difficulty to form more secure single bonds with other atoms. Thus ethylene reacts readily with chlorine to form dichloroethane:

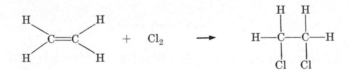

BENZENE

Carbon atoms can bond together to form closed rings as well as long chains. Benzene, C_6H_6, contains a particularly stable ring of six C atoms:

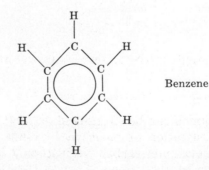

Benzene

The stability of the *benzene ring* is due to a bonding molecular orbital shaped like a pair of doughnuts above and below the ring itself. This orbital, which contains six electrons, is symbolized by the circle in the diagram of the benzene ring and supplements the single C—C bonds also present.

In *aromatic* compounds one or more of the H atoms in a benzene molecule are replaced by other atoms or atom groups. An example is toluene:

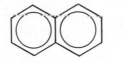
Toluene

The normal C and H atoms in a benzene ring are often omitted in representing the structures of aromatic molecules, as above. Some aromatic compounds contain two or more benzene rings fused together, as in naphthalene, $C_{10}H_8$:

Naphthalene

Organic compounds whose molecules contain no ring structures are called *aliphatic*.

ISOMERS

Isomers are compounds whose molecules have the same composition but different structures. For example, there are two isomers of butane, C_4H_{10}:

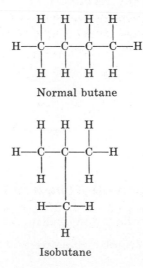
Normal butane

Isobutane

Isomers of a compound usually have different physical and chemical properties.

FUNCTIONAL GROUPS

Many organic compounds can be considered as *hydrocarbon derivatives* formed from a hydrocarbon by the substitution of *functional groups* for one or more of the H atoms. The chemical behavior of a hydrocarbon derivative is largely determined by the nature of the functional groups it contains. Some common functional groups and the classes of compounds characteristic of their presence are listed in Table 32-1.

Table 32-1.   Common Functional Groups

Name of Group	Structural Formula	Class of Compound	Example	
Hydroxyl	—O—H	Alcohol	H—C—O—H (with H above and below C)	Methyl alcohol
Ether	—O—	Ether	H—C—O—C—H (with H above and below each C)	Dimethyl ether
Aldehyde	—C(=O)—H	Aldehyde	H—C(=O)—H	Formaldehyde
Ketone	—C(=O)—	Ketone	H—C—C(=O)—C—H (with H above and below)	Acetone
Carboxyl	—C(=O)—O—H	Acid	H—C(=O)—O—H	Formic acid
Ester	—C(=O)—O—	Ester	H—C—C(=O)—O—C—H	Methyl acetate

Solved Problems

32.1. As a class, organic compounds have low melting and boiling points. Why?

Organic molecules are either nonpolar or nearly so, hence the van der Waals forces between them are weak.

32.2. Why is it preferable to use the symbol ⬡ for the benzene ring instead of the traditional symbol ⬡? What does the circle represent?

In a benzene ring there are single covalent bonds between each pair of carbon atoms and in addition six electrons are shared by the entire ring of carbon atoms. The presence of the latter electrons is indicated by the circle, so the symbol 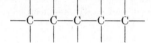 correctly represents the molecule. The traditional symbol suggests that three of the C—C bonds are double and three are single, whereas all the bonds are experimentally found to have the same character. Molecules with double bonds are quite reactive and form addition compounds, whereas benzene molecules are stable and form substitution compounds in which one or more H atoms are replaced by other atoms or atom groups.

32.3. Why are all aromatic compounds unsaturated?

Aromatic compounds contain one or more benzene rings in their molecules, and a benzene ring has six delocalized bonding electrons in addition to single bonds between adjacent carbon atoms.

32.4. The aliphatic compound pentene has the molecular formula C_5H_{10}. Are all of the bonds in the pentene molecule single?

Each C atom must participate in four bonds. We begin by drawing the C atoms with single bonds between them and lines to represent the remaining bonds they can form:

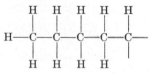

Now we distribute the ten H atoms among these bonds:

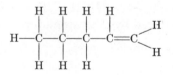

Two bonds are left over, which can be accounted for if one of the carbon—carbon bonds is a double bond. Thus the structural formula of one of the isomers of pentene is

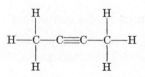

32.5. Each molecule of butyne, C_4H_6, has a triple bond between two of its carbon atoms. What is the structural formula of the butyne isomer in which the triple bond is in the middle of the molecule?

32.6. What is the structural formula of acetaldehyde, C_2H_4O? What is the nature of the bond between the C atoms?

Each C atom must participate in four bonds. We begin by drawing the C atoms with single bonds between them and lines representing the remaining bonds they can participate in:

Each H atom participates in a single bond and the O atom participates in a double bond, which takes care of all the bonds:

Since no bonds are left over, the bond between the C atoms is a single one. We recognize the

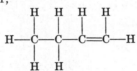 group as an aldehyde group.

32.7. There is a double bond between two of the C atoms in butene, C_4H_8. What are the structural formulas of the two isomers of butene?

The two isomers are

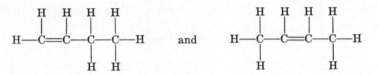   and

What might seem to be a third isomer,

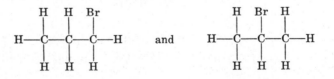

is really the first of the above structures reversed, which is not a true difference.

32.8. Propane has the formula $CH_3CH_2CH_3$. How many isomers does bromopropane, in which one H atom is replaced by a Br atom, have? What are their structural formulas?

There are two isomers of bromopropane,

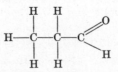   and

32.9. What are the possible names of the compound CH_3CH_2Br?

(*a*) Since CH_3CH_3 is ethane, the substitution of a Br atom for one of the H atoms yields bromoethane. (*b*) Since CH_3CH_2— is the ethyl group, CH_3CH_2Br can also be called ethyl bromide.

32.10. To what class of organic compounds does the compound belong whose structure is shown here?

The compound is an aldehyde (propionaldehyde).

32.11. The molecular formula of acetic acid, $HC_2H_3O_2$, is often written CH_3COOH to indicate that three H atoms are bonded to one of the C atoms. (a) What is the structural formula of acetic acid? (b) Why is one hydrogen ion more easily detached than the others when acetic acid is dissolved in water?

(a)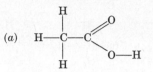

(b)   An O—H bond is weaker than a C—H bond.

32.12. The general formula for fatty acids (except formic acid, HCOOH) is $CH_3(CH_2)_nCOOH$. What is the structural formula of butyric acid, which corresponds to $n = 2$?

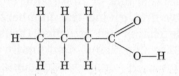

32.13. A compound in which an H atom in a hydrocarbon is replaced by an OH (hydroxyl) group is called an *alcohol*. Give the structural formula of ethyl alcohol (ethanol), which may be thought of as a derivative of ethane, C_2H_6.

The structural formula of ethane is

Replacing an H atom by an OH group gives ethyl alcohol:

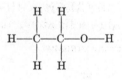

32.14. A compound in which the OH group of an organic acid is replaced by an NH_2 group is called an *amide*. Give the structural formula of acetamide, which may be thought of as a derivative of acetic acid.

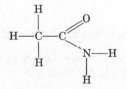

32.15. What is the structural formula of methyl ethyl ether?

In an ether, two hydrocarbon radicals are linked by an oxygen atom. Here the radicals are a methyl group (CH_3) and an ethyl group (CH_3CH_2), so the structure of methyl ethyl ether is

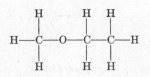

32.16. An alcohol and an acid react to form an ester and water by what may be **regarded** as the organic analog of an acid-base neutralization. Use structural formulas to show the formation of the ester methyl acetate from methyl alcohol and acetic acid.

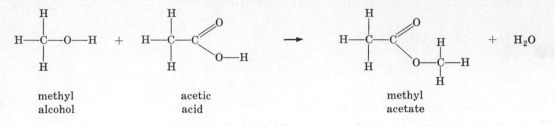

methyl   acetic   methyl
alcohol   acid   acetate

32.17. A hydrocarbon that contains one or more side chains of carbon atoms attached to a main chain is named by first giving the numbers of the carbon atoms in the main chain to which the side chains are attached (usually numbering from right to left), then the designation of the side chains, and finally the name of the main chain. What are the structural formulas of   (*a*) 2-methylpentane,   (*b*) 3,3-dimethylpentane, and (*c*) 2,4-dimethylpentane?

(*a*) The methyl group has the structural formula $\overset{\displaystyle H}{\underset{\displaystyle H}{-\overset{|}{\underset{|}{C}}-H}}$ and the pentane molecule has the

structural formula

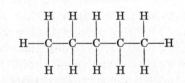

Hence in 2-methylpentane the methyl group is attached to the second C atom from the right:

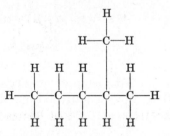

(*b*) In 3,3-dimethylpentane both methyl groups are attached to the same C atom:

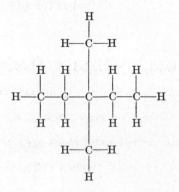

(*c*) In 2,4-dimethylpentane the methyl groups are attached to different atoms:

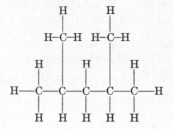

32.18. What is the structural formula of 1-methyl-3-ethylpentane?

Here a methyl group is attached to the first C atom in the pentane molecule and an ethyl group is attached to the third C atom:

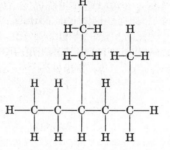

32.19. Benzene derivatives are named by first giving the numbers of the C atoms in the ring to which the atoms or atom groups that replace H atoms are attached (usually numbering clockwise from the top), then the designation of the atoms or atom groups, and finally the name of the basic compound. What are the structural formulas of (*a*) 1,3-dibromobenzene and (*b*) 2-bromo-5-chlorotoluene?

(*a*)

(*b*) Toluene has the structure so that 2-bromo-5-chlorotoluene has a Br atom attached to

the second C atom and a Cl atom attached to the fifth C atom:

Supplementary Problems

32.20. What is the principal bonding mechanism in organic compounds?

32.21. What properties of the carbon atom are responsible for the diversity of organic compounds?

32.22. Why are substances whose molecules contain triple carbon—carbon bonds relatively seldom found?

32.23. The compound heptane has the molecular formula C_7H_{16}. Are all of the carbon—carbon bonds in a heptane molecule single?

32.24. Which of the following are saturated hydrocarbons and which are unsaturated? C_3H_6, C_4H_6, C_4H_{10}, C_5H_{10}, C_6H_6.

32.25. The *alkanes* (or *paraffin hydrocarbons*) are saturated hydrocarbons with the general formula C_nH_{2n+2}. (a) Would you expect the *alkenes* (or *olefins*) whose general formula is C_nH_{2n} to be saturated or unsaturated? (b) Would you expect more, less, or about the same reactivity in the alkenes as in the comparable alkanes?

32.26. How many isomers does dichloroethane, $C_2H_4Cl_2$, have?

32.27. How many isomers does dichlorobenzene, $C_6H_4Cl_2$, have?

32.28. A molecule of xylene consists of a benzene ring with two of the H atoms replaced by methyl groups (CH_3). What are the structural formulas of the isomers of xylene?

32.29. A monohydric alcohol contains one OH group per molecule. The four simplest monohydric alcohols are methyl (CH_3OH), ethyl (C_2H_5OH), propyl (C_3H_7OH), and butyl (C_4H_9OH). Find the formula of a monohydric alcohol whose molecules contain n carbon atoms.

32.30. What is the name of the compound whose structural formula is shown here?

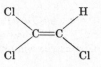

32.31. Ethylene glycol, which is widely used as an antifreeze, has the structural formula

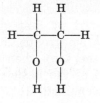

To what class of organic compounds does ethylene glycol belong?

32.32. To what class of organic compounds does the compound belong whose structure is shown here?

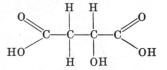

32.33. Each molecule of propene, C_3H_6, has a double bond between two of its carbon atoms. What is the structural formula of propene?

32.34. What is the structural formula of dimethyl ether?

32.35. What is the structural formula of methyl ethyl ketone?

32.36. Ethyl alcohol, CH_3CH_2OH, reacts with acetic acid, CH_3COOH, to give the ester ethyl acetate. What is the structural formula of ethyl acetate?

32.37. Propyl alcohol may be thought of as a derivative of propane, C_3H_8. What are the molecular and structural formulas of propyl alcohol?

32.38. The benzoic acid molecule consists of a benzene ring with one of the H atoms replaced by a carboxyl group. (a) What is the structural formula of benzoic acid? (b) Benzamide may be thought of as a derivative of benzoic acid (compare Problem 32.14). What is the structural formula of benzamide?

32.39. The ethyl group has the formula CH_3CH_2 and butane has the formula $CH_3CH_2CH_2CH_3$. What are the structural formulas of 2,2-diethylbutane and 1,3-diethylbutane?

32.40. Hexane has the formula $CH_3CH_2CH_2CH_2CH_2CH_3$. What is the structural formula of 2,2,5-trimethyl-hexane?

32.41. What is the structural formula of 2-ethyl-4,5-dimethylhexane?

32.42. The explosive trinitrotoluene (TNT) is prepared by mixing concentrated nitric and sulfuric acids with toluene. The sulfuric acid acts as a catalyst for a reaction between toluene and nitric acid in which three nitro (NO_2) groups are added to each toluene molecule. What is the structural formula of 2,4,6-trinitrotoluene?

Answers to Supplementary Problems

32.20. Covalent bonds that consist of shared electron pairs.

32.21. A carbon atom can form four covalent bonds with other atoms. Because these bonds can be with other carbon atoms, chains and rings of carbon atoms with other atoms or atom groups attached to them can exist. The number of possible combinations is therefore enormous.

32.22. Such molecules are extremely reactive.

32.23. Yes.

32.24. Only C_4H_{10} is saturated.

32.25. The alkenes are unsaturated with one $C\!=\!C$ bond per molecule and hence are more reactive than the alkanes.

32.26. two

32.27. three

32.28.

32.29. $C_nH_{2n+1}OH$

32.30. trichloroethylene

32.31. It is an alcohol.

32.32. The presence of  groups means that the compound is an acid. In fact it is malic acid.

32.33.
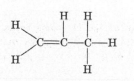

32.34.
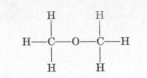

32.35.
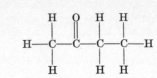

32.36.
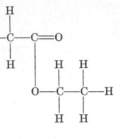

32.37. C_3H_7OH;
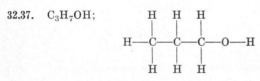

32.38. (a)
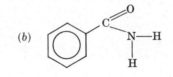

(b)
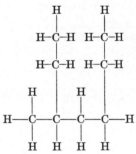

32.39.

2,2-diethylbutane:　　1,3-diethylbutane:
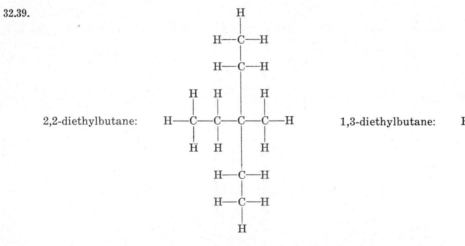

32.40.
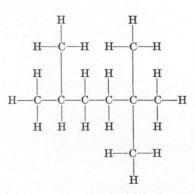

32.41.

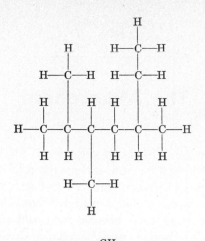

32.42.

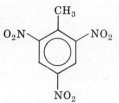

Chapter 33

The Atmosphere

COMPOSITION

The earth's atmosphere consists chiefly of nitrogen (78% by volume) and oxygen (21%). The remainder is largely argon (0.9%) and carbon dioxide (0.03%), plus traces of a number of other gases. Water vapor is present as well but to a variable extent, ranging from nearly 0 to 4%. The lower atmosphere also contains a considerable quantity of small, solid particles of different kinds (such as soot, bits of rock and soil, salt grains from the evaporation of sea water, and spores, pollen, and bacteria); these particles provide nuclei for the condensation of atmospheric water vapor to form clouds, fog, rain, and snow.

Nitrogen, oxygen, and carbon dioxide are important biologically. Nitrogen is a key ingredient of the amino acids of which all proteins consist, and certain bacteria are able to convert atmospheric nitrogen into nitrogen compounds which plants can utilize in manufacturing amino acids. Plants also convert water and atmospheric CO_2 into carbohydrates and oxygen in photosynthesis; animals obtain the carbohydrates and amino acids they need by eating plants. Plants and animals both derive energy by using atmospheric oxygen to convert carbon in their foods to CO_2.

High in the atmosphere solar X- and ultraviolet radiation split N_2 and O_2 molecules into atoms and into ions and electrons. One result is the formation of a small amount of *ozone*, O_3, from reactions between O_2 molecules and O atoms. Ozone is a very efficient absorber of solar ultraviolet at wavelengths longer than those absorbed by N_2 and O_2 and so prevents this potentially lethal radiation from reaching the earth's surface. The region of the upper atmosphere that contains ions and electrons is known as the *ionosphere;* long-range radio communication is possible because radio waves are channeled between the earth's surface and the ionosphere by reflection at both, instead of simply escaping into space.

STRUCTURE

The character of the atmosphere changes more or less abruptly at the *tropopause, stratopause,* and *mesopause,* which occur respectively at altitudes that average about 10 km (6 mi), 50 km (31 mi), and 80 km (50 mi). These surfaces divide the atmosphere into four regions, listed in order of increasing altitude:

1. The *troposphere* is the dense lower part of the atmosphere in which meteorological phenomena such as clouds and storms occur. Air temperature in the troposphere decreases with altitude until it reaches about −55 °C at the tropopause.

2. In the *stratosphere* the air is clear and dry. The temperature is constant in the lower part of the stratosphere but then rises because of heating from the absorption of solar energy by the ozone layer. The stratopause is defined by a temperature maximum, which is usually about 10 °C.

3. In the *mesosphere* the temperature falls steadily to about −80 °C at the mesopause. Air pressure at the mesopause is only about 3/100,000 of sea-level pressure.

4. In the *thermosphere* the absorption of solar X- and ultraviolet radiation results in high temperatures (1000 °C or more) and considerable ionization. The various layers of the ionosphere are found here. It must be kept in mind that the high temperatures of the thermosphere represent the average molecular energies there; because the number of molecules per m^3 is so small, the total energy per m^3 is also small despite the high temperature.

ENERGY BALANCE

Solar radiation reaches the top of the atmosphere at the rate of 20 kcal/min per m^2 of area perpendicular to its direction; about one-third of the arriving energy is reflected back into space, largely by clouds. Since the average temperature of the earth's surface does not change when reckoned on a long-term basis, the earth and its atmosphere reradiate away as much energy as they absorb. However, although intake and outgo are in balance for the earth as a whole, tropical regions receive more direct solar energy than they radiate and polar regions receive less; large-scale winds in the atmosphere and, to a lesser extent, currents in the oceans shift energy from the tropics to the high latitudes. The atmosphere receives the energy it carries around the globe from long-wavelength infrared radiation given off by the earth, which is absorbed by CO_2 and water vapor and communicated to the other atmospheric gases in molecular collisions.

MOISTURE

The atmosphere transports water as well as energy around the world. More water falls as rain and snow on land masses than is lost by them through evaporation, which compensates for the runoff of continental water to the oceans by rivers and streams; in turn the oceans lose more water by evaporation than they gain by precipitation.

Air which contains the maximum amount of water vapor that can evaporate at a given temperature is said to be *saturated*. The higher the temperature, the greater the concentration of water vapor at saturation. The *relative humidity* of a volume of air refers to its degree of saturation: relative humidities of 0, 50%, and 100% mean respectively that there is no moisture present, that the air contains half as much moisture as the maximum possible, and that the air is saturated. If a volume of air is cooled past the temperature at which it is saturated, the excess water vapor will condense on suitable nuclei into droplets of water or, in certain circumstances, into ice crystals.

Clouds form when rising air is cooled by its expansion. The water vapor usually condenses around salt particles in the air and forms tiny droplets or ice crystals small enough to remain suspended aloft indefinitely. Precipitation occurs when some of the water droplets or ice crystals in a cloud coalesce with others and grow heavy enough to fall.

Solved Problems

33.1. In what two important processes does the carbon dioxide in the atmosphere have essential roles?

   (a) In photosynthesis plants manufacture carbohydrates from atmospheric CO_2 and water, with oxygen as a by-product.

   (b) By absorbing infrared radiation emitted by the earth, CO_2 is an intermediary in the process by which solar energy is transferred to the lower atmosphere and carried around the earth by winds.

33.2. Illustrate the nitrogen cycle with a diagram.

See Fig. 33-1.

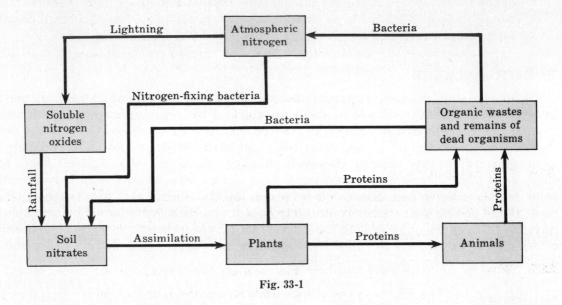

Fig. 33-1

33.3. Illustrate the oxygen-CO_2 cycle with a diagram.

See Fig. 33-2.

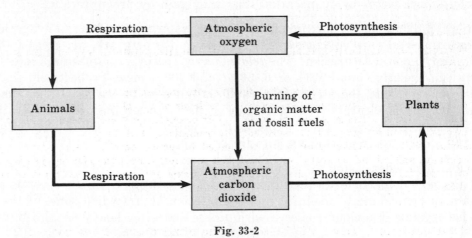

Fig. 33-2

33.4. What are the five chief classes of atmospheric pollutants?

Suspended particles (smoke, dust); carbon monoxide; unburned hydrocarbons; sulfur oxides (SO_2, SO_3); and nitrogen oxides (NO, NO_2).

33.5. (a) What is ozone? (b) Why is the ozone content of the atmosphere concentrated in a layer rather than being distributed uniformly at all levels? (c) Why is the ozone layer so important for terrestrial life?

(a) Ozone is a form of oxygen whose molecules consist of three O atoms, so that its formula is O_3. Ozone is less stable than O_2 and tends to break up into $O_2 + O$.

(b) The first step in the formation of an ozone molecule is the decomposition of an O_2 molecule into two O atoms. The energy required for this process is provided by photons of ultraviolet light from the sun. The second step is the attachment of an O atom to an O_2 molecule to form O_3. The rate of ozone production thus depends upon both the O_2 concentration and the intensity of solar ultraviolet light. At extremely high altitudes there are not enough O_2 molecules for an appreciable amount of O_3 to be formed. Between 15 and 35 km above the ground, however, the atmosphere is dense enough for the production of O_3 but not so dense that the ozone molecules undergo disruptive collisions too often. At lower altitudes the ultraviolet light has already been absorbed, so no ozone can come into being there except as a result of lightning strokes.

(c) Ozone is an excellent absorber of ultraviolet light of frequencies not absorbed by other atmospheric constituents, and thus prevents this light from reaching the earth's surface where it would harm most living things.

33.6. Why is the sky blue?

Sunlight is scattered by molecules and dust particles in the atmosphere. The shorter the wavelength, the more the scattering, hence blue light is scattered to a greater extent than light of other colors. What we see as "the sky" is scattered sunlight, which is therefore predominantly blue.

33.7. What is the ionosphere and how does it affect the propagation of radio waves?

The ionosphere is the region of the earth's atmosphere which contains an appreciable number of ions and electrons. The ionosphere extends through altitudes from about 30 miles to several hundred miles and is produced by the action of solar ultraviolet light and X-rays. During the day the ionosphere has four layers, D, E, F_1, and F_2 in order of ascending altitude. At night the D layer disappears, the E layer weakens, and the F_1 and F_2 layers coalesce into a single weak F layer. The D layer partially absorbs radio waves, the other layers reflect them and so make possible long-range radio communication.

33.8. What is a "temperature inversion" in the atmosphere and what is its connection with smog?

Ordinarily air temperature falls steadily with increasing altitude in the troposphere. Sometimes, however, a situation arises in which a layer of air aloft is warmer than the air below it; this constitutes a temperature inversion. (For example, on a clear summer night the earth's surface in a certain region may cool rapidly by radiation, which leads to a layer of cool air near the surface while the overlying air has not changed in temperature by very much.) Gases emitted by chimneys and vehicle exhausts cannot rise past a temperature inversion because when they reach it their density is greater than the density of the warm air layer. Hence the inversion acts to trap such gases, whose increased concentration is evident as smog.

33.9. Solar radiation arrives at the top of the atmosphere at the rate of 20 kcal/min per m^2 of area perpendicular to its direction. On the average 34% of this energy is reflected back into space and 19% is absorbed in the upper atmosphere, so 47% reaches the ground. A parabolic reflector on the ground, whose cross-sectional area at the open end is 1 m^2, is oriented perpendicular to the solar radiation. If all of the energy it intercepts is focused on 1 kg of water, how long will it take for the water, initially at 20 °C, to be heated to the boiling point?

The heat required to raise the temperature of 1 kg of water from 20 °C to 100 °C is

$$Q = mc\,\Delta T = 1 \text{ kg} \times 1 \text{ kcal/kg-°C} \times 80 \text{ °C} = 80 \text{ kcal}$$

The rate at which heat is supplied is 0.47×20 kcal/min = 9.4 kcal/min, so the time required is

$$t = \frac{80 \text{ kcal}}{9.4 \text{ kcal/min}} = 8.5 \text{ min}$$

33.10. What is the "greenhouse effect" and how is it related to the absorption of solar energy by the earth's atmosphere?

The interior of a greenhouse is warmer than the outside air because sunlight can enter through its windows but the infrared radiation that the warm interior gives off cannot escape through them. The carbon dioxide and water vapor contents of the atmosphere act as a one-way mirror of this kind for the earth as a whole. The atmosphere is transparent to visible light, which is absorbed by the earth's surface. The temperature of the surface is thereby increased, which in turn increases the rate at which it emits infrared radiation. The carbon dioxide and water vapor in the atmosphere absorb the infrared radiation, which leads to a warming of the lower atmosphere.

33.11. What are the various ways in which clouds can be formed in an air mass that was originally warm and moist?

For clouds to form in such an air mass, it must move upward, expand due to the lower pressure, and thereby become cool enough for some of its water vapor to condense. There are three processes which can accomplish the uplift of an air mass:

1. **Convection.** The warm air rises due to its buoyancy. Cumulus clouds are formed in this way.

2. **Synoptic cooling.** A warm air mass moving horizontally encounters a cooler mass and, being less dense, is forced upward on top of it. Stratus clouds are formed in this way.

3. **Orographic cooling.** A warm air mass moving horizontally encounters a land barrier such as a mountain range and rises. Both cumulus and stratus clouds can be formed in this way.

33.12. How are clouds classified?

The three basic types of clouds are *cirrus* (wispy or featherlike), *stratus* (layered), and *cumulus* (puffy or heaped up). A cloud that combines the characteristics of two of these types is designated accordingly, for instance cirrostratus. A cloud that occurs at a higher altitude than is normal for its type is given the prefix *alto*, as in altostratus. Clouds from which precipitation occurs have the word *nimbus* (Latin for rain) in their names, for instance nimbostratus.

33.13. Clouds are divided into four families depending upon the altitudes at which they normally occur, as follows: high ($> 7\,\text{km}$), middle (2 to 7 km), low ($< 2\,\text{km}$), and clouds with vertical development (base usually $< 2\,\text{km}$, top may be $> 7\,\text{km}$). Give examples of clouds in each family.

HIGH: cirrus, cirrostratus, cirrocumulus

MIDDLE: altocumulus, altostratus

LOW: stratus, stratocumulus, nimbostratus

CLOUDS WITH VERTICAL DEVELOPMENT: cumulus, cumulonimbus

33.14. Clouds are sometimes "seeded" with silver iodide to induce precipitation. Why silver iodide?

The crystal structure of silver iodide resembles that of ice, hence water molecules in a cloud can readily attach themselves to a silver iodide crystal. Silver iodide crystals are thus efficient condensation nuclei and so promote precipitation from a cloud.

Supplementary Problems

33.15. Why does the earth have an atmosphere whereas the moon does not?

33.16. Distinguish between the troposphere and the tropopause.

33.17. What characterizes (*a*) the tropopause, (*b*) the stratopause, and (*c*) the mesopause?

33.18.   What is the color of space to an astronaut above the earth's surface?

33.19.   What is insolation?

33.20.   The air in a closed container is saturated with water vapor at 20 °C. (a) What is its relative humidity? (b) What happens to the relative humidity if the temperature is reduced to 10 °C? (c) If the temperature is increased to 30 °C?

33.21.   Why does the air in a heated room tend to be dry?

33.22.   Why does dew form during clear, calm summer nights?

33.23.   What does "dew-point temperature" mean?

33.24.   How are clouds and fog related?

33.25.   What do high-altitude clouds consist of? Low-altitude clouds?

33.26.   What is the significance for weather phenomena of dust and salt particles in the atmosphere?

Answers to Supplementary Problems

33.15.   The gravitational field of the moon is not sufficient to prevent the escape of the rapidly-moving gas molecules in an atmosphere; the escape velocity of the moon is only 1.2 mi/s. The gravitational field of the more massive earth is great enough to retain an atmosphere; its escape velocity is 7.0 mi/s.

33.16.   The troposphere is the dense lower part of the atmosphere in which clouds, storms, and other weather phenomena occur. The tropopause is the upper limit of the troposphere and divides it from the clear, cold air of the stratosphere above.

33.17.   (a) a temperature minimum; (b) a temperature maximum; (c) a temperature minimum

33.18.   Since there is no atmosphere in space, there is no scattered sunlight, and space appears black.

33.19.   "Insolation" stands for incoming solar radiation and refers to the solar energy arriving at the top of the earth's atmosphere.

33.20.   (a) 100%; (b) the air remains saturated and so the relative humidity remains 100%, while the excess water vapor condenses out; (c) the relative humidity decreases

33.21.   The outside air has a low moisture content because it is cold, even though its relative humidity may be high. When this air is heated, its moisture content remains the same, hence its relative humidity decreases.

32.22.   On such a night the earth's surface cools by radiation. The air in contact with the surface cools also until it becomes saturated with water vapor, which then condenses into droplets of liquid water.

33.23.   The dew-point temperature is the lowest temperature to which a sample of air can be cooled without the condensation of water vapor. The lower the dew point, the smaller the concentration of water vapor in the air sample.

33.24.   A fog is a cloud at ground level.

33.25.   ice crystals; water droplets

33.26.   Dust and salt particles provide the nuclei around which water vapor condenses to form clouds, fog, rain, and snow.

Chapter 34

Weather

WINDS

Winds are horizontal movements of air that take place in response to pressure differences in the atmosphere. The greater the difference between the pressures in two regions, the faster the air between them will move. All pressure differences between places on the earth's surface can be traced, directly or indirectly, to temperature differences. If one region is warmer than its surroundings, for example, the air above it is heated and expands. The hot air rises, leaving behind a low-pressure zone into which cool air from the high-pressure neighborhood flows. The horizontal flow toward the heated region at low altitudes is balanced by a horizontal flow outward of air that has risen, which cools and sinks to replace the air that has moved inward. In this way *convection cells* come into being that convert temperature differences into pressure differences and thus cause winds to occur. On a large scale, the differences between the solar heating of the equatorial and polar regions are what power the general circulation of the lower atmosphere.

CORIOLIS EFFECT

The rotation of the earth influences the path of an object that moves above its surface as this path is seen by an observer on the surface. In the northern hemisphere a path that would be a straight line across a stationary earth appears instead to be curved to the right (as the object recedes from the observer); in the southern hemisphere the curvature is to the left. Only motion along the equator is not affected. This phenomenon is called the *Coriolis effect*.

The Coriolis effect is responsible for converting the north-south convection currents that would be caused on a nonrotating earth by the uneven distribution of solar heating into winds that have easterly (that is, out of the east) and westerly (out of the west) components (see Fig. 34-1).

GENERAL CIRCULATION OF THE ATMOSPHERE

When the winds of the world are averaged over a long period of time, transient fluctuations disappear to leave a large-scale *general circulation* of the lower atmosphere. The principal features of the general circulation are shown in Fig. 34-1. Conspicuous are the easterly winds of the polar regions, the *prevailing westerlies* of the middle latitudes, and the northeast and southeast *trade winds* of the tropics. An equatorial belt of low pressure, the *doldrums*, separates the trade winds of each hemisphere, and belts of high pressure, the *horse latitudes*, separate the trade winds from the prevailing westerlies; winds in these belts are weak and erratic.

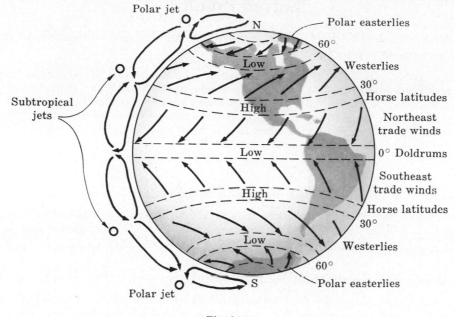

Fig. 34-1

With increasing altitude the regions of westerly winds broaden until almost the entire flow of air is west-to-east at the tropopause. The westerly flow aloft is not uniform but contains narrow cores of high-velocity winds called *jet streams*. The jet streams form wavelike zigzag patterns around the earth that change continuously and give rise to the variable weather of the middle latitudes by their effect on air masses closer to the surface.

CYCLONES AND ANTICYCLONES

Winds in the middle latitudes are associated with weather systems called *cyclones* and *anticyclones* that are several hundred to a thousand or more miles across and move from west to east. At the center of a cyclone the air pressure is low, and as air rushes in toward it the moving air is deflected toward the right in the northern hemisphere and toward the left in the southern, because of the Coriolis effect. As a result cyclonic winds blow in a counterclockwise spiral (as viewed from above) in the northern hemisphere and in a clockwise spiral in the southern hemisphere. An anticyclone is centered on a high-pressure region from which air moves outward. The Coriolis effect therefore causes anticyclonic winds to blow in a clockwise spiral in the northern hemisphere and in a counterclockwise spiral in the southern hemisphere. These spirals are conspicuous in cloud formations photographed from earth satellites.

As a rule, cyclones bring unstable weather conditions with clouds, rain, strong winds, and abrupt temperature changes. The weather associated with anticyclones, on the other hand, is usually settled and pleasant with clear skies and little wind.

Middle-latitude cyclones originate at the *front,* or boundary, between the cold polar air mass and the warmer air mass adjacent to it. It is common for a kink to develop in this front with a wedge of warm air protruding into the cold air mass. This produces a low-pressure region which moves eastward as a cyclone. The eastern side of the warm air wedge is a *warm front* since warm air moves in to replace cold air there: the western side is a *cold front* since cold air replaces warm air. After a few days the cold front, which moves faster, overtakes the warm front to force the wedge of warm air upwards, and soon afterward both it and the cyclone disappear.

Solved Problems

34.1. The earth is closest to the sun in January, but January is a winter month in the northern hemisphere. Why?

   The earth's axis of rotation is tilted with respect to its axis of revolution around the sun. As a result the daylight side of the northern hemisphere is tilted away from the sun in January, which means that sunlight strikes the northern hemisphere at a glancing angle and delivers less energy per square meter of surface than during the summer when the northern hemisphere is tilted toward the sun. The effect of the tilted axis is more than enough to counterbalance the relative closeness of the sun during the northern winter.

34.2. Distinguish between an isobar and a millibar.

   An isobar is a line on a weather map that joins points which have the same atmospheric pressure. A millibar (mb) is a unit of pressure; average sea-level atmospheric pressure is 1013 mb.

34.3. Why does a rising body of air become cooler?

   Atmospheric pressure decreases with altitude, hence a rising body of air expands. An expanding gas does work on its surroundings, and its loss of energy means a drop in temperature. Dry rising air cools at the rate of about 1 °C per 100 m.

34.4. Use a diagram to show the convection currents around a heated area on the earth's surface.

   See Fig. 34-2.

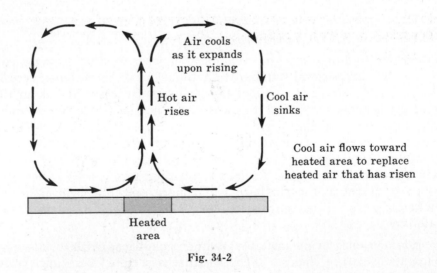

Fig. 34-2

34.5. (a) On summer days coastal regions often experience an onshore wind. What is the origin of such a sea breeze? (b) At night the sea breeze stops and is often replaced by an offshore wind. What is the origin of such a land breeze?

   (a) Sunlight causes the land to warm up fairly rapidly in the morning, since it is absorbed in a thin surface layer. The water temperature changes very little, partly because the incoming solar energy is shared by a thicker layer of water and partly because the specific heat capacity of water is large. The air over the warm land becomes warm in turn and rises by convection, whereupon cooler, denser air from the sea — the sea breeze — sweeps in to replace it.

(b) At night the land cools rapidly by radiation while the sea surface remains at about the same temperature as during the day because heat transfer is more efficient in water than in rock and soil. When land and sea are at the same temperature, the sea breeze stops. If the land cools still further, air warmed by the sea rises and cool air sweeps off the land to replace it.

34.6. The northeast and southeast trade winds meet in a belt called the doldrums. What is the characteristic weather of the doldrums?

The doldrums are at the equator, so it is quite warm there with considerable evaporation of water and thus high humidity. The air flow is largely upward, so surface winds are light and erratic. The rising currents of moist air lead to considerable rainfall.

34.7. Why are most of the world's deserts found in the horse latitudes, which separate the trade winds from the prevailing westerlies in both hemispheres?

The air flow in the horse latitudes is largely downward. Descending air is warmed by compression and its relative humidity decreases, hence there is little rainfall in the horse latitudes.

34.8. (a) When you face a wind associated with a cyclone in the northern hemisphere, in what approximate direction will the center of low pressure be? (b) In what direction will the center of low pressure be if you do this in the southern hemisphere?

(a) The circulation of air around a low-pressure region in the northern hemisphere is counterclockwise since air flowing toward the region is deflected to its own right by the Coriolis effect. Hence when you face the wind, the center of low pressure will be on your right.

(b) In the southern hemisphere the Coriolis effect causes moving air to be deflected to its own left, so the circulation around a low-pressure region is clockwise there. When you face the wind in this case, the center of low pressure will be on your left.

34.9. Sketch a typical mature cyclone as it might appear on a weather map of the northern hemisphere and identify its main features.

See Fig. 34-3.

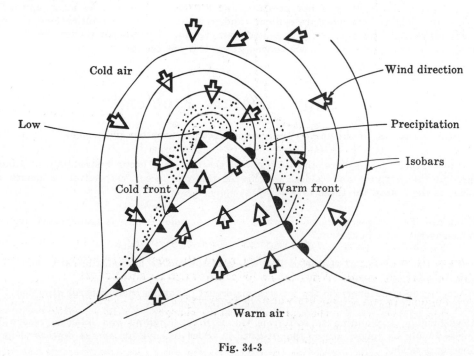

Fig. 34-3

34.10. What is an occluded front? At what stage in the evolution of a cyclone does it occur?

In a typical cyclone, a wedge of eastward-moving warm air penetrates a cold air mass. There is a warm front on the east and a cold front on the west of the wedge. The cold front moves faster than the warm front, and when it overtakes the warm front their intersection is lifted above the ground as the cold air burrows under the warm air of the wedge. The formation of such an occluded front is the last stage in the evolution of a cyclone, which soon afterward disappears.

34.11. What are the characteristic properties of the air in the following air masses: (*a*) continental polar; (*b*) continental tropical; (*c*) maritime polar; (*d*) maritime tropical?

(*a*) Cold, dry, stable. (*b*) Hot, dry, unstable. (*c*) Cool, moist, unstable in winter; cool, dry, stable in summer. (*d*) Warm, moist, unstable.

34.12. Distinguish between a hurricane and a tornado.

A hurricane is a large, violent tropical storm typically a hundred miles in diameter whose winds spiral inward and upward at velocities of 75 mi/hr or more around an "eye" of low pressure. Heavy rainfall accompanies the passage of a hurricane except in the eye, which may be 10 or 20 miles across. Most hurricanes occur on the western sides of the Pacific, Indian, and North Atlantic oceans during the late summer and early fall and their most violent phases last for a few days to a week or so. Hurricanes usually move at 10 to 30 mi/hr but may move faster or remain in one place for a day or more.

A tornado is a small, funnel-shaped rotational storm several hundred feet in diameter that appears to descend from a cumulonimbus (thunderstorm) cloud. Velocities in a tornado are apparently several hundred mi/hr, though the tornado itself moves at only 20 to 40 mi/hr; a tornado usually last for less than an hour. Tornados most often occur on central continental plains, notably those of the United States and Australia, but are also found at sea, where they are called *waterspouts* and are less violent.

34.13. Why is the sky in the vicinity of a cyclone cloudy whereas it is clear in the vicinity of an anticyclone?

A cyclone is a region of low pressure, and air flowing into it rises in an upward spiral. The rising air cools and its moisture content condenses into clouds. An anticyclone is a region of high pressure, and air flows out of it in a downward spiral. The descent warms the air and its relative humidity accordingly drops, hence condensation does not occur.

Supplementary Problems

34.14. What is the source of the energy that is manifested in weather phenomena?

34.15. In the northern hemisphere, the longest and shortest days occur respectively in June and December, but the warmest and coldest weather of the year occur respectively a month or two later. What is the reason for these time lags?

34.16. What are the two mechanisms by which energy of solar origin is transported around the earth? Which is the most important?

34.17. What is the direction of the prevailing winds of the middle latitudes in each hemisphere?

34.18. Where in the atmosphere do the jet streams occur? What is their general direction?

34.19. A yachtsman is planning to sail from the U. S. to England and later to return home. What routes should he follow across the Atlantic in order to have his course downwind as much of the time as possible?

34.20. How does the weather associated with a typical cyclone differ from that associated with a typical anticyclone?

34.21. What is the approximate sequence of wind directions when the center of a cyclone passes north of an observer in the northern hemisphere?

34.22. What is the approximate sequence of wind directions when the center of an anticyclone passes south of an observer in the northern hemisphere?

34.23. What is the difference between the rainfall that accompanies the passage of a warm front and that which accompanies the passage of a cold front?

34.24. What is the usual lifetime of a cyclone in the middle latitudes?

34.25. What is the direction of air flow in the central eye of a hurricane? Why is the sky clear over the eye?

Answers to Supplementary Problems

34.14. Solar radiation reaching the earth.

34.15. The heat capacities of land, sea, and air are very large, and when the rate of arrival of solar energy changes considerable energy must be absorbed or lost by a region before it reaches equilibrium. Since the difference between the rates of energy absorption and energy loss is always small, the temperature of the surface cannot change rapidly enough to keep pace with changes in the rate at which solar energy arrives, hence the time lags in seasonal weather conditions.

34.16. Winds and ocean currents carry energy around the earth in the forms of warm air and warm water respectively. Winds are more effective in energy transport than ocean currents.

34.17. Westerly (from the west) in both hemispheres.

34.18. The jet streams occur near the top of the troposphere and flow from west to east.

34.19. To England: northeast and then east in order to take advantage of the prevailing westerlies. Back to the U.S.: first south to the trade winds, then west, and finally north.

34.20. Anticyclonic weather is generally steady with a relatively constant temperature, clear skies, and light winds. Cyclonic weather is unsettled with rapid changes in temperature that accompany the passages of cold and warm fronts, cloudy skies, rain, and fairly strong, shifting winds.

34.21. From the southwest – from the west – from the northwest.

34.22. From the northwest – from the west – from the southwest.

34.23. Rainfall associated with the passage of a warm front is generally lighter and of longer duration than that associated with the passage of a cold front.

34.24. Three to five days.

34.25. The air flow is downward and, near the bottom, outward. As the air descends, it is compressed and therefore warmed, which decreases its relative humidity below saturation. Clouds come into being only in saturated air.

Chapter 35

The Oceans

OCEAN WATER

During the early history of the earth the gases in its initial atmosphere escaped into space. The constituents of the present atmosphere plus the water of the hydrosphere are believed to have emerged over a long period of time as a by-product of volcanic activity from the rocks of the earth's interior, where they were incorporated in various minerals.

The oceans cover about 71% of the earth's surface and contain about 97% of the water in the hydrosphere. The average salinity of ocean water is 3.5%; most of the ions are Na^+ and Cl^-, though many other ions are also present. Ocean water near the surface is saturated with atmospheric gases, whose concentrations decrease with depth.

The water in the top hundred meters of the ocean varies in temperature with location and season, and may be 20 °C or more. In the next km the temperature drops to a few °C and remains just above the freezing point to the ocean bottom, whose depth averages 3.7 km.

WAVES

Wind blowing across the surface of a body of water produces waves. The wave height depends upon the amount of energy transferred to the water and therefore increases with wind velocity, the time during which the wind has blown from the same direction, and the distance (called *fetch*) over which the wind has blown. When a wave passes a certain place in the ocean, water particles at and near the surface move in circular orbits. At the wave crest, the surface water moves in the direction of the wave; at the wave trough, it moves in the opposite direction. The orbital motion becomes negligible at a depth of about $\lambda/2$, where λ is the wavelength.

CURRENTS

The progress of a wave involves the transport of energy but not of water. However, a wind whose direction remains more or less constant will produce a net motion of surface water called a *current*. The wind patterns of the general circulation of the atmosphere lead to corresponding patterns of ocean currents, with the Coriolis effect playing a part. The latter patterns take the form of huge whirlpools, or *gyres*, in each ocean basin: gyres of clockwise flow in the North Atlantic and North Pacific Oceans, gyres of counterclockwise flow in the South Atlantic, South Pacific, and Indian Oceans. In addition, smaller counterclockwise gyres are present in the extreme northern parts of the North Atlantic and North Pacific. Eastward *equatorial countercurrents* flow in the doldrums of the Atlantic and Pacific between the gyres to the north and south, and a continuous eastward current flows around the entire Southern Ocean.

The westward-moving currents in the major gyres, which are propelled by the trade winds, are warmed by the tropical sun, and much of the added heat is retained in their further progress poleward and then eastward. Thus the North Equatorial Current is quite warm when it turns north as the Gulf Stream, and as the North Atlantic Drift it significantly moderates the climate of northern Europe by heating the westerly winds that blow across it.

THE TIDES

The twice-daily rise and fall of the tides can be traced to the gravitational forces exerted by the moon and, to a lesser extent, by the sun. Each of these bodies attracts different parts of the earth with slightly different forces. Because the moon is much closer to the earth than the sun is, the variations in the forces it exerts are greater than those in the forces exerted by the sun, even though the forces themselves are stronger in the case of the sun. For this reason the moon is chiefly responsible for the tides, with the solar influence limited to modifying the tidal range depending on the position of the sun with respect to the earth and moon.

The tidal bulges A and B in Fig. 35-1 are held in place by the moon as the earth rotates under them, thus causing the periodicity of the tides. Water at A is closest to the moon, and is heaped up by its gravitational pull. Water at B is farthest from the moon, and is least attracted by it; this water accordingly tends to be left behind, so to speak, as the earth is pulled away from

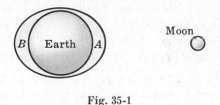

Fig. 35-1

under it due to the revolution of the earth and moon about their common center of mass. (The earth's pull on the moon, which causes it to move in an orbit, has a counterpart in the moon's pull on the earth, which also moves in an orbit, although a much smaller one. Earth and moon may be thought of as the two ends of a dumbbell that is rotating about its balance point, which happens to lie within the earth about 4700 km from the center. It is this center of mass that moves in an elliptical orbit around the sun.)

Solved Problems

35.1. List the chief reservoirs of the earth's water content in the order of the amount of water each contains.

>   Seas and oceans; icecaps and glaciers; ground water; lakes and rivers; atmospheric moisture.

35.2. Sketch the profile of a typical continental margin and identify its main features.

>   See Fig. 35-2.

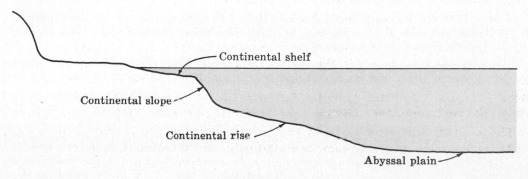

Fig. 35-2

35.3. Even in the intense cold of the polar regions, sea ice is seldom more than 3 m thick. (Icebergs are huge masses of ice that have broken off from icecaps or glaciers that formed from accumulated snow on a land base.) Why is sea ice so relatively thin?

There are several reasons. The chief one is that, as the water under a surface layer of ice is cooled, it becomes denser and sinks, to be replaced by warmer water from underneath. Another reason is that ice itself acts as an insulator. Also, as seawater freezes, its salt content is not incorporated in the ice crystals but stays behind to increase the salinity of the remaining water, whose freezing point is accordingly lowered.

35.4. What are plankton, where do they live, and why are they important to life on land as well as to aquatic life?

Plankton are minute oceanic plants (phytoplankton) and animals (zooplankton). Most plankton are found in the upper 10 m or so of the oceans where sunlight penetrates, and constitute the first step in the food chain of all other forms of aquatic life such as fish, shellfish, and mammals (whales, porpoises). The mass of phytoplankton in the oceans exceeds the mass of land vegetation, and phytoplankton are thus important in maintaining the global oxygen-carbon dioxide balance.

35.5. Why is marine life most abundant in the colder parts of the oceans?

Cold ocean water contains more of the dissolved CO_2 and O_2 required by marine plants and animals because gases are more soluble in cold water than in warm water. Another reason is that vertical mixing of ocean water is impeded by the presence of a layer of warm surface water, so organic debris that sinks to the bottom in warm parts of the ocean is never returned to the surface where it can provide nutrients (especially nitrates and phosphates) to living things. In cold parts of the ocean, on the other hand, the upwelling of bottom water helps to feed plants and animals near the surface where sunlight and dissolved gases are also available.

35.6. The salinity of seawater varies somewhat with location, for example being greater than average in warm, enclosed seas such as the Mediterranean where evaporation is high, and less than average where a river discharges into the ocean. However, the relative proportions of the various ions in solution are almost exactly the same everywhere regardless of local circumstances. What is the significance of the latter observation?

Because their waters must be thoroughly mixed in the course of time to obtain a uniform relative composition, the seas and oceans of the world cannot be static bodies but must contain large-scale currents, both vertical and horizontal.

35.7. (*a*) Under what circumstances is the range of the tides a maximum? (*b*) Under what circumstances is it a minimum? (*c*) What phases of the moon are associated with these extremes?

(*a*) When the sun and moon are in line with each other and with the earth, their tide-producing forces add together and high tides are at their highest, low tides are at their lowest. Such tides are called *spring tides*.

(*b*) When the sun-earth and moon-earth lines are perpendicular, the tide-producing forces of the sun and moon are in opposition, and high tides are at their lowest, low tides are at their highest. Such tides are called *neap tides*.

(*c*) Spring tides occur at new moon and full moon, neap tides occur at first quarter and last quarter (when half the moon's disk is illuminated).

35.8. The earth makes a complete rotation on its axis once every 24 hr. Why is the interval between successive high tides 12 hr 25 min instead of 12 hr?

The moon revolves around the earth in the same direction as the earth's rotation and takes 29.5 days to circle the earth relative to the sun; that is, 29.5 days elapse from new moon to new moon. Thus the moon is directly overhead a certain place on the earth 24 hr/29.5 days = 0.81 hr later each day, which is 50 min. There are two tides a day, and half of 24 hr 50 min is 12 hr 25 min.

Supplementary Problems

35.9. Which ions are predominant in seawater?

35.10. Why do icebergs float?

35.11. A wind begins to blow over the surface of a calm body of deep water. What factors govern the height of the waves that are produced?

35.12. Why do submarines submerge in storms?

35.13. Do ocean waves actually transport water from one place to another? If not, what if anything do such waves transport?

35.14. (a) If you were planning to drift in a raft across the North Atlantic from the U.S. to Europe by making use of ocean currents, what would your route be? (b) If you were planning to drift from Europe to the U.S., what would your route be?

35.15. England and Labrador are at about the same latitude on either side of the North Atlantic Ocean, but England is considerably warmer than Labrador on the average. Why?

Answers to Supplementary Problems

35.9. Na^+ and Cl^-.

35.10. Water expands when it freezes, and so has a lower density, because water molecules are farther apart in the regular structure of an ice crystal than they are in the irregular arrangements characteristic of liquid water.

35.11. (a) The greater the wind velocity, the higher the waves. (b) The longer the period of time during which the wind blows, the higher the waves. (c) The greater the distance (*fetch*) over which the wind blows across the water, the higher the waves. Each of the above factors ceases to have a strong effect on wave height after a certain point; for example, after a day or two the waves will have reached very nearly the maximum height possible for the wind velocity and fetch of a given situation.

35.12. Below a depth of about half the wavelength of an ocean wave, there is almost no disturbance of the water, hence a submarine at such depths is in no danger.

35.13. Water particles move in approximately circular orbits as a wave passes by, so no net transport of water is involved in wave motion. Water waves do transport energy, as all other waves do.

35.14. (a) At first northward with the Gulf Stream, then northeastward with the North Atlantic Drift. (b) At first southward with the Canary Current, then westward and finally northwestward with the North Equatorial Current and the Gulf Stream.

35.15. The North Atlantic Drift, which is fed by the Gulf Stream, brings warm water to the shores of northwestern Europe. The prevailing westerlies that blow over this warm water lead to a moderate climate for this part of Europe. In the case of Labrador, the westerly winds have blown over the cold land mass of northern Canada; in addition, the waters that bathe the Labrador coast originate in Baffin Bay off Greenland and are accordingly very cold.

Chapter 36

Earth Materials

THE EARTH'S CRUST

The *crust* of the earth is its outer shell of rock. The crust is typically 5 miles thick under the oceans and 20 miles thick under the continents. The most abundant elements in the crust are oxygen (47% by mass) and silicon (28%); then come aluminum, iron, calcium, sodium, potassium, and magnesium, which range from 8% to 2% in that order. Since the O^{--} ion is relatively large, over 90% of the volume of the crust is oxygen.

MINERALS

Rocks are aggregates of homogeneous substances called *minerals*. Some rocks, for instance limestone, consist of a single mineral only, but the majority consist of several minerals in varying proportions. Although over 2000 minerals have been identified, only a few are significant constituents of most rocks. Among the commoner minerals or mineral groups are the feldspars, quartz, the ferromagnesian minerals, the clay minerals, and mica, which are all silicates, and calcite, which is composed of calcium carbonate.

The basic structural element of all silicates is the SiO_4^{-4} tetrahedron, with each Si^{+4} ion surrounded by four O^{--} ions that may be thought of as occupying the corners of a tetrahedron centered on the Si^{+4} ion (Fig. 36-1). The SiO_4^{-4} tetrahedra may occur singly, as in the olivines where Fe^{++} and Mg^{++} ions then bond them together; in single or double chains in which adjacent tetrahedra share O^{--} ions, with the chains cross-linked by metal ions; in sheets in which each tetrahedron shares three O^{--} ions with its neighbors, with the sheets bonded together by metal ions; and in three-dimensional networks in which all the O^{--} ions of each tetrahedron are shared by its neighbors, as in quartz, SiO_2.

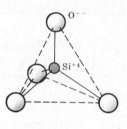

Fig. 36-1

IGNEOUS ROCKS

Rocks are classified as igneous, sedimentary, or metamorphic on the basis of their origins. *Igneous rocks* cooled from a molten state. Two-thirds of crustal rocks are igneous, with basalt constituting the bedrock under the oceans and granite the bedrock under the continents. *Intrusive* igneous rocks solidified beneath the surface where slow cooling resulted in large mineral grains; *extrusive* igneous rocks solidified after emerging from a volcano or other vent, and the rapid cooling resulted in small mineral grains.

SEDIMENTARY ROCKS

Most *sedimentary rocks* are composed of fragments of other rocks that have been eroded by the action of running water, glaciers, or wind. A few consist of material that either was precipitated from solution or was consolidated from the shells of marine organisms. Sediments are usually deposited in layers, and the layering is evident in the resulting rocks. Although sedimentary rocks make up only about 8% of the crust, three-quarters of surface rocks are of this kind.

METAMORPHIC ROCKS

Metamorphic rocks were formed from igneous and sedimentary rocks under the influence of heat and pressure below the earth's surface. Sometimes the changes are in the minerals themselves, which turn into other varieties more stable in the different environment of the earth's interior, sometimes the changes are in the characters of the mineral grains.

SOIL

Soil is a mixture of rock debris, largely clay minerals and quartz fragments, and organic matter. Much of the latter consists of microorganisms such as bacteria and fungi and their remains. *Podzol soils* occur in moist, fairly cool climates under coniferous forests; *latosol soils* develop in tropical rain forests; *chernozem soils* are extremely fertile and were formed under prairie grasses in temperate, subhumid climates; and *desert soils* are found under arid conditions and contain little organic matter but many soluble minerals absent from soils in regions of abundant rainfall.

Solved Problems

36.1. In the silicate minerals each Si^{+4} ion is always surrounded by four O^{--} ions, yet no mineral has the formula SiO_4. Why not?

An isolated SiO_4 tetrahedron has a net charge of -4 and so an assembly of such tetrahedra is electrically impossible. The only mineral composed only of silicon and oxygen is quartz, SiO_2, in which adjacent tetrahedra share all of the O^{--} ions. Minerals that contain isolated SiO_4^{-4} tetrahedra, such as olivine, also have positive metal ions in their crystal structures that bond these tetrahedra together, producing electrical neutrality.

36.2. The silicon-oxygen bond is a very strong one, hence the more oxygen atoms per SiO_4 tetrahedron that are shared by neighboring tetrahedra, the more stable the resulting mineral. On the basis of this information, which of the following minerals would you expect to be the most stable and which the least stable? Quartz, olivine, mica, feldspar.

In quartz all the oxygen atoms are shared between neighboring SiO_4 tetrahedra, so it is the most stable mineral of the ones listed. In olivine none of the oxygen atoms in the SiO_4 tetrahedra are shared, hence it is the least stable of these minerals.

36.3. The relative proportions of Fe^{++} and Mg^{++} in olivine are extremely variable. Why should this be so?

Fe^{++} and Mg^{++} ions have the same charge and nearly the same size (their radii are respectively 0.74×10^{-10} m and 0.66×10^{-10} m), and so are interchangeable in a given mineral structure. The relative proportions of iron and magnesium in a given olivine specimen depend upon the chemical and physical conditions under which the mineral was formed.

36.4. What property of Al^{+++} ions makes it possible for them to substitute for some of the Si^{+4} ions in many silicate minerals?

The two ions are nearly the same size (Al^{+++} has a radius of 0.51×10^{-10} m, Si^{+4} has a radius of 0.42×10^{-10} m) and so four O^{--} ions (radius 1.4×10^{-10} m) can fit as well around an Al^{+++} ion as around a Si^{+4} ion.

36.5. Certain silicate minerals, such as pyroxene, consist of chains of SiO_4 tetrahedra in which adjacent tetrahedra share an oxygen atom. (*a*) What is the ratio of silicon atoms to oxygen atoms in such minerals? (*b*) How is the requirement of electrical neutrality met by such minerals?

(*a*) Each SiO_4 tetrahedron shares two of its oxygen atoms with adjacent tetrahedra, so the chain contains three oxygen atoms per silicon atom.

(*b*) The SiO_3 chains are cross-linked by metal ions.

36.6. The feldspar minerals are abundant in the earth's crust. Describe the structures and compositions of the feldspars.

A feldspar consists of a three-dimensional assembly of SiO_4 tetrahedra in which some of the Si^{+4} are replaced by Al^{+++} ions plus other metal ions to maintain electrical neutrality. Thus in orthoclase each Al atom is accompanied by a potassium atom; the chemical formula of orthoclase is $KAlSi_3O_8$. In the plagioclase feldspars sodium or calcium atoms or both are present to supplement aluminum atoms in replacing some of the silicon atoms. The compositions of these feldspars range from $NaAlSi_2O_8$ through various combinations of Na and Ca to $CaAl_2Si_2O_8$.

36.7. In certain minerals, such as beryl, rings of six Si atoms occur. (*a*) How many O atoms are associated with each ring? (*b*) What is the net charge of each ring? (*c*) In beryl, the silicate rings are linked by Be^{++} and Al^{+++} ions in the ratio of three Be^{++} ions to two Al^{+++} ions. What is the chemical formula of beryl?

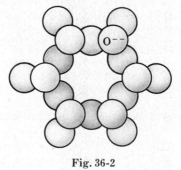

O^{--}

(*a*) There are 18 oxygen atoms in the ring configuration, with the six silicon atoms in the center of the six tetrahedra (Fig. 36-2).

(*b*) Since each ring consists of six Si^{+4} ions and 18 O^{--} ions, the total charge is −12.

(*c*) Each Be_3Al_2 group has a charge of +12, hence the chemical formula of beryl is $Be_3Al_2 \cdot Si_6O_{18}$.

Fig. 36-2

36.8. Igneous rocks rich in silicon are often called *silicic*, and rocks in which iron and magnesium are abundant are often called *mafic*. (*a*) Give examples of silicic and mafic rocks. (*b*) In general, silicic and mafic rocks are different in color. What is the difference? (*c*) Which are denser, silicic or mafic rocks?

(*a*) Silicic: granite and rhyolite; mafic: gabbro and basalt. (*b*) Silicic rocks are light in color, mafic rocks are dark in color. (*c*) Mafic rocks are denser.

36.9. Obsidian is a rock which resembles glass, in particular by sharing the property that its structure is closer to that of a liquid than to that of a crystalline solid. What does this observation suggest about the manner in which obsidian is formed?

To have the amorphous structure of a liquid, obsidian must have solidified so rapidly that crystals had no chance to develop. This can have occurred only by the cooling of a lava flow at the earth's surface.

36.10. Granite and rhyolite have similar compositions but granite is coarse-grained whereas rhyolite is fine-grained. What does the difference in grain size indicate about the environments in which each rock formed?

Large mineral grains can form only during slow cooling, hence granite must have solidified deep inside the crust. Small mineral grains occur when cooling is rapid, hence rhyolite must have solidified at or near the earth's surface.

36.11. Shale is a sedimentary rock that consolidated from mud deposits. What are the various metamorphic rocks that shale can become under progressively increasing temperature and pressure?

In order of increasing metamorphism, shale can become slate, schist, and gneiss.

36.12. (a) What is the origin of limestone? (b) What rock does limestone metamorphose into?

(a) Limestone is produced both by consolidation of shell fragments and by precipitation of $CaCO_3$ from solution. (b) Marble.

36.13. The mineral grains of many metamorphic rocks are flat or elongated and occur in parallel layers. (a) What is this property called? (b) How does it originate?

(a) *Foliation.* (b) Foliation occurs when the minerals of a rock recrystallize under great pressure, which causes them to grow out perpendicular to the direction of the stress.

36.14. What happens to the density of a rock that undergoes metamorphism?

The density increases because the pressures under which metamorphism occurs lead to more compact rearrangements of the atoms in the various minerals.

Supplementary Problems

36.15. What is the relationship between rocks and minerals?

36.16. What is the most abundant mineral in the earth's crust?

36.17. What gives asbestos its characteristic fibrous structure?

36.18. In quartz each SiO_4 tetrahedron shares all of its oxygen atoms with adjacent tetrahedra. What is the ratio between the numbers of oxygen and silicon atoms in quartz?

36.19. Olivine is a silicate mineral in which the SiO_4 tetrahedra do not share oxygen atoms with each other but are bonded together by iron and magnesium ions. (a) From this information would you expect olivine to exhibit cleavage? (b) Would you expect olivine to be light or dark in color?

36.20. The silicate mineral mica is readily broken apart into thin flakes. What does this observation indicate about the crystal structure of mica?

36.21. Arrange the three classes of rock in order of their abundance in the crust.

36.22. In what type of rocks are fossils found?

36.23. Diorite is an intrusive rock and andesite, whose composition is similar, is an extrusive rock. How can they be distinguished apart?

36.24. What are the three most common types of cemented-fragment (or *clastic*) sedimentary rocks? What distinguishes them from one another?

36.25. What is the origin of chert and why is it so resistant to chemical and mechanical attack?

36.26. Limestone and marble have the same composition. What is the difference in structure that meta-morphism produces when limestone becomes marble?

36.27. Gneiss is formed at much greater depths than slate. Which rock would you expect to have the greater density?

36.28. In what type of rocks is foliation found?

Answers to Supplementary Problems

36.15. A rock is an aggregate of grains of one or more minerals.

36.16. feldspar

36.17. Chains of SiO_4 tetrahedra linked by the sharing of O^{--} ions are responsible for the fibrous structure of asbestos.

36.18. There are two O atoms per Si atom in quartz.

36.19. (a) No cleavage; when struck, olivine breaks irregularly. (b) The presence of iron leads to a dark color.

36.20. The SiO_4 tetrahedra in mica are bonded into flat sheets, which comes about because each tetrahedron shares three of its O^{--} ions with neighboring tetrahedra in the same plane.

36.21. Igneous, metamorphic, sedimentary.

36.22. Fossils are found in sedimentary rocks.

36.23. Diorite is coarse-grained and andesite is fine-grained.

36.24. Conglomerate, sandstone, and shale; they are distinguished according to grain size, from large (> 2 mm) through medium (1/16 mm to 2 mm) to small (< 1/16 mm) in the above order.

36.25. Chert is a chemical precipitate. It consists largely of microscopic quartz crystals and hence is hard and durable.

36.26. The grains become larger.

36.27. gneiss

36.28. Metamorphic rocks.

Chapter 37

Erosion and Sedimentation

WEATHERING

The gradual disintegration of exposed rocks is called *weathering*. The chief mechanism of *mechanical weathering* is the freezing of water in crevices in rocks; water expands as it turns into ice, and considerable forces can be developed in this way. Surface water, which is slightly acid due to the formation of carbonic acid from dissolved CO_2 and to the presence of organic acids from decaying plant and animal matter, is the principal agent of *chemical weathering*. The weakly acid water turns feldspar into clay minerals, dissolves the calcite of limestone, and otherwise attacks many of the minerals of common rocks; quartz is the mineral that is least susceptible to chemical weathering.

STREAMS

The most powerful agent of erosion is the running water of streams and rivers, whose work is aided by the pebbles and stones carried in their flow. When the slope of a stream is steep, it carves a narrow V-shaped valley, with the debris being transported away by the swiftly moving water. At lower elevations where the slope is more gradual, the stream tends to widen rather than deepen its bed, and some of its load of *alluvium* (erosional debris) is deposited on the broad *flood plain* of the resulting valley. A *delta* may be formed from alluvium where a stream empties into a lake, sea, or ocean. Deposition is also common where a stream leaves a steep valley and slows down as it enters a plain; the resulting cone of sand and gravel pointing upstream is called an *alluvial fan*. The most widespread accumulations of sediments are found in the shallow parts of the oceans adjacent to the continents.

SEDIMENTS

The physical characteristics of sediments often reveal something about their origin. A smoothly flowing stream tends to deposit first the coarse fragments it carries, with finer and finer particles settling out farther and farther along its path. Thus a *well-sorted* sediment, which consists of particles of very nearly the same size, indicates that such a stream was responsible. On the other hand, a turbulent stream usually leaves a *poorly-sorted* sediment composed of particles of all sizes mixed together.

The type of *bedding*, or layering, of a sediment also provides clues as to the conditions under which it was formed. Slight changes in the composition or grain size of a sediment produce layers of different appearance, which may lie parallel to one another or at different angles. The latter situation, called *cross-bedding*, can arise from turbulent flow which produces patterns of ripples on the channel bottom; when the angles between the beds are steep, the deposits are more likely to have been laid down by winds in the form of dunes than by streams.

GLACIERS

A *glacier* is a large mass of ice that has formed from the recrystallization under pressure of accumulated snow. *Valley glaciers* occur in mountain valleys originally cut by streams, and flow slowly downhill to melt at their lower ends. A valley glacier grinds out a characteristic U-shaped trough with a round, steep-walled *cirque* at its head.   The piled-up debris at the foot of a glacier is called a *moraine*.

Continental glaciers, or *icecaps,* that are thousands of feet thick cover most of Greenland and Antarctica; motion in such an icecap is from the center outward, and icebergs are fragments that have broken off the edges into the sea.   Similar sheets of ice extended across Canada and northern Eurasia in relatively recent geological history.

WIND AND WAVES

In desert regions wind acts as an erosional agent by virtue of the sand it carries along the surface, but its effects are minor compared with those of the flash floods caused by the occasional rain storms.   Wind is significant in shaping certain landscapes for another reason, its ability to transport fine silt for considerable distances while leaving behind dunes of heavier sand particles.   Deposits of windblown silt are called *loess.*

Ocean waves battering a coastline gradually wear it away to form beaches and cliffs, but, as in the case of winds, the more important effect of their activity lies in the redistribution of sediments produced by other agents.

GROUNDWATER

The outer part of the earth's crust is permeable to water, and a considerable fraction of the rainwater reaching the ground is absorbed.   The *water table* is the upper surface of the *saturated zone* in which the pore spaces of the rocks are filled with water.   Groundwater moves slowly in the saturated zone and emerges in hillside *springs* where channels exist to the saturated zone and in streams, lakes, and swamps where the water table is above ground level.   A *well* is a hole dug deep enough to reach to penetrate the water table and so make groundwater accessible.

The carbonate minerals of limestone and dolomite are soluble in groundwater made slightly acid by dissolved CO_2, and underground caverns are produced by the action of groundwater in regions containing such rocks.   The roof of a cavern near the surface may collapse to form a *sinkhole;* a distinctive *Karst topography* occurs in places where underground channels, caverns, and sinkholes are abundant.

Groundwater contains traces of minerals in solution besides calcium carbonate, such as silica and ferric oxide, all of which, when precipitated, act as cementing materials in the hardening of sediments to form rock.   Other processes also enter into *lithification,* such as the compaction of sediments by the pressure of overlying deposits and the recrystallization of certain of the minerals they contain.   *Veins* are formed by the precipitation of various substances from groundwater in rock fissures.

Solved Problems

37.1.   What is the source of energy that makes possible the erosion of landscapes?

   The ultimate source of the energy that goes into erosion is the sun.   Solar energy evaporates surface water, some of which subsequently falls as rain and snow on high ground.   The gravitational potential energy of the latter water turns into kinetic energy as it flows downhill, and some of the kinetic energy becomes work done in eroding landscapes along its path.

37.2.   Why are igneous and metamorphic rocks in general more susceptible to chemical weathering than sedimentary rocks?

Igneous and metamorphic rocks were formed under conditions of heat and pressure very different from those at the earth's surface, and minerals stable under the former conditions are not necessarily stable under the latter. Most sedimentary rocks, on the other hand, consist of rock debris that has already undergone chemical weathering, and so are relatively resistant to further attack. The chief exception is limestone, which is soluble in water that contains carbon dioxide.

37.3. Granite consists of feldspars, quartz, and ferromagnesian minerals. (*a*) What becomes of these minerals when granite undergoes weathering? (*b*) What kinds of sedimentary rocks can the weathering products form?

(*a*) The weathering of granite produces quartz fragments, clay minerals, and ferric oxides, plus K^+, Na^+, Mg^{++}, and Ca^{++} ions in solution.

(*b*) The quartz fragments can become sandstone or, when mixed with clay minerals, shale; the Ca^{++} and Mg^{++} ions can become limestone or dolomite.

37.4. Trace the evolution of the landscape of an initially uplifted region in which stream erosion is the chief geological process. Why is this cycle no longer believed to be widely applicable in its simplest form?

In the young landscape, streams have steep gradients and cut narrow, deep valleys in the predominantly high land mass. As time goes on, tributary streams develop and the region is carved into an intricate pattern of ridges and valleys. This mature landscape is eventually worn down into a series of broad flood plains covered with alluvium and separated by low hills. With old age the landscape becomes a low, rolling *peneplain* near sea level.

The main reason the above cycle, though useful in classifying landscapes, has fallen out of favor with geologists is that there are very few regions in which geological processes involving uplift do not occur at the same time as stream erosion. Thus most actual landscapes are the result of a complex of different factors, and reflect a balance among them rather than the action of stream erosion alone.

37.5. Material deposited by a glacier is called *till*. What are the distinctive properties of till?

Till is a mixture of debris ranging from fine, claylike material to large boulders. Till is neither sorted in size nor deposited in distinct layers as other sediments are. Some of the boulders exhibit flattened, scratched faces where they were scraped against bedrock.

37.6. Why do flash floods occur in deserts and why are they so important in forming desert landscapes despite their infrequency?

In a desert region there is little soil or vegetation to absorb water from a rainfall, so most of it remains on the surface to flood channels that are dry at other times. The rapid flood waters are effective erosional agents and carry considerable loads of sediments, which are often deposited in alluvial fans at the feet of desert mountain ranges since no permanent watercourses exist to carry them to the sea. The floodwaters themselves eventually evaporate. Wind is much less able to affect landscapes than running water, hence desert landscapes reflect the action of flash floods to a greater extent than that of winds.

37.7. Why are clay minerals and quartz particles abundant in sediments which have not been chemically deposited?

Quartz is resistant to chemical attack and so survives weathering and erosion. Feldspar, the most common mineral, is converted into clay minerals by the "carbonic acid" of surface waters.

37.8. What properties would you expect to find in the sediments deposited as an alluvial fan where a steep stream emerges onto level ground?

The deposited material will consist of irregularly bedded sand and gravel or clay, along with rounded pebbles and stones.

37.9. What is the distinction between the *porosity* and the *permeability* of a rock bed?

Porosity refers to the total volume of pore space in a rock bed; the greater the porosity of a bed, the more water it can contain. *Permeability* refers to the relative ease with which water can move through a rock bed; the greater the permeability of a bed, the faster water can flow through it.

37.10. Draw a cross-sectional view of a region underlain by permeable rock and show the water table, a well, a lake, and a spring.

See Fig. 37-1.

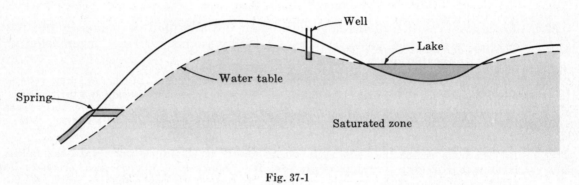

Fig. 37-1

37.11. What is "hard" water? How can it be made "soft"?

"Hard" water contains dissolved minerals which prevent soap from forming suds, react with soap to produce a precipitate, and form insoluble deposits in boilers. Calcium and magnesium ions are usually responsible for hard water. Groundwater often contains these ions through the solvent action of water containing dissolved CO_2 on rocks such as limestone. To soften water, the Ca^{++} and Mg^{++} ions must be removed, which can be done in a variety of ways. In one common method, hard water is passed through a column containing a mineral called zeolite, which absorbs Ca^{++} and Mg^{++} ions into its structure while releasing an equivalent number of Na^+ ions. Since Na^+ ions do not affect soap, nor do sodium compounds precipitate out from hot water, the water is now soft.

37.12. What is *evapotranspiration* and why is it important in the water budget of a region?

Evapotranspiration refers to the return to the atmosphere of water from the land by direct evaporation and by transpiration by plants, which is the process by which water is drawn into a plant through its roots and given off to the atmosphere through its leaves. If annual evapotranspiration in a region exceeds the water deposited annually by precipitation, the water needed for normal plant growth will have to be provided by irrigation.

37.13. When CO_2 is dissolved in water, some of it reacts with the water to form H^+ and HCO_3^- ions. The resulting slightly acid solution is commonly referred to as a "carbonic acid solution," although it is unlikely that any H_2CO_3 exists as such. Calcite ($CaCO_3$) reacts with "carbonic acid" to form the much more soluble calcium bicarbonate, $Ca(HCO_3)_2$. What is the formula for the overall reaction by which calcite goes into solution as calcium and bicarbonate ions?

$$CaCO_3 + H_2O + CO_2 \longrightarrow Ca^{++} + 2HCO_3^-$$

Supplementary Problems

37.14. In what way is the weathering of rock important to terrestrial life?

37.15. What common rocks are almost immune to chemical weathering?

37.16. What agent of erosion produces valleys with a V-shaped cross-section?   A U-shaped cross-section?

37.17. Where would you expect to find Karst topography?

37.18. Under what circumstances does a glacier form?

37.19. Glaciers are observed to wear down bedrock that is harder than glacial ice. How can this happen?

37.20. Glaciers grind away rock with far more force than rivers or streams, yet running water has had more influence in shaping landscapes around the world than glaciers have. Why?

37.21. What is the eventual site of deposition of most sediments?

Answers to Supplementary Problems

37.14. The rock debris produced by weathering is the principal constituent of soil.

37.15. Quartz is highly resistant to chemical attack, hence rocks which are largely quartz, such as chert and many quartzites, are not subject to chemical weathering.

37.16. streams; glaciers

37.17. Karst topography is found in regions of abundant rainfall that are underlain by carbonate rocks such as limestone and dolomite.

37.18. A glacier forms when the average annual snowfall in a region exceeds the annual loss by evaporation and melting.

37.19. Embedded in glaciers are stones and boulders, some of which are hard enough to erode the bedrock.

37.20. Glaciers were insignificant or absent during most of the earth's history except for relatively brief "ice ages"; even today glaciers are active over only a small (10% or so) proportion of the earth's land area.

37.21. The ocean floor.

Chapter 38

Vulcanism and Diastrophism

VOLCANOES

Molten rock is called *magma* when it is underground and *lava* when it is on the surface. A *volcano* is an opening in the earth's crust through which lava, rock fragments, and hot gases emerge. Since lava cools and hardens relatively near the opening, a characteristically conical mountain is usually built up. Most volcanoes are only intermittently active. Explosive eruptions occur when the magma is highly viscous and has a large gas content; a less viscous magma with a small gas content flows out more or less quietly.

Volcanic rocks are fine-grained because lava cools too rapidly for large mineral crystals to form, and often they contain holes where gas bubbles were trapped; *pumice* is an extreme example of the latter effect. In explosive eruptions bits of magma are blown out which solidify into fragments of various sizes. Deposits of the finer fragments may consolidate into *tuff*, and deposits of the coarser ones into *breccia*.

The majority of today's volcanoes occur in a band that encircles the Pacific Ocean and in another that extends from the Mediterranean region across Asia to join the Pacific band in the Indonesian archipelago.

PLUTONS

A *pluton* is a body of magma that has risen through the crust and hardened while still underground. Plutons consist of coarse-grained igneous rocks. A *batholith* is a large pluton that may be several miles thick and extend over thousands of square miles; most of the rock in batholiths is granite. Batholiths are always associated with mountain ranges, either past or present. *Sills* and *laccoliths* are smaller plutons that have intruded parallel to existing strata; *dikes* are wall-like plutons that have intruded into fissures that cut across existing strata.

DIASTROPHISM

Diastrophism refers to movements of the solid rock of the earth's crust. An example is the relative displacement of the rocks on both sides of a fracture; a fracture along which such a slippage has occurred is called a *fault*. The commonest cause of earthquakes is the sudden dislocation of rocks along a fault. Sometimes a crustal segment subjected to a stress may flow instead of faulting; thus folds in rock strata result from gradual yielding to a horizontal compression.

MOUNTAIN BUILDING

Most of the major continental mountain ranges, such as the Rockies, the Alps, and the Himalayas, are the result of large-scale folding. Smaller mountain ranges have often been produced by faulting which elevated crustal blocks; volcanic action is also responsible for some mountain ranges.

The sedimentary layers in fold mountains typically increase in thickness toward the center of the range, which suggests that where the range now stands was once a large basin that gradually sank as sediments accumulated in it. Such a sinking basin is called a *geosyncline*. Eventually diastrophic activity, probably connected with continental drift (Chapter 40), led to the folding and uplifting of the sedimentary deposits, which were later intruded by granite batholiths and other plutons. It is thought that the granite came into being when the base of the geosyncline descended far enough to melt and the granitic magma then rose to intrude the overlying strata.

Solved Problems

38.1. A *caldera* is a large craterlike depression up to several miles across that is associated with former volcanic activity. How is a caldera formed?

　　　A caldera can be formed in two ways, or by a combination of both: a violent explosion that ejects a vast quantity of material to leave behind a depression; and the collapse of a volcanic cone into an underground cavity left by the loss of magma, either through being blown out or by seeping away.

38.2. What is the chief factor that determines the viscosity of a magma, that is, how readily it flows? What kinds of landscapes are produced by volcanoes whose lavas have relatively high and relatively low viscosities?

　　　The greater the silicon content of a magma, the higher its viscosity and the less readily it flows. Highly viscous lavas (such as andesite) usually produce steep conical mountains and, in general, a rugged landscape; less viscous lavas (such as basalt) spread out to produce more even landscapes.

38.3. What is the chief constituent of volcanic gases?

　　　Steam comprises most of the gases emitted by a volcano. Some of the steam comes from groundwater heated by magma, some comes from the combination of hydrogen in the magma with atmospheric oxygen, and some (called *juvenile water*) was formerly incorporated in rocks deep in the crust and is carried upward by the magma to be released at the surface.

38.4. What is the origin of the glasslike rock obsidian?

　　　Obsidian forms from a rhyolitic lava, which is rich in silica, that cools so rapidly that crystals do not have time to grow. The resulting rock, like glass, has a structure that is essentially that of a liquid.

38.5. What kinds of rocks are likely to be found in (*a*) a batholith and (*b*) a dike?

　　(*a*) Because magma cools slowly in a batholith, coarse-grained igneous rocks such as granite, diorite, and gabbro are likely to be found.

　　(*b*) Magma may cool slowly or rapidly in a dike, depending on the circumstances. Hence both coarse- and fine-grained igneous rocks may be found: granite, diorite, gabbro, rhyolite, andesite, and basalt, for instance.

38.6. What effect does the intrusion of a batholith have on the nature of nearby rocks?

　　　The initial high temperature of a batholith, which persists for a long time since it cools slowly owing to its great mass, produces changes in nearby rocks that constitute *contact* (or *thermal*) *metamorphism*. In general, mineral grains grow larger in such rocks and become interlocked. These changes are especially marked in soft sedimentary rocks: thus sandstone becomes quartzite, shale becomes hornfels, and limestone becomes marble. Contact metamorphism produces unfoliated rocks since only heat is involved and not pressure as well.

38.7. Show the relationships among sedimentary, igneous, and metamorphic rocks, unconsolidated sediments, and magma by means of a diagram.

     See Fig. 38-1.

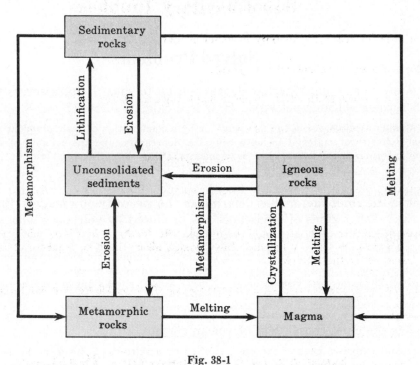

Fig. 38-1

38.8. What are the principal classes of faults?

     A *normal fault* is an inclined surface along which a rock mass has slipped downward. A normal fault is the result of tension in the crust and increases the area covered by the rocks involved.

     A *thrust* (or *reverse*) *fault* is an inclined surface along which a rock mass has moved upward to override the neighboring mass. A thrust fault is the result of compression in the crust and decreases the area covered by the rocks involved.

     A *strike-slip fault* is a surface along which one rock mass has moved horizontally with respect to the other. A strike-slip fault is the result of oppositely-directed forces in the crust that do not act along the same line, so that the result is a distortion of the crust rather than a change in area.

38.9. What topographical features are associated with faults?

     Both normal and thrust faults produce cliffs called *fault scarps*. A strike-slip fault is often marked by a *rift*, which is a trench or valley caused by erosion of the disintegrated rock produced during the faulting.

38.10. Distinguish between a *syncline* and an *anticline* in folded rock strata.

     A *syncline* is a troughlike fold that is concave upward; an *anticline* is an archlike fold that is convex upward.

38.11. Why is it believed that the region where the Rocky Mountains now stand was once near or below sea level?

    The Rocky Mountains contain thick layers of sedimentary rocks that can only have been formed from sediments deposited over a long period of time. Hence the region must once have been low enough for rivers and streams containing erosional debris to flow into it.

Supplementary Problems

38.12. What kinds of rocks are likely to be found in lava flows?

38.13. What is the most common volcanic rock?

38.14. What is the cause of the holes found in many volcanic rocks?

38.15. Distinguish between a dike and a vein.

38.16. Why are metamorphic rocks often found near plutons?

38.17. The energy source of erosional processes is the sun. Where does the energy involved in diastrophic activity come from?

38.18. Masses of igneous rock are found to intrude the folded sedimentary and metamorphic rocks of large mountain ranges. What does this suggest about the time sequence of the various events in the formation of these ranges?

38.19. What geological process is chiefly responsible for the topography of a mountain range?

Answers to Supplementary Problems

38.12. rhyolite, andesite, basalt, obsidian

38.13. basalt

38.14. Such holes were produced by bubbles of gas trapped in lava as it solidified.

38.15. A dike consists of molten rock that has intruded into a fissure and hardened there; a vein consists of material that has precipitated in a fissure from solution in groundwater.

38.16. The intruded magma that solidifies into a pluton is very hot, and thus nearby rocks often undergo thermal metamorphism.

38.17. The earth's interior.

38.18. The first phase was the deposition of sediments in a geosyncline and their hardening into rocks, then the folding and raising of the sedimentary layers, and finally the intrusion of plutons.

38.19. erosion

Chapter 39

The Earth's Interior

SEISMIC WAVES

The majority of earthquakes are caused by the sudden displacement of crustal blocks along a fault, which relieves stresses that have built up over a period of time. An earthquake gives rise to waves of three kinds:

1. *Primary* (or *P*) *waves*, which are longitudinal waves that involve back-and-forth vibrations of particles of matter in the same direction as that in which the waves travel. P waves are essentially pressure waves, like sound waves.

2. *Secondary* (or *S*) *waves*, which are transverse waves that involve vibrations of particles of matter perpendicular to the direction in which the waves travel. The waves produced by shaking a stretched string are similar to S waves. Both P and S waves are body waves that travel through the earth's interior. S waves cannot occur in a liquid, though P waves can.

3. *Surface* (or *L*) *waves*, which are analogous to water waves, involve orbital motions of particles of matter and are limited to the earth's surface.

The three kinds of seismic waves travel with different velocities and so arrive at a distant observer at different times after an earthquake; P waves are the first, next S waves, and finally L waves. The velocities of P and S waves increase with depth, so that refraction causes them to travel along curved paths through the earth. In addition, both refraction and reflection occur at boundaries between regions with different physical properties. The analysis of seismic waves received at observatories around the world has led to the identification of three principal regions within the earth.

INTERIOR STRUCTURE

The earth consists of a central *core* 2160 mi (3500 km) in radius; a surrounding *mantle* 1800 mi (2900 km) thick; and a *crust* whose thickness ranges from typically 5 mi under the oceans to an average of 20 mi under the continents.

The core, which constitutes 19% of the earth's volume, is thought likely to be composed largely of molten iron with some nickel and perhaps traces of sulfur and silicon. The inability of S waves to travel through the core is strong evidence for its liquid nature, since transverse waves can only be transmitted by solid materials. The inner part of the core behaves differently from the rest and is probably solid; this *inner core* has a radius of about 780 mi (1300 km).

The mantle, which constitutes 80% of the earth's volume and about 67% of its mass, is solid and is thought likely to be composed of ferromagnesian silicate minerals such as olivine, pyroxene, and garnet. The crust consists of a global layer several miles thick of simatic (basaltic) rock with thicker layers of sialic (granitic) rock under the continents. The crust-mantle boundary is called the *Mohorovičić discontinuity*.

LITHOSPHERE AND ASTHENOSPHERE

The crust and the outermost part of the mantle comprise a rigid shell of rock 50 to 100 km thick called the *lithosphere*. A layer about 100 km thick in the mantle below the lithosphere, known as the *asthenosphere,* is apparently capable of plastic flow, unlike the rigid lithosphere and the rest of the mantle. Stresses applied to the asthenosphere over a long period of time, such as the weight of a continental block or the horizontal forces exerted during continental drift, cause the asthenosphere to flow gradually. Short-period stresses, such as those produced by an earthquake, are transmitted in the same way as in a rigid material, though with a reduced velocity.

The lithosphere in essence floats on the asthenosphere, with irregularities such as the continental blocks being supported by their buoyancy in the denser material of the asthenosphere. This concept, called *isostasy*, is supported by the observation that the higher a mountain range is, the deeper are its roots.

GEOMAGNETISM

The earth's magnetic field closely resembles the field that would be produced by a giant bar magnet located near the earth's center and tilted by 11° from the axis of rotation. Since the core is molten, no such magnet can actually exist there. Instead, the field is believed to arise from coupled fluid motions and electric currents in the liquid iron of the core; a current in the form of a loop is surrounded by a magnetic field of the same form as that of a bar magnet. The geomagnetic field cannot originate in the mantle because it is a nonconductor, hence no electric currents can exist there, and because it is too hot for a ferromagnetic substance such as iron to retain its magnetism.

Measurements of the magnetization of crustal rocks indicate that the geomagnetic field has often reversed its direction in the past. Such reversals seem consistent with the hypothesis that the field is due to electric currents in the core, since changes may well occur in the patterns of flow of the liquid iron there from time to time.

Solved Problems

39.1. The travel times of seismic waves depend only on the distance between earthquake and observing station and do not vary around the earth for the same such distance. What does this indicate about the uniformity of the material in the earth's interior?

Because travel times depend only on distance and not on location, any variations in the material of the interior can only occur along a radius and not transversely. Thus the division of the interior into an inner core, an outer core, and a mantle, which form concentric shells, is consistent with the above observation, but a difference between the material in, say, the northern and southern hemispheres is ruled out.

39.2. Seismic S waves are never detected beyond about 7000 miles (measured along the earth's surface) from an earthquake. P waves also disappear at this distance, but reappear at distances greater than about 10,000 miles. How do these observations fit in with the hypothesis that the earth has a liquid core?

The presence of a liquid core affects both S and P waves: the S waves cannot travel through it at all, and the P waves travel in it with a different velocity than in the mantle. The latter fact means that P waves entering the core change their directions due to refraction, as shown in Fig. 39-1. As a result a "shadow zone" occurs in a band around the earth in which P waves are not found, and there is a still larger region in which S waves are absent since they are absorbed in the core. (The velocities of P and S waves vary with depth, which causes their paths to be curved due to refraction, in addition to the sharp change in direction of P waves at the core-mantle boundary.)

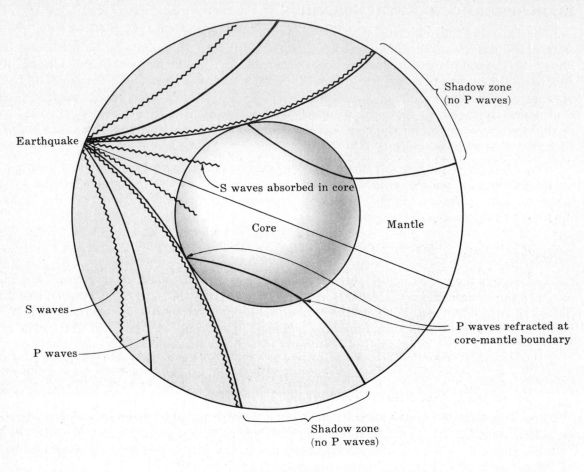

Fig. 39-1

39.3. The earth's radius is 6.4×10^6 m and its mass is 6.0×10^{24} kg (see Problem 4.11). Calculate the average density of the earth as a whole and compare it with the average density of crustal rocks, 2.7 g/cm³. What does this comparison indicate about the composition of the earth's interior?

The volume of a sphere of radius r is $\frac{4}{3}\pi r^3$, hence the earth's volume is

$$V = \frac{4}{3}\pi r^3 = \frac{4}{3} \times \pi \times (6.4 \times 10^6 \text{ m})^3 = 1.1 \times 10^{21} \text{ m}^3$$

and its density is

$$d = \frac{\text{mass}}{\text{volume}} = \frac{6.0 \times 10^{24} \text{ kg}}{1.1 \times 10^{21} \text{ m}^3} = 5.5 \times 10^3 \frac{\text{kg}}{\text{m}^3} = 5.5 \frac{\text{g}}{\text{cm}^3}$$

The average density of the earth as a whole is about twice the average density of crustal rocks, which signifies that the materials of the earth's interior are much denser than those of its surface.

39.4. What is the difference between the earth's crust and its lithosphere?

The crust is distinguished from the mantle beneath it by a sharp difference in seismic wave velocity, which suggests a difference in the composition of the minerals involved, or in their crystal structure, or in both. The lithosphere is distinguished from the asthenosphere beneath it by a difference in their behaviors under stress: the lithosphere is rigid whereas the asthenosphere is capable of plastic flow.

39.5. List some of the considerations that have led to the belief that the material of the mantle is similar to such ultramafic (mainly composed of ferromagnesian minerals) igneous rocks as eclogite and peridotite.

(1) Only dense rocks such as these transmit seismic waves in the manner exhibited by the mantle. (2) These rocks are similar in composition to the basalt of the lower crust but are sufficiently different to be able to give rise to the observed Mohorovičić discontinuity in seismic wave transmission between crust and mantle. (3) Most volcanic magmas come from the upper mantle, and their composition is consistent with an origin in eclogite or peridotite. (4) Diamonds can form only under conditions of high temperature and pressure, such as are found in the mantle but not in the crust, and eclogite and peridotite are common in diamond-bearing rock formations which must have originated in the mantle. (5) Stony meteorites consist chiefly of the minerals olivine and pyroxene, just as these rocks do, and it is an attractive notion that both the mantle and stony meteorites had the same origin in the early solar system.

39.6. (*a*) Why is it considered likely that the earth's outer core is liquid? (*b*) Why is the liquid thought to be largely iron? (*c*) Why is nickel believed to be present as well?

(*a*) Transverse waves cannot propagate through a liquid, and it is observed that seismic S waves, which are transverse, are unable to pass through the core although P waves, which are pressure waves, are able to. In support of the idea that the core is a liquid is the observation that the earth's magnetic field, which originates in its interior, fluctuates in both magnitude and direction, which is hard to explain if the interior is solid but easy to explain if some of the interior is an electrically conducting liquid.

(*b*) Iron is a fairly abundant element in the universe; its density is just about right for an iron core to account for the total mass of the earth, given the ferromagnesian silicate composition of the mantle; and iron is a good conductor of electricity, which is necessary in order to explain the origin of the earth's magnetic field.

(*c*) Meteorites that contain iron always contain a small proportion of nickel as well, which suggests that these two metals occur together in the solar system.

39.7. What evidence is there in favor of the idea that the earth's interior is very hot?

(1) Measurements made in mines and wells indicate that temperature increases with depth. (2) Molten rock from the interior emerges from volcanoes. (3) The outer core is liquid, which means it must be at a high temperature.

39.8. What origins are possible for the high temperatures found in the earth's interior?

The decay of radioactive isotopes inside the earth evolves considerable energy which is manifested as heat. Moreover, when the earth was formed, the initial matter came together and contracted under the influence of gravity into a tightly packed aggregate. In this process the initial gravitational potential energy became heat, and some of this heat may still be present since the mantle is a poor conductor.

39.9. The earth's magnetic field is very nearly the same as that which would be produced by a loop of electric current located in the interior of the earth. Assuming that the field is exactly the same as that of a current loop, what is the simplest procedure you could use to find the locations of the geomagnetic poles, which are those points where the axis of the loop intersects the earth's surface?

A magnetized needle suspended so it can move freely will point vertically downward at the geomagnetic poles since the direction of the magnetic field at these points is along the axis of the loop.

Supplementary Problems

39.10. What are the similarities between seismic P and S waves? What are the differences?

39.11. How is it possible to tell from the seismograph record of the waves from an earthquake how far away the earthquake occurred?

39.12. Where is the earth's crust thinnest? Where is it thickest?

39.13. What evidence is available from seismic wave studies that supports the existence of the asthenosphere?

39.14. How is a plastic asthenosphere possible with a rigid lithosphere above it and a rigid mantle below it?

39.15. How does the radius of the earth's core compare with the radius of the earth as a whole? How does the mass of the core compare with the mass of the earth as a whole?

39.16. Why does a compass needle usually not point due north?

39.17. Why is it unlikely that the earth's magnetic field originates in a huge bar magnet located in its interior?

Answers to Supplementary Problems

39.10. P and S waves both can travel through a solid medium, but only P waves can travel through a liquid. P waves are longitudinal and, like sound waves, consist of pressure fluctuations; S waves are transverse and are analogous to waves in a stretched string. P waves are faster than S waves in the same medium.

39.11. The P and S waves of an earthquake require different lengths of time to go from their source to a particular location on the earth's surface. The time interval between arrivals of each kind of wave at an observatory can be compared with the known travel-time-versus-distance data to find the distance that corresponds to this interval.

39.12. The crust is thinnest under the oceans and thickest under the continents.

39.13. The velocity of seismic waves is lower in the asthenosphere than above or below it in the mantle, which suggests that its physical properties are different. The difference is consistent with a plastic rather than a rigid character for the asthenosphere material.

39.14. The asthenosphere is plastic because its material is close to its melting point under the conditions of temperature and pressure found in that region of the mantle. Above the asthenosphere the temperature is too low, and below it the pressure is too high, for the material of the mantle to be plastic.

39.15. The core's radius is about 55% of the radius of the earth; the core's mass is about one-third the mass of the earth.

39.16. The axis of the earth's magnetic field is inclined with respect to its axis of rotation, hence the geomagnetic and geographic north poles are not in the same place.

39.17. Ferromagnetic materials lose their magnetic properties at high temperatures, and sufficiently high temperatures exist throughout all of the earth's interior except near the surface of the crust. Also, both the direction and strength of the field are observed to vary, and in fact the field has reversed its direction many times in the past, which cannot be reconciled with the notion of a permanent magnet in the interior.

Chapter 40

Continental Drift

THE OCEAN FLOORS

The earth's crust is subject not only to vertical changes, with entire regions being thrust upward and other regions subsiding, but also to horizontal changes, with the continents continually shifting their positions around the earth. The evidence for the latter events is recent but conclusive. Some of the major findings concern the ocean floors:

1. The ocean floors are relatively recent in origin; the oldest sediments date back only about 135 million years, in contrast to continental rocks which date back as much as 4000 million years. Many parts of the ocean floor are much younger still, so that about one-third of the earth's surface has come into existence in 1.5% of the earth's history.

2. A worldwide system of narrow *ridges* and somewhat broader *rises* runs across the oceans. An example is the Mid-Atlantic Ridge, which virtually bisects the Atlantic Ocean from north to south; Iceland, the Azores, and Ascension Island are some of the higher peaks in this ridge.

3. The direction of magnetization of ocean-floor rocks is the same along strips parallel to the midocean ridges, but the direction is reversed from strip to strip going away from a ridge on either side.

4. A system of *trenches* several km deep occurs around the rim of the Pacific Ocean; it coincides with the belt in which most current earthquakes and volcanoes occur. The trenches have *island arcs* on their landward sides that consist of volcanic mountains projecting above sea level.

PLATE TECTONICS

The preceding observations are accounted for by the theory of *plate tectonics*. According to this theory, the lithosphere is divided into seven very large *plates* plus a number of smaller ones, all of which float on the plastic asthenosphere. Three kinds of events may occur at a boundary between adjacent plates:

1. **Plate creation.** The plates move apart at several cm per year and molten rock rises to form new ocean floor on either side. A midocean ridge occurs at this boundary and consists of the latest rock to be deposited. As the new rock hardens, it is magnetized in the same direction as the geomagnetic field at the time; since this field reverses its direction several times per million years on the average, the result will be strips with alternate magnetization on both sides of the ridge, as observed.

2. **Plate destruction.** One plate slides underneath the other and melts when it reaches the mantle. An oceanic trench is produced at such a *subduction zone*. Volcanoes and island arcs are formed where the less dense components of the descending plate margin melt and rise to the surface. When continental blocks on adjacent plates are pressed together in a subduction zone, they are too light relative to the underlying material for either to be forced under, and instead they buckle to produce a mountain range.

3. **Plate motion.** At some boundaries the adjacent plates are simply sliding past each other without colliding or moving apart. These boundaries, where only horizontal motion occurs, are called *transform faults*. An ideal plate would thus have one edge growing at a ridge, the opposite edge disappearing into the mantle, and the other two edges sliding past the edges of adjoining plates.

CONTINENTAL DRIFT

As the various lithospheric plates shift around the earth, growing at some margins and being destroyed at others, the continents shift with them. Geological, magnetic, climatic, and biological evidence have made possible the reconstruction of past arrangements of the continents and the prediction of future ones. It seems likely that about 200 million years ago there was only a single supercontinent, Pangaea, and a single ocean, Panthalassa. After some millions of years Pangaea began to break apart into Gondwanaland (South America, Africa, Antarctica, India, and Australia) and Laurasia (Eurasia, Greenland, and North America); the Tethys Sea came into being between them. South America and Africa then broke off as a unit from the rest of Gondwanaland, and later they separated as the South Atlantic Ocean came into being. By about 65 million years ago the Atlantic Ocean had completed its extension northward, Australia had separated from Antarctica, and India had begun to drift toward Asia.

Some tens of millions of years from now the Atlantic Ocean will be wider than it is today and the Pacific will be narrower. Australia will have moved north. California will have been detached from the rest of North America, the Arabian peninsula will be attached to Asia, the Mediterranean will be much smaller, and East Africa will have broken away from the rest of Africa.

Solved Problems

40.1. Why are the ocean floors so much younger than the continents?

Owing to their low density and consequent buoyancy, the continental blocks are not forced down into the mantle in subduction zones but remain as permanent features of the lithospheric plates they are part of. The ocean floors, on the other hand, are continually being destroyed in such zones, as new ocean floors are deposited at midocean ridges.

40.2. When continental drift was first proposed half a century ago, it was assumed that the continents move through soft ocean floors. Why is this hypothesis no longer considered valid? How does continental drift actually occur?

The ocean floors are extremely rigid, so it is not possible for the continental blocks to move through them. Instead, the continents are each part of a lithospheric plate whose motion is accomplished by the destruction of the plate at one margin and the formation of new plate at the opposite margin.

40.3. How can observations of the magnetization of rocks provide information on continental drift?

When a sediment that includes iron-containing minerals is deposited, or when a magma with iron-containing minerals hardens, it becomes weakly magnetized by the earth's magnetic field. The direction of magnetization is the same as that of the lines of force of the earth's field at the time the rock was formed, hence it indicates the latitude at which the rock was then located and the positions of the magnetic poles then. Under the assumption that the magnetic and geographic axes have never been far apart—which is supported both experimentally and theoretically—it is possible to correlate such paleomagnetic evidence from rocks of different ages and from different locations into a picture of continental drift that agrees with the picture derived from geological findings.

40.4. There is strong evidence that today's continents were once parts of a single super-continent, Pangaea. What can be said about the likelihood that continental drift occurred before Pangaea came into existence?

> The motions of lithospheric plates are on such a huge scale that it is hard to believe they suddenly began only 200 million years ago, which is relatively recent in the geological history of the earth. Hence it is likely that continental drift was taking place even before Pangaea was formed, and in fact there is some evidence (such as the existence of the Ural Mountains) that Pangaea was the result of the coming together of three earlier continents, Gondwanaland, Asia east of the Urals, and a land mass consisting of North America, Greenland, and Europe.

40.5. What kind of biological evidence supports the notion that all the continents were once part of a single supercontinent?

> Until perhaps 180 million years ago fossils indicate that living things of the same kinds occurred everywhere that suitable habitats existed, whereas since that time many plants and animals have evolved differently in different continents.

40.6. Give examples of climatological findings that support the concept of continental drift.

> Glacial deposits and evidence of glacial erosion are found in tropical regions of South America, Africa, and India, which suggests that these regions were once much farther south. Coal is formed from plant debris that accumulates in swamps, and its presence in Antarctica suggests that this continent was once much farther north.

40.7. (a) What mountain ranges of today were once part of the Tethys Sea? (b) What kind of evidence would indicate that the region where these mountains are present was once below sea level?

> (a) The Pyrenees, Alps, and Caucasus of Europe; the Atlas Mountains of North Africa; and the Himalayas of Asia.
>
> (b) Thick deposits of sedimentary rocks; fossils of sea creatures.

40.8. The east coast of South America is a good fit against the west coast of Africa. What sort of evidence would you look for to confirm that the two continents had once been parts of the same land mass?

> If South America and Africa were once joined together, there should be similar geological formations and fossils of the same kinds at corresponding locations along their respective east and west coasts. This is indeed found for material deposited up to about 100 million years ago, which is when these continents must have begun to separate.

40.9. The distance between the continental shelves of the east coast of Greenland and the west coast of Norway is about 800 miles. If Greenland separated from Norway 65 million years ago and their respective plates have been moving apart ever since at the same rate, find the average velocity of each plate.

> Each plate must have moved 400 mi in 65×10^6 years, and so, since $1\,\text{mi} = 1.61\,\text{km} = 1.61 \times 10^5\,\text{cm}$, its velocity is
>
> $$ v = \frac{\text{distance}}{\text{time}} = \frac{400\,\text{mi} \times 1.61 \times 10^5\,\text{cm/mi}}{65 \times 10^6\,\text{years}} = 1\,\text{cm/year} $$

40.10. List the most plausible suggestions that have been made for the mechanism by which lithospheric plates are moved across the earth's surface. Are any of these ideas widely accepted?

(1) Convection currents in the asthenosphere.

(2) The sinking of the edge of a lithospheric plate in a subduction zone pulls the rest of the plate across the earth's surface.

(3) The upwelling magma in a midocean ridge pushes the adjacent plates apart.

(4) The elevated material in a midocean ridge forces the adjacent plates apart by virtue of its weight, much as in the case of a man standing with his feet in two adjacent rowboats.

Each of these ideas is open to serious objections and none is at present widely accepted by geologists. It is possible that several mechanisms contribute to plate motion.

Supplementary Problems

40.11. Which of today's continents were once part of Laurasia? Of Gondwanaland?

40.12. Madagascar is about 2700 million years old whereas Providence Island, which is not far away from it, is 36 million years old. What do these ages suggest about the origins of these islands?

40.13. North America, Greenland, and Eurasia fit quite well together in reconstructing Laurasia, but there is no space available for Iceland. Why is the omission of Iceland from Laurasia reasonable?

40.14. How does the origin of the Himalayas differ from that of the oceanic mountains that constitute the Mid-Atlantic Ridge?

40.15. The San Andreas Fault in California is a strike-slip fault that lies along the boundary between the Pacific and American plates. What does this indicate about the nature of the boundary?

40.16. How would you expect the ages of the South Pacific islands far away from the East Pacific Rise to compare with those near this rise?

40.17. The oldest sediments found on the floor of the South Atlantic Ocean 1300 km west of the axis of the Mid-Atlantic Ridge were deposited about 70 million years ago. What rate of plate movement does this finding suggest?

Answers to Supplementary Problems

40.11. Laurasia: North America, Greenland, Eurasia (except India). Gondwanaland: South America, Africa, Antarctica, Australia, India.

40.12. Madagascar must have broken off from the African continent, whereas Providence Island must have a volcanic origin.

40.13. Iceland is less than 70 million years old, much younger than North America, Greenland, and Eurasia, and was formed after the breakup of Laurasia from magma rising through the rift in the Mid-Atlantic Ridge.

40.14. The Himalayas were thrust upward by the collision of the Indian plate and the Eurasian plate. The Mid-Atlantic Ridge was formed by the upwelling of molten rock.

40.15. The San Andreas Fault is part of a transform fault; the Pacific plate is moving northwestward relative to the American plate.

40.16. The East Pacific Rise marks the rift through which magma wells to the surface to form new ocean floor, hence islands far from the rise were formed earlier than those near the rise.

40.17. 1.9 cm/year

Chapter 41

Earth History

RELATIVE TIME

Before the development of radioactive methods in this century, the geological events that have shaped the earth's surface could only be placed in historical order. The principles of historical geology, the chief of which are listed below, permit a relative time scale to be determined for many such events.

1. The geological processes that occurred in the past are the same as the ones that are occurring today, although not necessarily in the same places or to the same extent. This is the *principle of uniform change,* or *uniformitarianism.*

2. In a sequence of sedimentary rocks, the lowest bed is the oldest and the highest bed is the youngest. The beds were originally deposited in horizontal layers.

3. Folding and faulting occurred later than the youngest bed affected.

4. An igneous rock is younger than the youngest bed it intrudes.

FOSSILS

Fossils are the remains or other traces of living things of the past that are found in rocks. Organisms have evolved continuously from simple to complex forms and have also changed in response to environmental changes, and the fossil record mirrors this progression. Fossils are accordingly useful in historical geology. For instance, a series of rock beds can be arranged in the sequence of their formation on the basis of the fossils they contain, which may not be possible from purely geological considerations. In addition, beds deposited at the same time but in different places can be correlated from the presence in them of fossils of the same kinds. The type of fossil found in a particular bed also reveals something about the local environment at the time the organism was alive: whether the region was dry land or was covered by fresh or salt water, whether the climate was warm or cold, and so forth.

RADIOACTIVE DATING

Methods based on radioactive decay make it possible to establish the ages of many rocks on an absolute rather than a relative time scale. As mentioned in Chapter 19, every radioactive nuclide has a characteristic *half-life* during which half of an initial quantity decays. Half of the remainder decays in the next half-life, and so on. The ratio between the amounts of a certain nuclide and its stable daughter product in a sample therefore indicates the age of the sample; the greater the proportion of the daughter product, the older the sample.

Four radioactive nuclides found in common minerals are especially useful in dating igneous and metamorphic rocks:

Parent nuclide	Stable daughter nuclide	Half-life
Potassium 40	Argon 40	1.3×10^9 years
Rubidium 87	Strontium 87	47×10^9 years
Uranium 235	Lead 207	0.7×10^9 years
Uranium 238	Lead 206	4.5×10^9 years

Potassium 40 and rubidium 87 each decay to a stable daughter in a single step, but both uranium isotopes undergo several successive decays before becoming stable lead isotopes. The carbon 14 method (see **Problem 19.6**) can be used for dating sedimentary deposits that contain fossil carbon up to an age of about 40,000 years.

GEOCHRONOLOGY

Although the evolution of living things is a continuous process, the fossil record shows that three especially marked changes in the patterns of plant and animal life have taken place in the past. These times of change divide the most recent 570 million years of the earth's history into three *eras*: *Paleozoic* ("ancient life"), *Mesozoic* ("intermediate life"), and *Cenozoic* ("recent life"). The nearly 4 billion years from the earth's formation to the start of the Paleozoic era are lumped together into *Precambrian time*. Figure 41-1 shows the currently-accepted division of the eras into periods; the various periods are further subdivided into epochs, of which only those of the Cenozoic era are shown.

Millions of years before the present	Era	Period	Epoch	Duration in millions of years	The biological record	
65	Cenozoic	Quaternary	Recent	0.01	Man becomes dominant	
225			Pleistocene	2.5	Rise of man; large mammals abundant	
570		Tertiary	Pliocene	4.5	Flowering plants abundant	Age of Mammals
			Miocene	19	Grasses abundant; rapid spread of grazing mammals	
			Oligocene	12	Apes and elephants appear	
			Eocene	16	Primitive horses, camels, rhinoceroses	
			Paleocene	11	First primates	
	Mesozoic	Cretaceous		71	First flowering plants; dinosaurs die out	Age of Reptiles
		Jurassic		54	First birds; dinosaurs at their peak	
		Triassic		35	Dinosaurs and first mammals appear	
	Paleozoic	Permian		55	Rise of reptiles; large insects abundant	
		Pennsylvanian } Carboniferous		45	Large nonflowering plants in enormous swamps	
		Mississippian }		20	Large amphibians; extensive forests; sharks abundant	
		Devonian		50	First forests and amphibians; fish abundant	
4000	Oldest rocks	Silurian		35	First land plants and coral reefs	
		Ordovician		70	First vertebrates (fish) appear	
		Cambrian		70	Marine shelled invertebrates (earliest abundant fossils)	
	Precambrian time	Late Precambrian			Marine invertebrates, mainly without shells	
4500		Early Precambrian			Marine algae (primitive one-celled plants)	

Fig. 41-1

Solved Problems

41.1. An *angular unconformity* is an irregular surface that separates tilted lower rock strata from horizontal upper ones. How does such an unconformity originate?

An angular unconformity is a buried surface of erosion that involves at least four events: (1) deposition of the oldest strata; (2) diastrophic movement that raises and tilts the existing strata; (3) erosion of the elevated strata to produce an irregular surface that cuts across their exposed edges; (4) a new period of deposition that buries the eroded surface.

41.2. In Fig. 41-2, beds *A* to *F* consist of sedimentary rocks formed from marine deposits and *G* and *H* are granite. What sequence of events must have occurred in this region?

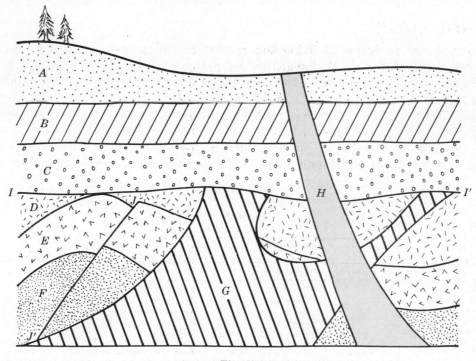

Fig. 41-2

(1) Deposition of beds *E* and *F* when the region was below sea level.

(2) Diastrophic movement that produced the fault *JJ'* and the folds in beds *E* and *F*.

(3) Deposition of bed *D*.

(4) Intrusion of the granite pluton *G*.

(5) Erosion that produced the irregular surface *II'*, which is an unconformity. The region must have been elevated above sea level for this erosion to have occurred.

(6) The region subsided below sea level and beds *A*, *B*, and *C* were deposited.

(7) Intrusion of the granite pluton *H*.

(8) Re-elevation of the region above sea level and the renewed erosion of the surface.

41.3. What is the biological basis for the division of geological time into eras and periods?

The fossil record shows that there were a number of occasions when animal and plant life became sharply reduced in both number and variety, to be followed in each case by the rapid evolution of new types. The intervals of extinction are used to divide geological history into periods; during a typical period there is an expansion of living things followed by a time in which biological change is more gradual, then an interval of extinction ends the period. The division into eras is based on exceptionally marked, worldwide extinctions and subsequent expansions of plant and animal life.

41.4. The earth's history is sometimes divided into two *eons, Cryptozoic* ("hidden life") and *Phanerozoic* ("visible life"), with the first corresponding to Precambrian time and the second extending from the beginning of the Paleozoic era to the present day. What is the reason for this division?

Abundant fossils exist from the Phanerozoic eon, which permit tracing the evolution of living things during this span of time. Few fossils exist from the Cryptozoic eon, making it difficult to determine the forms of life that were present then and how they developed.

41.5. List some of the various kinds of fossils.

(1) Actual plant or animal tissues, usually of a hard nature such as teeth, bones, hair, and shells. Entire insects have been found preserved in amber.

(2) Plant tissues that have become coal through partial decay but which retain their original forms.

(3) Tissues that have been replaced by a mineral (such as silica) from groundwater; petrified wood is an example. Sometimes a porous tissue such as bone will have its pore spaces filled with a deposited mineral.

(4) Impressions that remain in a rock of plant or animal structures that have themselves disappeared.

(5) Footprints, worm holes, or other cavities produced by animals in soft ground that have later filled with a different material and so can be distinguished today.

41.6. Why are most fossils found in beds that were once the floors of shallow seas?

Plant and animal life is abundant in such seas, and dead organisms sink to the bottom where they are soon buried in sediments that protect them from decay. Land organisms rarely leave fossils unless they fall into a swamp or lake, because their remains are subject to chemical and bacterial decay and the attacks of scavengers.

41.7. What are the two basic conditions that must be met by a radioactive nuclide in order that it be useful in dating a particular kind of rock?

The nuclide must occur in at least one of the minerals found in the rock, and it must have a half-life that is roughly comparable with the age of the rock (within a factor of 10 to 100, depending upon the details of the situation).

41.8. The early atmosphere of the earth probably consisted of carbon dioxide, water vapor, and nitrogen, with little free oxygen. What is believed to be the source of the oxygen in the present-day atmosphere? What bearing has this question on the relatively rapid development of varied and complex forms of life that marks the start of the Paleozoic era?

Most of the oxygen in the atmosphere was probably produced by photosynthesis. In Precambrian time photosynthesis was carried out by the blue-green algae that were abundant then; these algae do not require free oxygen, unlike plants. When the oxygen content of the atmosphere and oceans grew large enough, more complex, oxygen-dependent forms of life could develop.

41.9. Precambrian rocks include sedimentary, igneous, and metamorphic varieties. What does this suggest about the geological activity in Precambrian time?

Precambrian geological activity must have been similar to that of today.

41.10. Paleozoic sedimentary rocks derived from marine deposits are widely distributed in all the continents. What does this indicate about the height of the continents relative to sea level in the Paleozoic era?

Much of the area of the continents must have been near or below sea level during at least part of the Paleozoic since shallow seas must have been widespread on their surfaces then.

41.11. Why are fossils still useful in dating rock formations despite the development of radioactive methods?

Fossils are found in sedimentary rocks, and radioactive dating is not generally useful in such rocks. Dating by means of fossils is usually much easier than radioactive dating, which requires elaborate apparatus. Since the geological periods that correspond to specific fossil types are well established, fossil dating is reasonably accurate.

41.12. Under what circumstances is coal formed? During what geological periods were such conditions widespread?

Coal is formed from plant material that accumulates in an environment where partial decay occurs in which most of the hydrogen and oxygen is removed to leave a residue that is largely carbon. The residue consolidates into coal under the pressure of sediments deposited later. Swamps are especially favorable for the formation of coal, since there is an abundance of plant life and the decay of plant remains underwater leaves carbon residues. Most coal deposits were laid down during the Mississippian and Pennsylvanian periods, often jointly called the Carboniferous period.

41.13. The same reptiles were found on all continents during the Mesozoic era, but the mammals of the Cenozoic era are often different on different continents. Why?

During the Mesozoic era today's continents were joined together so the animal populations (which were largely reptiles) could move freely among them. During the Cenozoic era the continents were split apart, and the evolution of some of the mammals that replaced the reptiles proceeded differently on the various land masses.

41.14. What were the Ice Ages? When did they occur?

The Ice Ages involved the formation of ice sheets that covered large areas of the earth's surface. Ice advanced across the continents during four major episodes, which were separated by interglacial periods during which the ice retreated poleward. The Ice Ages took place during the past two million years, that is, in the Pleistocene epoch of the Quaternary period of the Cenozoic era. The most recent large-scale glaciation covered much of Canada and northeastern United States and began to recede only about 20,000 years ago. The origin of the worldwide climatic changes that produced the Ice Ages is not known, though a number of possible mechanisms have been proposed.

41.15. The Scandinavian land mass is rising at the rate of about 1 cm per year. What is believed to be the reason?

When the thick sheet of ice that covered this region during the most recent glacial period melted, the continental block became lighter and its buoyancy provided an upward force that has been raising it toward a level of isostatic equilibrium.

Supplementary Problems

41.16. What is an unconformity?

41.17. Precambrian rocks are exposed over a large part of eastern Canada. What does this observation suggest about the geological history of this region since the end of Precambrian time?

41.18. What conspicuous difference is there between Precambrian sedimentary rocks and those of later eras?

41.19. What are the chief kinds of organisms that have left traces in Precambrian sedimentary rocks?

41.20. Why are fossils never found in igneous rocks and only seldom in metamorphic rocks?

41.21. Why is it believed that large parts of the United States were once covered by shallow seas?

41.22. About 200 million years ago today's continents were all part of the supercontinent Pangaea. During what geological era did Pangaea break apart into Laurasia and Gondwanaland? During what era did Laurasia break up into North America, Greenland, and Eurasia?

41.23. What major change in land animal life occurred between the late Mesozoic and early Cenozoic era?

41.24. What is believed to be the origin of petroleum? In the rocks of what era are petroleum deposits most common?

41.25. During what geological era did birds develop? From what type of animal did they evolve?

41.26. What are some of the chief differences between reptiles and mammals?

41.27. Minnesota has a great many shallow lakes. How did they originate?

Answers to Supplementary Problems

41.16. An unconformity is an eroded surface buried under rocks deposited subsequently.

41.17. The region must have been above sea level for most of the 570 million years since the end of Precambrian time or else it would be covered with sedimentary rocks.

41.18. Precambrian sedimentary rocks contain few if any fossils, whereas later sedimentary rocks usually contain abundant fossils.

41.19. Bacteria and algae.

41.20. Igneous rocks have hardened from a molten state, and no fossil could survive such temperatures. Metamorphic rocks have been altered under conditions of heat and pressure severe enough to distort or destroy most fossils.

41.21. Sedimentary rocks that contain the fossil shells of marine organisms are found in many parts of the United States.

41.22. Mesozoic era; Cenozoic era

41.23. Reptiles, the dominant form of land animals during the Mesozoic, declined and were superseded by mammals, which are dominant in the Cenozoic.

41.24. Petroleum is thought to have originated in the remains of marine animals and plants which became buried under sedimentary deposits. Petroleum is usually found in Cenozoic rocks.

41.25. Mesozoic era; reptiles

41.26. Reptiles are cold-blooded (their body temperatures vary with the ambient temperature), lay eggs, have relatively small brains and scaly skins, and the teeth of each individual are all nearly the same. Mammals are warm-blooded (their body temperatures are constant), bear live offspring which are suckled, have relatively large brains and usually hairy coats and subcutaneous fat (both of which provide thermal insulation), and the teeth of each individual are of several different kinds.

41.27. The Pleistocene glaciation in that region left many depressions which subsequently filled with water to form lakes.

Chapter 42

Earth and Sky

PTOLEMAIC AND COPERNICAN SYSTEMS

Until the 16th century the earth was believed to be the center of the universe. According to the *Ptolemaic system,* which was a detailed picture of this idea, the sun and moon revolve around the earth in circular orbits while the planets each travel in a series of loops called *epicycles.* The stars are supposed to be fixed to a crystal sphere that turns once a day.

In 1543 Copernicus published the hypothesis that the sun is the center of the solar system, with the moon revolving around the earth. The stars are far away in space, and the earth rotates daily on its axis. In the *Copernican system* the planet nearest the sun is Mercury and next in order are Venus, the earth, Mars, Jupiter, Saturn, Uranus, Neptune, and Pluto. (The last three planets were not known in the time of Copernicus.)

Kepler modified the Copernican system by showing that the planetary orbits are ellipses rather than circles and discovered two other significant regularities obeyed by the planets. His three laws, which contributed to the discovery by Newton of the law of gravitation, are as follows:

1. The orbits of the planets around the sun are ellipses.

2. Each planet moves so that a line drawn from the planet to the sun sweeps out equal areas in equal times. Thus the planet moves most rapidly when it is closest to the sun and least rapidly when it is farthest from the sun.

3. The ratio between the square of the time required by a planet to revolve around the sun and the cube of its average distance from the sun has the same value for all the planets.

MOTIONS OF THE EARTH

The earth rotates on its axis and revolves around the sun. The *day* is the time required for a complete rotation and the *year* is the time required for a complete revolution. The axis of rotation is tilted by 23.5° with respect to a perpendicular to the plane of the orbit. The angle of tilt is constant but its direction in space changes very slowly; thus the earth resembles a wobbling top. The wobble is called *precession* and its period is about 26,000 years.

The *seasons* occur because, as a result of the tilt of the earth's axis, for half of each year one hemisphere receives more sunlight than the other, and in the other half of the year it receives less sunlight. On 22 June (the *summer solstice*) the noon sun is at its highest in the sky in the northern hemisphere and the period of daylight is longest; on 22 December (the *winter solstice*) the noon sun is at its lowest and the period of daylight is shortest. In the southern hemisphere the situation is reversed. On 21 March (the *vernal equinox*) and 23 September (the *autumnal equinox*) the sun is directly overhead at noon on the equator and the periods of daylight and darkness are equal everywhere.

LATITUDE AND LONGITUDE

Locations on the earth's surface are specified in terms of its axis of rotation. A *great circle* is any circle on the earth's surface whose center is the earth's center. The *equator* is a great circle midway between the North and South Poles. A *meridian* is a great circle that passes through both poles, and it forms a right angle with the equator. The *prime meridian* passes through Greenwich, England. The *longitude* of a point on the earth's surface is the angular distance between a meridian through this point and the prime meridian; the prime meridian is assigned the longitude 0°, and longitudes are given in degrees east or west of the prime meridian. Thus a longitude of 60° W identifies a meridian 60° west of the prime meridian.

The *latitude* of a point on the earth's surface is the angle between a line drawn from the earth's center to it and another line drawn from the earth's center to a point on the equator on the same meridian. Thus a latitude of 60° N identifies a circle (smaller than a great circle) 60° north of the equator. The latitude and longitude of a place on the earth's surface specify its location. (See Fig. 42-1.)

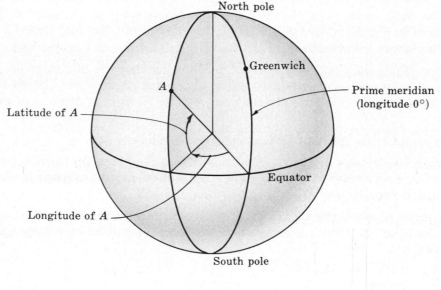

Fig. 42-1

TIME

A *solar day* is the period of the earth's rotation with respect to the sun; a *sidereal day* is its period with respect to the stars. Because the earth revolves around the sun, the sidereal day is about 4 min shorter than the solar day. Ordinary timekeeping is based on the average solar day, which is divided into 24 hr with each hour further divided into minutes and seconds.

From the point of view of a stationary observer on the earth, the sun moves around the earth in a westward direction once every 24 hr, which is 15° of longitude per hour. This means that noon (or any other time reckoned with respect to the sun) occurs 1 hr later at a longitude 15° west of a particular place and 1 hr earlier at a longitude 15° east of it. For convenience, the world is divided into 24 *time zones*, each about 15° of longitude wide and each keeping time 1 hr ahead of the zone west of it and 1 hr behind the zone east of it.

The *international date line*, which follows the 180° meridian except for deviations to avoid going through Alaska and island groups in the Pacific, separates one day from the next. A person traveling eastward around the world sets his watch ahead at each successive time zone, and when he crosses the date line he subtracts a day from his calendar in order to compensate. A person traveling westward sets his watch behind at each successive time zone, and when he crosses the date line he adds a day to his calendar.

Solved Problems

42.1. Did Copernicus prove that his system is correct and the Ptolemaic system is incorrect? If not, why is the Copernican system accepted today?

The Copernican system, by referring the planetary motions to the sun rather than to the earth, was much simpler than the Ptolemaic system and, when modified by Kepler, was more accurate in describing these motions. However, the Ptolemaic system could also be modified to achieve as much accuracy as the Copernican system, though in a much more complicated way. The Copernican system, when it was proposed, was better than the Ptolemaic system because of its simplicity, but this simplicity did not make the Ptolemaic system wrong since all it involved was a shift in the choice of the reference point for reckoning planetary motions. The Copernican system is today considered correct (and the Ptolemaic system incorrect) because there is direct experimental evidence for the motions of the planets around the sun; for example, the change in apparent position of nearby stars relative to distant stars as the earth revolves around the sun.

42.2. Why are Mercury and Venus always seen either around sunset or around sunrise?

Mercury and Venus are closer to the sun than the earth is, hence an observer on the earth always sees them in the vicinity of the sun. When one of them is east of the sun, it disappears below the horizon after the sun and is visible in the early evening; when it is west of the sun, it rises above the horizon before the sun and is visible in the early morning.

42.3. The *astronomical unit* (AU) is the average radius of the earth's orbit; it is equal to 1.495×10^{11} m, which is about 93,000,000 miles. (a) Find the value of the constant ratio in Kepler's third law in terms of AU and years. (b) The average radius of Jupiter's orbit is 5.2 AU. How long does Jupiter require to complete a revolution around the sun?

(a) If the average radius of a planet's orbit is R and its period of revolution is T, Kepler's third law states that T^2/R^3 has the same value for all the planets. For the earth, $T = 1$ yr and $R = 1$ AU, so $T^2/R^3 = C = 1$ yr²/AU³.

(b) For Jupiter, $R = 5.2$ AU, and so

$$ T^2 \;=\; CR^3 \;=\; 1\,\frac{\text{yr}^2}{\text{AU}^3} \times (5.2\ \text{AU})^3 \;=\; 141\ \text{yr}^2 $$

$$ T \;=\; \sqrt{141\ \text{yr}^2} \;=\; 11.9\ \text{yr} $$

42.4. Neptune's period of revolution is 165 yr. (a) What is the average radius of Neptune's orbit in AU? (b) In miles?

(a) According to Kepler's third law, $T^2/R^3 = C$, hence $R^3 = T^2/C$. With $T = 165$ yr and $C = 1$ yr²/AU³ we have

$$ R^3 \;=\; \frac{T^2}{C} \;=\; \frac{(165\ \text{yr})^2}{1\ \text{yr}^2/\text{AU}^3} \;=\; 2.72 \times 10^4\ \text{AU}^3 $$

$$ R \;=\; \sqrt[3]{2.72 \times 10^4\ \text{AU}^3} $$

To find this cube root, we begin by rewriting 2.72×10^4 as 27.2×10^3 since $\sqrt[3]{10^3} = 10$. Because $\sqrt[3]{\text{AU}^3} = \text{AU}$ as well, we obtain

$$ R \;=\; \sqrt[3]{2.72 \times 10^4\ \text{AU}^3} \;=\; \sqrt[3]{27.2 \times 10^3}\ \text{AU} \;=\; \sqrt[3]{27.2} \times \sqrt[3]{10^3}\ \text{AU} \;=\; \sqrt[3]{27.2} \times 10\ \text{AU} $$

For an approximate result, we note that $3 \times 3 \times 3 = 27$, so $\sqrt[3]{27.2} \approx 3$. Numerical tables give the more exact result

$$ R \;=\; 3.01 \times 10\ \text{AU} \;=\; 30.1\ \text{AU} $$

(b) Since $1\ \text{AU} = 9.3 \times 10^7$ mi, $R = 30.1\ \text{AU} \times 9.3 \times 10^7$ mi/AU $= 2.8 \times 10^9$ mi.

42.5. (*a*) How are the North and South Poles and the equator defined? (*b*) What are their respective latitudes?

(*a*) The North and South Poles are the points where the earth's axis of rotation intersects its surface. The equator is an imaginary line around the earth midway between the poles.

(*b*) The latitude of the North Pole is 90° N, that of the South Pole is 90° S, and that of the equator is 0°.

42.6. (*a*) How are the Arctic and Antarctic Circles defined? (*b*) What are their latitudes?

(*a*) On 22 December, the shortest day of the year in the northern hemisphere, the 23.5° tilt of the earth's axis means that no sunlight reaches any point within 23.5° of the North Pole. The Arctic Circle is the boundary of this region of darkness. On the same day, which is the longest day of the year in the southern hemisphere, there are 24 hours of daylight at all points within 23.5° of the South Pole, and the Antarctic Circle is the boundary of this region of daylight. On 22 June the situations in the two hemispheres are reversed.

(*b*) The latitude of the North Pole is 90° N, hence that of the Arctic Circle is 90° N − 23.5° = 66.5° N. Similarly the latitude of the Antarctic Circle is 66.5° S.

42.7. (*a*) How are the Tropics of Cancer and Capricorn defined? (*b*) What are their latitudes?

(*a*) The Tropic of Cancer is the most northerly latitude in the northern hemisphere at which the sun is ever directly overhead at noon. The Tropic of Capricorn is the corresponding latitude in the southern hemisphere.

(*b*) On 22 June, when the North Pole is tilted closest to the sun and hence is the day of maximum sunlight in the northern hemisphere, the noon sun is directly overhead 23.5° north of the equator; hence the latitude of the Tropic of Cancer is 23.5° N. Similarly the latitude of the Tropic of Capricorn is 23.5° S; the South Pole is tilted closest to the sun on 22 December, when the noon sun is directly overhead at this latitude.

42.8. Why does the duration of the solar day vary slightly throughout the year?

The earth's orbit is an ellipse, and its velocity is greater when it is near the sun than when it is far from the sun. Since the solar day depends on both the earth's rotation and its orbital motion, the duration of the solar day varies during the year.

42.9. Why does the sky not become dark as soon as the sun sets below the horizon?

Refraction and scattering of sunlight by the earth's atmosphere enable some sunlight to reach the earth's surface for an hour or more after the sun has set.

42.10. *Greenwich mean time* (GMT) is local time at the prime meridian (0° longitude) when variations in the earth's orbital velocity are averaged out. When GMT is 0800 (8:00 AM), what is the local time at (*a*) New York City (longitude 74° W) and (*b*) Venice (longitude 12° E)?

(*a*) Since a longitude difference of 15° leads to a time difference of 1 hr, 1° of longitude difference is equivalent to a 4 min time difference. The longitude of New York City is 74° W, hence the time difference with respect to GMT is

$$\Delta T = 74° \times 4 \text{ min/}° = 296 \text{ min} = 4 \text{ hr } 56 \text{ min}$$

Since New York City is west of the prime meridian, the time difference must be subtracted from GMT (it is earlier in New York City than in Greenwich), so local time in New York City at 0800 GMT is

$$T = \text{GMT} - \Delta T = 0800 - 0456 = 0760 - 0456 = 0304$$

which is 3:04 AM.

(*b*) The longitude of Venice is 12° E, hence the time difference is

$$\Delta T = 12° \times 4 \text{ min/}° = 48 \text{ min}$$

Since Venice is east of the prime meridian, the time difference must be added to GMT (it is later in Venice than in Greenwich), so local time in Venice at 0800 GMT is

$$T \;=\; \text{GMT} + \Delta T \;=\; 0800 + 0048 \;=\; 0848$$

which is 8:48 AM.

42.11. What is the GMT of local noon at Lisbon, longitude $9° \, \text{W}$?

Since $1°$ of longitude is equivalent to 4 min of time, the time difference between $0°$ and $9°$ longitude is

$$\Delta T \;=\; 9° \times 4 \text{ min/}° \;=\; 36 \text{ min}$$

Lisbon is west of the prime meridian, so when it is $T = 1200$ at Lisbon the GMT is

$$\text{GMT} \;=\; T + \Delta T \;=\; 1200 + 0036 \;=\; 1236$$

which is 12:36 PM.

42.12. When the local time in Moscow (longitude $38° \, \text{E}$) is 1000 (10:00 AM), what is the local time in New York City (longitude $74° \, \text{W}$)?

The longitude difference between Moscow and New York City is $38° + 74° = 112°$ since they are on opposite sides of the prime meridian. The corresponding time difference is

$$\Delta T \;-\; 112° \times 4 \text{ min/}° \;=\; 448 \text{ min} \;=\; 7 \text{ hr } 28 \text{ min}$$

Since New York City is west of Moscow, local time in New York City is earlier than local time in Moscow, and the New York City local time that corresponds to 1000 Moscow local time is

$$T \;=\; 1000 - 0728 \;=\; 0960 - 0728 \;=\; 0232$$

which is 2:32 AM.

42.13. What is your longitude if local noon occurs at 1440 GMT (2:40 PM)?

The time difference is $2 \text{ hr } 40 \text{ min} = 160 \text{ min}$, and the corresponding longitude difference is

$$\frac{160 \text{ min}}{4 \text{ min/}°} \;=\; 40°$$

Since local noon occurs *after* 1200 GMT, your location must be west of the prime meridian, so the longitude is $40° \, \text{W}$.

42.14. Why are leap years necessary?

The length of the year is 365 days 5 hr 48 min 46 s. Since the difference between the actual length of the year and 365 days is very nearly 6 hr, which is $\frac{1}{4}$ day, adding an extra day to February every four years (namely those years evenly divisible by 4) enables the seasons to recur at very nearly the same dates each year. (The discrepancy of 11 min 14 s per year that remains adds up to a full day after 128 years. To remove most of this discrepancy, century years not divisible by 400 are not leap years; thus 1900 was not a leap year, but 2000 will be one. This step makes the calendar accurate to 1 day per 3300 years. A further modification leaves 4000, 8000, 12,000 and so on as 365-day years rather than leap years; the resulting calendar is accurate to 1 day per 20,000 years, which is adequate for the time being.)

Supplementary Problems

42.15. The average radius of Mercury's orbit is 0.387 AU. (*a*) How many meters is this? (*b*) Use Kepler's third law to find Mercury's period of revolution around the sun.

42.16. The period of revolution of Uranus is 84 yr. Use Kepler's third law to find the average radius of its orbit in AU.

42.17. As seen from the earth, the sun drifts eastward relative to the stars; that is, at sunset or sunrise on a given day, the sun appears eastward of its position the previous sunset or sunrise. Through approximately what angle does the sun move eastward each day relative to the stars?

42.18. The earth is closest to the sun in December and farthest from the sun in June, yet in the northern hemisphere December is a winter month and June is a summer month. Why?

42.19. (*a*) If the earth's axis were tilted by 30° instead of by 23.5°, would the seasons be more or less pronounced than they now are? (*b*) What would the latitudes of the Arctic Circle and the Tropic of Cancer be?

42.20. (*a*) In May, does the length of the day (that is, the period between sunrise and sunset) change when one travels north from the Tropic of Cancer? If so, does it become longer or shorter? (*b*) Does the length of the day change when one travels west from the prime meridian? If so, does it become longer or shorter?

42.21. What must be your location if the stars move across the sky in a circle centered directly overhead?

42.22. Where on the earth can the entire sky be observed during the course of a year?

42.23. What is the local time in Tokyo (longitude 140° E) when it is local noon in Moscow (longitude 38° E)?

42.24. When GMT is 1430 (2:30 PM), what is the local time in Chicago, longitude 88° W?

42.25. What is GMT when it is 1800 (6:00 PM) local time in Tokyo, longitude 140° E?

42.26. What is your longitude if local noon occurs at 0800 GMT (8:00 AM)?

42.27. Can you think of any evidence in favor of the earth's rotation that does not involve any reference to astronomical bodies outside the earth?

Answers to Supplementary Problems

42.15. 5.79×10^{10} m; 0.241 years

42.16. 19 AU

42.17. 360°/365 days $\approx$ 1°/day

42.18. The tilt of the earth's axis produces a much greater annual variation in the amount of sunlight reaching a point on the earth than does the eccentricity of its orbit.

42.19. more pronounced; 60°, 30°

42.20. The day becomes longer; no change.

42.21. At the North or South Pole.

42.22. Between the Tropics of Cancer and of Capricorn, which are the limiting latitudes that intersect the plane of the earth's orbit.

42.23. 1848 (6:48 PM)

42.24. 0838 (8:38 AM)

42.25. 0840 (8:40 AM)

42.26. 60° E

42.27. The equatorial bulge; the pattern of winds in the general circulation of the atmosphere.

Chapter 43

The Solar System

THE PLANETS

Like the earth, the other planets rotate on their axes and revolve around the sun, and all except Mercury, Venus, and Pluto have satellites. Most planet and satellite orbits lie near the same plane, and most of the various rotations and revolutions are in the same direction (counterclockwise as seen looking down from above the earth's North Pole). The planets and their satellites are visible by virtue of the sunlight they reflect.

The inner planets — Mercury, Venus, the earth, and Mars — are considerably smaller, less massive, and denser than the outer planets — Jupiter, Saturn, Uranus, and Neptune; also, the inner planets rotate much more slowly than do the outer planets. Pluto is a special case, with properties closer to those of the inner planets despite its distance from the sun. The outer planets (except Pluto) apparently consist largely of hydrogen and hydrogen compounds such as methane and ammonia, which accounts for their low densities.

Thousands of minor planets called *asteroids* orbit the sun between Mars and Jupiter. The largest asteroid is not quite 500 mi in diameter; most are much smaller.

THE MOON

The moon revolves around the earth once every 27.3 days. Because the moon rotates on its axis with the same period, the same lunar hemisphere always faces the earth.

As the moon revolves around the earth, the extent of its illuminated hemisphere visible from the earth changes, which accounts for the *phases* of the moon. When the moon is on the opposite side of the earth from the sun, sunlight reaches the entire lunar hemisphere facing the earth, and the resulting bright disk is called *full moon*. At *new moon* the moon is between the earth and the sun, and the hemisphere facing the earth receives no sunlight and appears dark. At intermediate positions in the lunar orbit different portions of the illuminated part of the moon's surface are visible. Owing to the motion of the earth around the sun, the period between full moons is 29.5 days, which is longer than the 27.3-day orbital period of the moon.

The plane of the moon's orbit is slightly tilted relative to the plane of the earth's orbit, which is why sunlight normally is able to reach the moon when it is on the opposite side of the earth from the sun, and why the moon normally does not obstruct the sun at new moon. On certain occasions, however, earth, sun, and moon lie along a straight line, and *eclipses* then take place. During a *lunar eclipse* the earth is between the sun and the moon and its shadow falls on the moon; during a *solar eclipse* the moon is between the sun and the earth and its shadow falls on the earth. Total eclipses occur because, although their actual diameters and their distances from the earth are different, the apparent diameters of both sun and moon as seen from the earth are the same at certain times.

COMETS

Comets are members of the solar system whose orbits around the sun are very long, narrow ellipses. Most comet orbits are so large that their periods range up to a million years or more; a few have periods short enough to be seen regularly from the earth, for instance Halley's comet whose period is 76 years.

When it is far from the sun, a comet is a compact aggregate of frozen water, ammonia, and methane with some particles of metallic and stony characters probably present as well. Near the sun the H_2O, NH_3, and CH_4 vaporize and spread out to form a cloud of thin gas tens or hundreds of thousands of miles across. This cloud is excited by solar ultraviolet radiation and reradiates visible light; a small part of the glow of a comet is reflected sunlight. Such clouds often exhibit tails that always point away from the sun; they are the result of pressure exerted both by sunlight and by the *solar wind* of protons and electrons that continually stream outward from the sun.

METEORS

Meteoroids are particles of matter, usually small, that travel through the solar system in orbits around the sun. When a meteoroid enters the atmosphere, it is heated by friction and glows brightly: the visual phenomenon is called a *meteor*. *Meteorites* are the remains of meteoroids that reach the ground. Most meteoroids occur in swarms and produce meteor showers at regular intervals. Such meteoroids are thought to be the debris of comets, and in some cases their orbits can be identified as those of former comets. Meteorites are divided into two main classes: *iron meteorites*, which are largely iron with some nickel also present, and *stony meteorites*, which consist chiefly of silicate minerals in a characteristic structure. A few meteorites are composed of mixtures of iron and stony material.

Solved Problems

43.1. (a) How is it possible to distinguish the planets from the stars by observations with the naked eye? (b) By observations with a telescope?

(a) When viewed over a period of time, a planet will be seen to change its position in the sky relative to the stars. (This is the reason for the name "planet," which is Greek for "wanderer".)

(b) Seen through a sufficiently powerful telescope, the planets appear as disks whereas the stars, which are much more distant, appear as points of light.

43.2. Why is Venus a brighter object in the sky than Mars?

Venus is closer to the sun than Mars and hence receives more sunlight to reflect. It is larger than Mars, so the reflecting surface is greater in area. Venus is surrounded by clouds whereas Mars has none, and these clouds constitute a better reflector of sunlight than the Martian surface; the white polar caps on Mars are too small to make much difference in this respect. As a result of all these factors, Venus is not only brighter than Mars but is also at times the brightest object in the sky after the sun and moon.

43.3. Which planets would you expect to show phases like those of the moon?

Mercury and Venus, because their orbits are closer to the sun than that of the earth.

43.4. What is the nature of Saturn's rings?

The rings of Saturn consist of large numbers of small particles which orbit the planet like miniature satellites.

43.5. Stars, planets, and satellites are all very nearly spherical in form, but some of the smaller asteroids apparently have irregular shapes. Explain these observations.

Gravitational forces in a large object are strong enough to prevent more than minor departures from sphericity (except for equatorial bulges due to rotation). If a significant protuberance were to occur, the gravitational pull of the rest of the object on it would create such overwhelming pressures that the most rigid underlying material would flow and the protuberance would be drawn down until almost level with the rest of the surface. In the case of a small asteroid, which might be only a few miles across, the gravitational forces are much smaller and the rigidity of their material may well be sufficient for an irregular shape to be maintained.

43.6. As seen from the earth, the moon drifts eastward relative to the stars; that is, on a given night the moon appears eastward of its position the night before at the same time. Through what angle does the moon move eastward each day relative to the stars?

The moon circles the earth in 27.3 days relative to the stars, hence it travels through 360° in 27.3 days or

$$\frac{360°}{27.3 \text{ days}} = 13°/\text{day}$$

43.7. How long a time elapses between the moon at first quarter (when it appears as a half moon) and the full moon?

From first quarter to full moon is $\frac{1}{4}$ of a lunar cycle. Since the complete cycle takes 29.5 days, here we have $\frac{1}{4} \times 29.5$ days = 7.4 days.

43.8. Why is it believed that the moon's interior is different in composition from the earth's interior?

The average density of the moon is 3.3 g/cm³ whereas that of the earth is 5.5 g/cm³. Part of the reason for the smaller density of the moon is its smaller total mass, which means that pressures in its interior are less than those in the earth's interior. However, this factor is not sufficient to account for the large difference in densities. Possible reasons include a smaller proportion of iron in the moon's interior than in the earth's, and the presence of large quantities of low-density substances such as graphite.

43.9. List some theories of the moon's origin. Are any of them free from serious objections?

(1) The moon was initially part of the earth and split off from it to become an independent body. (2) The moon was formed elsewhere in the solar system and later was captured by the earth's gravitational field. (3) The moon and the earth came into being together as a double-planet system.

There are serious objections to each of these theories.

43.10. When a comet is close enough to the sun to be seen from the earth, stars are visible through both the comet's head and tail. What does this imply about the danger to the earth from a collision with a comet?

The density of a comet is extremely low when it is in the vicinity of the earth, and in a collision the comet material would simply be absorbed in the upper atmosphere.

43.11. Some meteor showers recur each year at about the same time, for instance the Perseid shower that appears early every August. Does this mean that the orbits of the meteoroids in the Perseid swarm all have periods of exactly one year?

No. The annual occurrence of a meteor shower at a particular date simply means that the earth's orbit intersects the common orbit of a meteoroid swarm at the particular point corresponding to that date. (The Perseid meteoroids are believed to be the remnants of a comet whose period was 105 years.) If the number of meteors seen is about the same each year, the meteoroids must be spread out along their common orbit; if the number varies considerably, the meteoroids must be bunched together.

43.12. Over ninety percent of the meteorites found after a known fall are stony, yet most of the meteorites in museums are iron. Why?

Stony meteorites resemble ordinary rocks whereas iron ones are conspicuously different; also, stony meteorites are more readily eroded than iron meteorites.

Supplementary Problems

43.13. Which of the planets are readily visible with the naked eye?

43.14. On which planet is the length of the year shortest? On which planet is it longest?

43.15. According to *Bode's law,* which was discovered two centuries ago by Titius, the mean orbital radii of the planets in AU are given by the formula $R_n = 0.4 + 0.3 \times 2^n$, where $n = 0$ for Venus, $n = 1$ for the earth, and so on; $R = 0.4$ for Mercury. Compare the predictions of Bode's law with the actual orbital radii in AU. Include the asteroids, whose mean orbital radii average 2.9 AU.

43.16. Approximately how many days elapse between new moon and full moon?

43.17. (*a*) In what phase must the moon be at the time of a solar eclipse? (*b*) At the time of a lunar eclipse?

43.18. If the moon were smaller than it is, would total eclipses of the sun still occur? Would total eclipses of the moon still occur?

43.19. Would you expect the rings of Saturn to revolve with the same period, like parts of the same phonograph record?

43.20. If the earth had no atmosphere, would comets still be visible from its surface? Would meteoroids?

43.21. Why do comets have tails only in the vicinity of the sun? Why do these tails always point away from the sun, even when the comet is receding from it?

Answers to Supplementary Problems

43.13. Mercury, Venus, Mars, Jupiter, and Saturn.

43.14. Mercury; Pluto

43.15.

Planet	Predicted Distance	Actual Distance
Mercury	$0.4 + 0.0 = 0.4$	0.39
Venus	$0.4 + 0.3 = 0.7$	0.72
Earth	$0.4 + 0.6 = 1.0$	1.00
Mars	$0.4 + 1.2 = 1.6$	1.52
Asteroids	$0.4 + 2.4 = 2.8$	2.90 (average)
Jupiter	$0.4 + 4.8 = 5.2$	5.20
Saturn	$0.4 + 9.6 = 10.0$	9.54
Uranus	$0.4 + 19.2 = 19.6$	19.18
Neptune	$0.4 + 38.4 = 38.8$	30.06
Pluto	$0.4 + 76.8 = 77.2$	39.44

43.16. 15 days   **43.17.** new moon; full moon   **43.18.** no; yes

43.19. No. Each particle in the rings pursues its own orbit around Saturn, and by Kepler's third law the inner particles must have shorter periods than the outer particles.

43.20. yes; no

43.21. The tail of a comet is formed by pressure exerted by solar emissions. Sunlight is partly responsible but the chief influence is the "solar wind" of protons and electrons. Far from the sun the effect of these emissions is too small to produce a tail. A comet's tail always points away from the sun since the direction of the solar emissions is always radially outward from the sun.

Chapter 44

The Sun

SOLAR ENERGY

The sun is a typical star composed largely of hydrogen with a substantial proportion of helium and small amounts of most of the other elements. The sun's energy originates in the fusion of hydrogen nuclei (protons) to form helium nuclei (see Problem 44.6). The conversion of hydrogen to helium can take place under the conditions believed to exist in the sun's interior in two different ways. In one of them, the *proton-proton cycle*, collisions of protons result in the formation of heavier nuclei whose collisions in turn yield helium nuclei. The *carbon cycle* is a sequence of steps in which carbon nuclei absorb a succession of protons until they ultimately disgorge helium nuclei to become carbon nuclei once more. In the sun the proton-proton cycle predominates; in hotter stars, the carbon cycle predominates.

SOLAR ATMOSPHERE

The light that reaches us from the sun is emitted by its *photosphere,* a relatively thin layer of gas whose temperature is about 6000 K. Below the glowing surface of the photosphere negative hydrogen ions are present which absorb radiation from the sun's interior. Energy reaches the photosphere through this opaque region by means of convection, with columns of hot gas bubbling upward to give the photosphere a mottled, granular appearance.

The atoms in the upper part of the photosphere and in the lower part of the *chromosphere* that lies above it absorb light of certain characteristic frequencies and so produce the dark-line spectrum of the sun. (Absorption spectra are described in Chapter 20.) This region is sometimes referred to as the *reversing layer.* The chromosphere itself emits reddish light that can be seen during solar eclipses and in special telescopes. Large flamelike *prominences* frequently extend outward from the chromosphere. The chromosphere gradually merges into the *corona,* a very hot (over 10^6 K in its outer part) cloud of ionized gas that extends for millions of miles outward from the sun. The density of the corona is very low, so despite its high temperature it radiates much less light than the photosphere.

Ions, largely protons and electrons, continuously stream outward in the corona and constitute the *solar wind* that helps cause comet tails to point away from the sun. The faster ions from the sun produce the *aurora* in the upper atmosphere of the earth by exciting the atoms and molecules there to radiate.

SUNSPOTS

Sunspots are dark markings on the solar surface that persist for a few hours to a week or more. A sunspot appears dark because its temperature is about 1000 K cooler than the rest of the photosphere. The number of sunspots varies through an 11-year cycle; at sunspot minimum, few if any spots are visible, whereas at sunspot maximum 100 or more spots may be present at one time. It seems possible that a number of terrestrial phenomena are correlated with the sunspot cycle, in particular certain periodic weather variations. Strong magnetic fields are associated with sunspots; because the polarities of these fields are reversed in successive 11-year cycles, the true period of the sunspot cycle is 22 years.

A *solar flare* is a sudden release of a large amount of energy from the solar surface. In a flare, a region of the chromosphere becomes much brighter than usual, and ultraviolet light, X-rays, and streams of fast ions are emitted copiously. Flares occur most often in the neighborhood of sunspot groups and are most frequent at sunspot maximum. The various radiations associated with flares affect the earth's ionosphere and magnetic field, produce exceptional auroral activity, and are dangerous to astronauts.

Solved Problems

44.1. How is the sun's mass determined?

The sun's mass can be found from the characteristics of the earth's orbital motion around the sun. For simplicity we will assume a circular orbit. The centripetal force on the earth as it revolves around the sun is mv^2/R, where m is the earth's mass, v its orbital velocity, and R the radius of its orbit. This force equals the gravitational force GMm/R^2 the sun exerts on the earth, where G is the universal gravitational constant and M is the sun's mass. Hence

$$\text{Gravitational force} = \text{centripetal force}$$

$$\frac{GMm}{R^2} = \frac{mv^2}{R}$$

$$M = \frac{v^2R}{G}$$

Since v, R, and G are known, M can be calculated.

44.2. The sun is 1.5×10^{11} m from the earth and its angular diameter as seen from the earth is $0.53°$. Find the diameter of the sun.

The ratio between the sun's diameter d and the circumference of a circle of radius $r = 1.5 \times 10^{11}$ m is equal to the ratio between the angle $0.53°$ and the angle $360°$ that corresponds to a complete revolution (see Fig. 44-1). Hence

$$\frac{d}{2\pi r} = \frac{0.53°}{360°}$$

$$d = \frac{2\pi r \times 0.53°}{360°}$$

$$= \frac{2\pi \times 1.5 \times 10^{11} \text{ m} \times 0.53°}{360°}$$

$$= 1.4 \times 10^9 \text{ m}$$

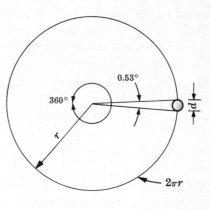

Fig. 44-1

44.3. (a) State two methods for determining the rate at which the sun rotates. (b) Is the rotation rate uniform?

(a) Sunspots appear to move across the sun's disk because of solar rotation. From the velocity of this motion the rate at which the sun rotates can be found. Another approach is to measure the Doppler shift in the spectral lines of radiation from the limbs (edges) of the sun's disk. At one limb the spectral lines are found to be shifted toward the blue end of the spectrum, which signifies motion toward the observer, and at the other limb the spectral lines are found to be shifted toward the red end of the spectrum, which signifies motion away from the observer. From the amounts of each shift the rotation rate can be determined.

(b) The period of solar rotation is about 25 days at the equator and increases to about 35 days near the poles.

44.4.   How can the composition of the sun be determined?

The presence of the spectral lines of a particular element in the solar spectrum means that this element must be present in the sun.

44.5.   Why is it impossible for combustion to be the source of the sun's energy?

Combustion involves the reaction of oxygen with another substance to form an oxygen-containing compound. Combustion cannot occur in the sun because the high temperatures there would prevent the formation of any such compound. Also, even if combustion could occur, the energy liberated would be totally inadequate to account for the observed rate at which the sun radiates energy.

44.6.   The proton mass is 1.673×10^{-27} kg and the mass of the $^4_2$He nucleus is 6.642×10^{-27} kg. (a) Find the energy liberated in the formation of a helium nucleus from four protons. (b) In the sun about 4 billion kg of matter is converted to energy each second. Find the power output of the sun in watts.

(a) Four protons combine to form a $^4_2$He nucleus in either the proton-proton or carbon cycle; in either case two positrons (positive electrons) are given off, which conserves electric charge. The mass that disappears in the formation of a $^4_2$He nucleus (neglecting the masses of the positrons) is

$$\Delta m = 4 \times m_p - m_{\text{He}} = 4 \times 1.673 \times 10^{-27}\ \text{kg} - 6.642 \times 10^{-27}\ \text{kg}$$
$$= 0.050 \times 10^{-27}\ \text{kg} = 5.0 \times 10^{-29}\ \text{kg}$$

and the energy liberated is

$$E = (\Delta m)c^2 = 5.0 \times 10^{-29}\ \text{kg} \times (3 \times 10^8\ \text{m/s})^2 = 4.5 \times 10^{-12}\ \text{J}$$

(b) Since 4 billion kg $= 4 \times 10^9$ kg and 1 W $= 1$ J/s, the power radiated by the sun is

$$P = \frac{E}{t} = \frac{mc^2}{t} = \frac{4 \times 10^9\ \text{kg} \times (3 \times 10^8\ \text{m/s})^2}{1\ \text{s}} = 3.6 \times 10^{26}\ \text{W}$$

44.7.   How is it possible for helium to have been discovered in the sun before it was found on the earth?

Helium is abundant in the sun, and its spectral lines appear as part of the solar spectrum. These spectral lines at first could not be identified as those of any terrestrial element, so it was assumed they corresponded to a new element that was called helium after *helios*, which is Greek for "sun." Later a new gas was discovered on the earth, and its spectrum turned out to be the same as the helium spectrum.

44.8.   Why does the sun appear brightest in the center of its disk and dimmest at the edge?

Light originating deep in the photosphere, where the temperature is highest, can emerge at the center of the disk because the path length through the absorbing H$^-$ ions of the photosphere is a minimum. Toward the edge of the disk the light reaching the earth must pass obliquely through the rest of the photosphere and so light from the lower part of the photosphere is more likely to be absorbed. Hence the light reaching the earth from near the edge of the sun comes from the outer, cooler part of the photosphere, and the center of the sun's disk accordingly appears brighter.

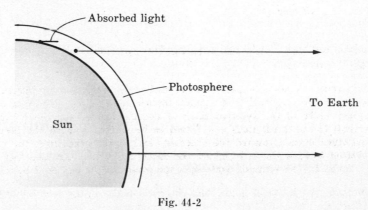

Fig. 44-2

44.9. Why are auroras most common near the polar regions?

The earth's magnetic field deflects the streams of solar ions that cause the aurora so that they usually reach the atmosphere in doughnut-shaped zones centered about the north and south geomagnetic poles.

44.10. What is the evidence for the belief that strong magnetic fields are associated with sunspots?

Spectral lines from atoms in a magnetic field are each found to be split into two or more component lines, with the spacing between the component lines varying with the strength of the field. This phenomenon is called the *Zeeman effect* after its discoverer. The presence of split lines in the spectra of sunspots signifies the presence of a magnetic field, and the strength of the field can be determined from the separation of the component lines.

Supplementary Problems

44.11. Information about the various parts of the solar atmosphere is obtained by analyzing their respective spectra. Why can this direct approach not be used to obtain information about the sun's interior?

44.12. (a) What aspect of the formation of helium from hydrogen results in the evolution of energy by the process? (b) Do all nuclear reactions evolve energy?

44.13. The sun's mass is 2×10^{30} kg, and it currently is losing about 4×10^9 kg of its mass per second as its hydrogen is converted into helium. If the sun has been radiating energy at the same rate as at present during the 4.5-billion-year existence of the earth, what fraction of its original mass has been lost? What does this suggest about the possibility that the solar radiation rate has indeed been approximately constant during most of this period of time, as is suggested by geological evidence?

44.14. Stars whose interior temperatures exceed 10^8 K obtain part of their energy from the fusion of helium nuclei to form carbon nuclei. Given that the mass of the $^4_2$He nucleus is 6.642×10^{-27} kg and that of the $^{12}_6$C nucleus is 19.918×10^{-27} kg, find the energy liberated each time a $^{12}_6$C nucleus is formed.

44.15. The photosphere of the sun is at a temperature of 6000 K, whereas the temperature of much of the corona exceeds 1,000,000 K. Why is the photosphere rather than the corona the source of most solar radiation?

44.16. What is the reason for the belief that there is a region between the photosphere and the corona whose temperature is less than that of either?

44.17. Why are the chromosphere and the corona ordinarily not visible? How do we know they exist?

44.18. Why do sunspots appear dark if their temperatures are typically 5000 K?

44.19. Intense auroral activity is observed in the earth's atmosphere about a day after a solar flare occurs. What is the average velocity of the ions emitted during the flare?

Answers to Supplementary Problems

44.11. The photosphere is opaque by virtue of the negative hydrogen ions it contains, hence no light from the sun's interior can emerge to be analyzed.

44.12. (a) A helium nucleus has less mass than the total mass of the four hydrogen nuclei that combine to form it, and the "missing" mass appears as energy. (b) No.

44.13.   The fraction of the sun's mass that has been lost in this period if its radiation rate has been un-changed is 2.8×10^{-4}, which is 0.028%. So little of the sun's mass has been converted to energy since the earth was formed that it is entirely possible that the sun's radiation rate has not changed appreciably during this period.

44.14. 7.2×10^{-13} J

44.15.   The corona has an extremely low density, so the amount of energy it radiates is small despite its high temperature.

44.16.   The solar spectrum consists of dark lines on a bright background, which can only be produced by absorption in a cool gas above the radiating surface.

44.17.   The photosphere is so much brighter than the chromosphere and corona that they cannot be seen unless the photosphere is masked, which is done by the moon during a total solar eclipse and also in special telescopes called *coronagraphs*.

44.18.   Sunspots appear dark only by comparison with the rest of the photosphere, whose temperature is 6000 K.

44.19. 1.7×10^6 m/s

Chapter 45

The Stars

APPARENT MAGNITUDE

The brightness of a star is expressed on a scale derived from that of the Greek astronomer Hipparchus, who classified stars into magnitudes ranging from 1st (the brightest) to 6th (the faintest visible to the naked eye). In the modern scale, a difference of five magnitudes is assumed to represent a brightness ratio of 100, so each magnitude step represents a brightness ratio of $\sqrt[5]{100} = 2.512 \approx 2.5$. Thus a 1st magnitude star is about 2.5 times brighter than a 2nd magnitude star, $2.5 \times 2.5 = 6.3$ times brighter than a 3rd magnitude star, and so on. Sirius, the brightest star in the sky, has an apparent magnitude of -1.4 because it is much brighter than what is today considered a 1st magnitude star; at its brightest, Venus has the apparent magnitude -4; a good pair of binoculars can make 10th magnitude stars just visible; and the limit of the largest telescope is magnitude 20 for visual observation and 23.5 for photographic observation.

ABSOLUTE MAGNITUDE AND STELLAR DISTANCES

The apparent brightness of a star as seen from the earth depends upon two factors, its actual brightness (or *luminosity*) and its distance from the earth. Apparent brightness decreases as $1/r^2$, where r is the distance from the earth, because the area illuminated by the light from a star increases fourfold for each doubling of the distance from the star. Two units are used to express stellar distances. The *light year* is the distance traveled by light in one year and is equal to 9.46×10^{15} m or 5.88×10^{12} mi. The *parsec* is the distance at which a star will change its apparent position in the sky back and forth through an angle of $1''$ in the course of a year, during which the earth makes a complete circuit of its orbit. ($1° = 60' = 60$ min and $1' = 60'' = 60$ sec, so $1° = 3600''$.) One parsec = 3.26 light years = 3.086×10^{16} m. The *absolute magnitude* of a star is the apparent magnitude it would have if it were exactly 10 parsecs away. The absolute magnitude of the sun is about +5.

Distances to stars up to 100 parsecs or so can be directly determined by observing their *parallax*, or change in position in the sky during the course of a year. Distances to stars farther away can often be found indirectly since certain properties of many stars, such as their spectral type, vary with absolute magnitude and these properties can be determined by observation. A knowledge of both the absolute and apparent magnitudes of a star then gives its distance at once.

HERTZSPRUNG-RUSSELL DIAGRAM

Stars are classified into seven types according to the nature of their spectra, as in Table 45-1; the sun is a G-type star. The change in properties from one type to the next is gradual. Since the color of a hot object depends upon its temperature, going from red through yellow and white to blue with increasing temperature, surface temperatures can be established for stars on the basis of their spectral type.

Table 45-1.   Star Classes

Class	Color	Surface Temperature, K	Spectral Characteristics
O	Blue	Over 25,000	Lines of ionized helium and other ionized elements, hydrogen lines weak
B	Blue-white	11,000—25,000	Hydrogen and helium prominent
A	White	7500—11,000	Hydrogen lines very strong
F	Yellow-white	6000—7500	Hydrogen lines weaker, lines of ionized metals becoming prominent
G	Yellow	5000—6000	Lines of ionized and neutral metals, especially calcium, prominent
K	Orange	3500—5000	Lines of neutral metals and band spectra of simple compounds present
M	Red	Under 3500	Band spectra of many compounds present

The *Hertzsprung-Russell diagram* (Fig. 45-1) is a graph in which the absolute magnitude of a star is plotted against its spectral type.   Most stars belong to the *main sequence,* which extends from the upper left to the lower right of the H-R diagram.   The absolute magnitude of a main-sequence star increases in a regular way with its mass, and the *mass-luminosity relationship* permits the masses of such stars to be established once their absolute magnitudes are known.   Since the spectral type of a star is related to its temperature, the H-R diagram can equally well be considered as a graph of mass versus temperature for main-sequence stars.   Thus stars at the upper left of the main sequence are large, much more massive than the sun, hot, and bright; stars at the lower right are small, much less massive than the sun, cool, and faint.   The sun is near the middle of the main sequence, a rather average star.

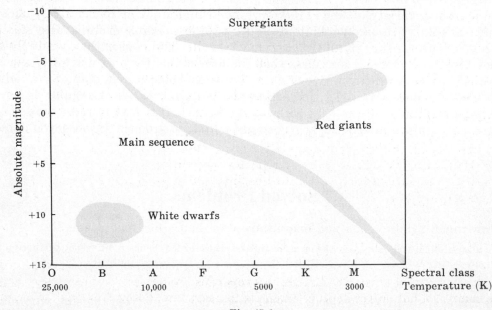

Fig. 45-1

At the upper right of the H-R diagram are *red giants,* which are large, massive, and very luminous but cool; the *supergiants* are even larger, more massive, and brighter.   At the lower left are *white dwarfs,* which are small and dim but hot.   The outstanding property of white dwarfs is their high density — a typical one has the mass of the sun but the size of the earth. The atoms in a white dwarf have collapsed, and their nuclei and electrons are packed closely together.

STELLAR EVOLUTION

Stars are believed to originate in gas clouds in space, which consist largely of hydrogen. Local concentrations occur from time to time in such clouds, and if the density of one of them is sufficiently great, gravitation will both attract more gas and cause the accumulation to contract. The contraction liberates potential energy which heats the *protostar* and causes it to glow. Such a very young star appears among the cooler giants and supergiants in the H-R diagram. As the protostar continues to contract, its luminosity decreases, and it moves downward on the H-R diagram. After some thousands or millions of years more, the protostar's temperature rises, shifting it to the left on the H-R diagram, and eventually it becomes hot enough for nuclear reactions to occur that convert its hydrogen to helium. Now the star ceases to contract and becomes a stable member of the main sequence, where it remains for a period that depends on its mass: the greater the mass, the more rapidly it consumes its hydrogen content, and the shorter its period of stability.

When the hydrogen in a star's interior is sufficiently depleted, it contracts, which releases gravitational potential energy and, by raising the temperature near the center, increases the rate of nuclear reactions there. Some of these reactions now involve the building-up of nuclei larger than that of helium, for instance in the combination of three helium nuclei to form a carbon nucleus. The outer part of the star is heated and expands, but the expansion then causes a cooling so the result is a very large, cool star with a hot core. Such a star is a red giant; the shift from the main sequence to the upper right of the H-R diagram is relatively rapid. When the sun becomes a red giant about 5 billion years hence, it will expand until it swallows up some or all of the inner planets.

It seems likely that most stars end their lives as white dwarfs, though the transition from red giant to white dwarf is not well understood. White dwarfs cannot exceed 1.2 solar masses, but a variety of ways are known by which stars more massive than this can eject sufficient matter to become white dwarfs. Another possible destiny for a star is to become a *neutron star*, a sort of superdwarf much smaller and denser than white dwarfs and with such enormous internal pressures that the only stable form of matter is the neutron. Calculations show that a neutron star of one solar mass would have a diameter of only 10 km or so. *Pulsars*, which emit brief, intense bursts of radio waves at regular intervals, are believed to be rotating neutron stars with magnetic fields that lead to radio emission in narrow beams; as a pulsar rotates, its beams swing with it to produce the observed fluctuations.

Solved Problems

45.1. How much brighter is a 2nd magnitude star than a 5th magnitude star?

Each magnitude step represents a brightness ratio of 2.5, hence a difference of three magnitudes means a ratio of $2.5 \times 2.5 \times 2.5 = 16$.

45.2. On the earth we receive about 400,000 times more light from the sun, whose apparent magnitude is -26.5, than from the full moon. What is the approximate magnitude of the full moon?

Since a brightness ratio of 100 is represented by a magnitude difference of 5, a brightness ratio of $100 \times 100 = 10,000$ is represented by a magnitude difference of $5 + 5 = 10$. The brightness ratio between sun and full moon is 400,000 and 400,000/10,000 − 40, so the magnitude difference is more than 10 by the number of steps needed to give a further brightness ratio of 40. Each magnitude step is a brightness ratio of 2.5 and $2.5 \times 2.5 \times 2.5 \times 2.5 \approx 40$, hence the sun is $10 + 4 = 14$ magnitudes brighter than the full moon. Thus the apparent magnitude of the full moon is $-26.5 + 14 = -12.5$.

45.3. Sirius has an apparent magnitude of −1.4 and an absolute magnitude of +1.4. Is it closer to or farther from the earth than 10 parsecs?

Absolute magnitudes are based on a distance of 10 parsecs. Since the absolute magnitude of Sirius is a larger number than its apparent magnitude, it would be less bright if 10 parsecs away than it appears at present, hence it is less than 10 parsecs from the earth. (Sirius is actually 2.7 parsecs away.)

45.4. Why is the sun considered to be a star?

The sun's size, mass, temperature, luminosity, spectrum, and position on the H-R diagram are those of a typical star.

45.5. How are stellar masses determined?

A direct measurement of the mass of a star is only possible when it is a member of a binary system. Such a system consists of two stars that revolve around their common center of mass. If the period of revolution and the separation of the stars are both known, the sum of their masses can be calculated; if the relative position of the center of mass can also be established from the motions of the stars, their individual masses can be inferred. The mass of a single main-sequence star can be indirectly inferred from its absolute magnitude by using the mass-luminosity relationship.

45.6. How can the motion of a star be detected?

A star moving toward or away from the earth will show a Doppler shift in its spectrum toward the blue or the red end respectively. A star moving across our line of sight will change its position relative to other stars, as revealed by photographs taken at different times.

45.7. The spectrum of a certain star shows a Doppler shift that varies periodically between indicating motion toward the earth and motion away from the earth. What kind of a star is this?

The star must be a member of a binary system (a "double star"), and the periodic Doppler shifts occur as it revolves around the center of mass of the system.

45.8. Why are most stars part of the main sequence on the H-R diagram?

A star on the main sequence is in an equilibrium condition with its tendency to expand owing to its high temperature exactly balanced by its tendency to contract gravitationally. The condition lasts until the star's hydrogen content decreases beyond a certain proportion, which requires a relatively long time compared with its earlier and later phases. Therefore most stars are members of the main sequence simply because this is the longest stage in a star's evolution.

45.9. Main-sequence stars are supposed to evolve into red giant stars, but relatively few stars lie between the main sequence and the group of giant stars on the H-R diagram. Why?

When the hydrogen content of a main-sequence star has become sufficiently depleted by nuclear reactions so that it is no longer in stable equilibrium, it becomes a red giant relatively rapidly and so at any time few stars are in the midst of this transition.

45.10. Why do all main-sequence stars have masses that lie between about 0.1 and 100 times the mass of the sun?

A star whose mass is less than about 1/10 the mass of the sun is not able by gravitational contraction to reach a temperature high enough for nuclear reactions to occur in its interior; thus it can never attain the stable configuration of a main-sequence star. On the other hand, a star whose mass is more than about 100 times the mass of the sun would become so hot as a result of accelerated nuclear reactions that the outward pressure would exceed the inward force of its gravitation, and it would not be stable either.

45.11. A giant star is much redder than a main-sequence star of the same absolute magnitude. How does this observation indicate that the giant star is larger than the main-sequence star?

The color of a star depends upon its temperature, with the hottest ones appearing blue and the coolest appearing red. Hence a giant star must be cooler than a main-sequence star of the same absolute magnitude. The amount of light radiated by an object increases with both its temperature and its surface area. Since a giant star is brighter for its temperature than a main-sequence star, it must be larger than a main-sequence star of the same absolute magnitude.

45.12. The brightness of certain stars fluctuates. What are the three chief classes of variable stars?

(1) **Eclipsing binaries.** These consist of double stars of which one member of the pair periodically moves across the face of the other to produce a change in apparent brightness. Strictly speaking, an eclipsing binary is not a true variable since the luminosity of each member of the pair does not change.

(2) **Pulsating variables.** Such a star expands and contracts in a more or less regular rhythm, and its light output fluctuates in the same rhythm. Apparently a change in temperature accompanying each change in size is primarily responsible for the variation in luminosity, with the change in size a secondary factor. Most pulsating variables are giant or supergiant stars. The *Cepheid variables*, which fluctuate with periods of a few days, are significant because their periods are related to their absolute magnitudes. Thus they provide a way to find the distances of star groups so remote that spectrum analysis is impossible: from the period of a Cepheid in such a star group the absolute magnitude of the Cepheid can be established, and comparison with its apparent magnitude then yields the distance of the star and so of the group.

(3) **Eruptive variables.** Most such stars exhibit sudden increases in brightness, notably the *novae* and *supernovae*; a few decrease in brightness. A nova is a small, hot star that abruptly flares up to thousands or tens of thousands of times its normal luminosity as it ejects a shell of gas that expands rapidly. A nova may take a day or less to attain its maximum brightness, then it gradually declines to its former brightness over a longer period, up to a year or more. A supernova flares up to hundreds of millions of times its normal luminosity, reaching an absolute magnitude of −14 to −20 and sometimes outshining the entire galaxy of which it is part. In a supernova the ejected material is a substantial part of the original star, much more than in the case of a nova. The *Crab nebula* is the expanding cloud of gas from a supernova observed in 1054. A pulsar is at the center of this nebula, which suggests that during the supernova explosion the core of the original star was imploded sufficiently to produce a neutron star.

45.13. What is a "black hole"?

According to the general theory of relativity, light waves are affected by gravitational fields. For example, starlight passing near the sun is deflected to a small but measurable extent by the sun's gravity. The smaller an object and the more massive it is, the stronger the gravitational field at its surface; if this field is strong enough, light will be unable to escape from the object, and it is a "black hole" in space. Conceivably some stars whose mass exceeds 1.2 solar masses ultimately contract to become black holes instead of shedding enough matter to become white dwarfs. (The upper limit to the mass of a white dwarf is 1.2 solar masses.)

Supplementary Problems

45.14. The apparent magnitudes of the stars Polaris, Aldebaran, and Sirius are respectively +2.0, +1.0, and −1.4. (*a*) Which star is brightest and which is faintest? (*b*) What is the ratio in brightness between Polaris and Aldebaran?

45.15. At their brightest, Uranus, Neptune, and Pluto have the apparent magnitudes +5.6, +7.9, and +14.9 respectively. Which (if any) of them can be seen with the naked eye? Through a good pair of binoculars?

45.16. The star Vega has an absolute magnitude of +0.5 and is 8 parsecs away. Is its apparent magnitude larger or smaller than +0.5?

45.17. Which varies more, the masses of the stars or their luminosities?

45.18. List three ways to determine the distance of a star from the earth. Can these methods be used with every star?

45.19. Most stars exhibit absorption spectra. What does this indicate about their structures?

45.20. Which stars have the highest densities? The lowest?

45.21. Which two properties of a star must be known in order to estimate its diameter?

45.22. Which of the following types of star is most common in the sky? Least common? Main-sequence stars, white dwarfs, red giants, supernovae, double stars, Cepheid variables.

45.23. The stars Betelgeuse and Deneb have similar absolute magnitudes but Betelgeuse is red and Deneb is white. (*a*) Which is larger? (*b*) Which has the greater density? (*c*) Which is hotter?

45.24. The sun occupies a position near the middle of the main sequence. Where would a star be located in the main sequence if its mass is ten times that of the sun? Would it remain in the main sequence for a longer or shorter time than the sun? Would it be hotter or cooler than the sun?

45.25. As a star evolves, does its position on the main sequence shift diagonally downward to the right, diagonally upward to the left, or does it remain more or less in the same place until it starts to become a red giant?

45.26. After a very long time, a white dwarf will cool down and become a "black dwarf." What will the corresponding evolutionary path of the star be on the H-R diagram?

45.27. A main-sequence star of the same absolute magnitude as a white dwarf star is much redder than the dwarf. How does this fact indicate that the dwarf star is smaller than the main-sequence star?

Answers to Supplementary Problems

45.14. (*a*) Sirius is brightest and Polaris is faintest. (*b*) Aldebaran is 2.5 times brighter than Polaris.

45.15. Uranus can be seen with the naked eye, and it and Neptune can be seen through a good pair of binoculars.

45.16. Its apparent magnitude is smaller because it would appear less bright if 10 parsecs away.

45.17. All stars have masses between about 0.1 and 100 solar masses, a range of 10^3, but luminosities vary between 10^{-6} and 10^6 times the sun's luminosity, a range of 10^{12}.

45.18. (*a*) By its parallax, which is its apparent shift in position in the sky as the earth moves around in its orbit. This method is suitable for stars within about 100 parsecs of the earth. (*b*) By using its spectral type to determine its absolute magnitude, and then comparing the latter with the apparent magnitude. This method is suitable for main-sequence stars for which satisfactory spectra can be obtained. (*c*) In the case of a Cepheid variable, the period indicates its absolute magnitude, and again comparison with the apparent magnitude yields the distance. This method is suitable for stars that are members of the same group in space as a Cepheid.

45.19. Such a star has a hot interior with a cooler atmosphere around it.

45.20. White dwarf and neutron stars have the highest densities, and giant stars have the lowest densities.

45.21. The star's absolute magnitude and its temperature.

45.22. main-sequence stars; supernovae

45.23. Betelgeuse; Deneb; Deneb

45.24. Farther to the left and higher in the main sequence; shorter time; hotter.

45.25. A star does not change its position in the main sequence by very much until it begins to become a red giant, when it leaves the main sequence entirely.

45.26. Diagonally downward (since the star's luminosity will decrease) and to the right (since its temperature will decrease).

45.27. The dwarf's color indicates that its temperature is higher than that of the main-sequence star. If both radiate the same amount of energy, the dwarf must be smaller since a hot object radiates more energy than a cool one of the same size.

Chapter 46

The Universe

MILKY WAY GALAXY

The stars are not uniformly distributed in space but occur in aggregates called *galaxies*. The stars that make up the Milky Way are part of the galaxy that includes the sun. Most of the stars in our galaxy are concentrated in a relatively thin disklike region which has a thicker central nucleus, much like a fried egg. The disk is about 100,000 light years in diameter and the stars in it are concentrated in two spiral arms that extend from the nucleus. The sun is located about 30,000 light years from the center of the galaxy and, like the other stars in the galaxy, revolves around the center; the sun's period of revolution is about 200 million years.

Associated with our galaxy are a number of huge *globular clusters* of stars that form a sort of halo or corona around the central disk. Thus the true form of the galaxy as a whole is roughly spherical. The stars in the spiral arms, called *Population I* stars, are of different ages, including very young ones still in the process of formation; those in the rest of the galaxy, called *Population II* stars, are all very old. Presumably all the matter of the galaxy was originally a spherical cloud of gas and dust that gradually concentrated in the spiral arms, leaving behind those stars that had already come into being which comprise Population II. New stars continued to form from the gas and dust in the spiral arms, and these younger stars comprise Population I.

COSMIC RAYS

High-energy atomic nuclei, largely protons, continually rain down on the earth from space. These *primary cosmic rays* circulate throughout our galaxy, to which they are confined by magnetic fields; a few exceptionally energetic protons probably originate outside the galaxy, since the intergalactic magnetic fields are not strong enough to trap them. It seems likely that primary cosmic rays consist of nuclei originally ejected during supernova explosions which were subsequently accelerated by the same magnetic fields that prevent their escape from the galaxy.

When a primary cosmic ray arrives at the earth's atmosphere, it disrupts atoms in its path to produce a shower of *secondary cosmic rays* which are what reach the surface. The secondaries consist of neutrons and protons, mesons and other unstable particles, and electrons and gamma rays from the decay of the latter.

OTHER GALAXIES

Most galaxies are either spiral or elliptical in character. Like our galaxy, other spiral galaxies consist of a nucleus and corona of Population II stars with spiral arms of Population I stars. Elliptical galaxies have no spiral arms and consist entirely of Population II stars; they range in shape from spheres to fairly flat ellipsoids. The most common galaxies are dwarf elliptical ones that contain only a few million stars.

Just as individual stars are always found to be grouped in galaxies, galaxies themselves seem to be grouped in clusters. Such a cluster contains from tens to thousands of galaxies in a region a few million light years across. The *Local Group* of galaxies, to which the Milky Way galaxy belongs, seems to have about 19 members; the uncertainty arises from the faintness of dwarf galaxies, which makes their detection difficult.

THE EXPANDING UNIVERSE

The spectral lines of all galaxies are found to be shifted toward the red by an amount that increases with distance. The only explanation in accord with existing experimental and theoretical knowledge is that the red shifts originate in the Doppler effect and signify motion away from the earth; the proportionality between recession velocity and distance means that the entire universe must be expanding, so that an observer anywhere finds that everything else is moving away from him.

The *big-bang* cosmological model, which is widely accepted today, holds that the present universe came into being perhaps 10 billion years ago in a gigantic explosion. As the primeval matter spread out in space, local concentrations formed that became galaxies. The fastest galaxies traveled farthest since the big bang occurred, which accounts for the correlation between red shift and distance. A modification of the big-bang model suggests that gravitation will slow down and eventually reverse the expansion, so that the universe will contract into a single mass again in the distant future. This mass will then explode in another big bang, starting a new expansion. Such an oscillating universe might have a period of 80 billion years for each cycle.

QUASARS

A *quasar* ("quasi-stellar radio source") appears in a telescope as a point of light, as a star does, but it is a far more powerful source of radio waves than any known star. The spectra of quasars show large red shifts; if quasar red shifts are related to their distances in the same way as galactic red shifts are, quasars are not only the most distant objects in the universe but are also giving off energy at prodigious rates. The light and radio outputs of quasars sometimes change in periods of a few weeks, so they cannot be more than a few light weeks across, which makes the mechanism of their energy emission even more difficult to explain. On the other hand, if quasars are actually nearby objects located in our galaxy, the energy problem is less severe, but now the origin of their red shifts has no explanation in current knowledge. The nature of quasars is a major challenge to astrophysics.

Solved Problems

46.1. Do all the stars in our galaxy revolve around its center with the same period?

The stars in the inner part of the galactic disk revolve around the center with about the same period, much as if they were part of a solid wheel. The outer stars behave like planets revolving around the sun: the more distant they are from the galactic center, the longer they take to complete a revolution. Star clusters in the corona revolve around the center in elongated orbits at higher velocities than do stars in the disk.

46.2. How can the rotation of a spiral galaxy be experimentally determined?

If the galaxy is so oriented that we see its disk edge-on or nearly so, the stars on one side of the center are moving toward us and those on the other side are moving away from us. These motions produce Doppler shifts in the spectrum of the galaxy that can be detected.

46.3. **What are the properties of globular clusters? Why are they not considered to be a type of galaxy?**

A typical globular cluster is an assembly of hundreds of thousands of Population II stars that are relatively close together — several light months apart in the densest regions. Globular clusters are found in all galaxies; in spiral galaxies, most of them are located in the corona outside the central disk, and they move at high velocities in elongated orbits about the galactic center. Since globular clusters are much smaller than galaxies and are always found as members of them, they cannot be considered as being themselves galaxies.

46.4. **Why is it unlikely that most cosmic rays come from the sun?**

A solar origin would produce a day-night variation in cosmic-ray intensity, but no such variation is observed. Also, no mechanism is known by which the sun could accelerate cosmic-ray primaries to the highest energies they are found to have, though such mechanisms are likely to exist in the galaxy as a whole. Some low-energy primaries, however, probably have a solar origin.

46.5. **Is cosmic-ray intensity the same everywhere on the earth?**

Cosmic-ray primaries are atomic nuclei and hence are charged particles. The earth's magnetic field is not strong enough to influence the paths of the more energetic primaries by very much, and they arrive equally often everywhere, but the slower ones are deflected so that fewer of them reach the equatorial regions than the polar regions.

46.6. **Do radio telescopes actually magnify something? If not, why are larger and larger ones being built?**

Radio telescopes are giant antennas that detect radio waves from space; they do not magnify anything. The larger a radio telescope is, the better able it is to respond to weak radio signals, and the more accurately it can identify the direction from which a given signal is coming.

46.7. **According to *Hubble's law,* the greater the distance of a galaxy, the greater its velocity of recession; it is this proportionality that led to the concept of the expanding universe. A recent approximate value for the velocity of recession is 25 km/s per million light years of distance. Use this value to estimate the age of the universe.**

Since   1 light year $= 9.5 \times 10^{15}$ m   and   1 km $= 10^3$ m,

$$s \;=\; 10^6 \text{ light years} \times 9.5 \times 10^{15} \text{ m/ly} \;=\; 9.5 \times 10^{21} \text{ m}$$

$$v \;=\; 25 \text{ km/s} \times 10^3 \text{ m/km} \;=\; 2.5 \times 10^4 \text{ m/s}$$

A galaxy 10^6 light years away has a velocity of about 25 km/s, so the time during which it has been moving (assuming that its velocity has always been the same) is about

$$t \;=\; \frac{\text{distance}}{\text{velocity}} \;=\; \frac{s}{v} \;=\; \frac{9.5 \times 10^{21} \text{ m}}{2.5 \times 10^4 \text{ m/s}} \;=\; 3.8 \times 10^{17} \text{ s}$$

There are 3.2×10^7 s in a year, hence the age of the universe in years must be in the neighborhood of

$$t \;=\; \frac{3.8 \times 10^{17} \text{ s}}{3.2 \times 10^7 \text{ s/year}} \;=\; 1.2 \times 10^{10} \text{ years}$$

which is 12 billion years.

46.8. **What is the *steady-state* theory of the universe?**

The basis of this theory is the belief that the universe is unchanging in its basic character, so that it has no beginning and no end. In order that the density of galaxies in space remain the same despite their observed spreading apart, the steady-state theory postulates that new galaxies are forming in empty space all the time, with the necessary matter coming into being by a process of spontaneous creation. The theory is attractive philosophically in its conception of an eternal universe, always evolving and yet always the same. However, there are serious objections to the notion of the spontaneous creation of matter, and present experimental evidence is not consistent with various predictions of the steady-state theory.

46.9. The most distant galaxies are very faint in even the largest of today's telescopes and reliable data on them is rare. If adequate such data were available, what information would you look for to decide whether the big-bang, oscillatory, or steady-state theory of the universe is nearest the truth?

(1) **Apparent ages of distant galaxies.** Galaxies form, mature, and age in an evolutionary sequence. The light reaching us from a distant galaxy left it long ago; any information we can obtain about a galaxy say 5 billion light years away represents its physical state 5 billion years ago. According to the big-bang and oscillatory theories, all galaxies came into being at about the same time soon after the expansion of the universe began. Thus these theories predict that distant galaxies will appear younger than nearby galaxies. According to the steady-state theory, nothing fundamental ever changes in the universe, so distant galaxies will appear the same as nearby ones.

(2) **Density of galaxies.** According to the big-bang and oscillatory theories, the galaxies long ago were closer together than they are today. Thus distant galaxies, which we see as they were in the past, should be closer together in space than nearby galaxies, which we see as they were more recently. According to the steady-state theory, the density of galaxies should not vary, since as they move apart new ones come into being.

(3) **Rate of expansion.** The big-bang theory predicts that the universe will continue to expand indefinitely, though its rate will slow down due to gravitation; the oscillatory theory predicts that gravitation will eventually cause the universe to contract; and the steady-state theory predicts that the expansion will continue forever at a constant rate. Thus determining exactly how the rate of expansion varies with distance for the farthest galaxies will help establish which of the three theories is correct — if indeed any of them is correct in its present form.

46.10. If the universe originated in a big bang, in its early moments it must have been a very hot, dense *primeval fireball*. Because of its high temperature, radiation from the fireball would have been in the high-frequency region of the spectrum, chiefly in the form of X-rays. Today space is pervaded by a sea of low-frequency radio waves which have been identified as the remnants of the fireball radiation. (This is one of the pieces of evidence that contradict the steady-state theory.) What caused the change in frequency?

The expansion of the fireball produced Doppler shifts in the radiation that decreased the frequencies present in it, just as the light from distant galaxies is found to exhibit red shifts.

Supplementary Problems

46.11. Why is the sun thought to be part of the central disk of our galaxy?

46.12. State two differences between Population I and Population II stars.

46.13. Where is most of the interstellar gas in our galaxy located? What is its chief constituent?

46.14. Would you expect elliptical galaxies to contain abundant gas and dust?

46.15. Cosmic-ray primaries are mostly protons, but few protons are in the cosmic rays that reach the earth's surface. Why?

46.16.   What effect, if any, would the disappearance of the earth's magnetic field have on the distribution of cosmic rays around the earth?

46.17.   According to current ideas, where are elements heavier than hydrogen formed?

46.18.   The spectra of quasars exhibit red shifts, never blue shifts. Why does this suggest that quasars are not members of our galaxy?

Answers to Supplementary Problems

46.11.   The Milky Way is composed of stars in the spiral arms of our galaxy and so defines its central disk. Since the earth is close to the plane of the Milky Way, the sun must be part of the central disk of the galaxy.

46.12.   Population I stars are found in the spiral arms of spiral galaxies and are of all ages, including very young stars. Population II stars are found in the coronas and nuclei of spiral galaxies and as the sole constituents of elliptical galaxies, and are all very old.

46.13.   in the spiral arms; hydrogen

46.14.   Since elliptical galaxies contain only Population II stars, which are very old, they cannot contain much gas and dust since young stars would then be present.

46.15.   Cosmic-ray primaries lose their initial energies in collisions with the nuclei of atoms in the atmosphere, and the cosmic rays that reach the earth are almost all secondaries produced in these collisions.

46.16.   In the absence of the geomagnetic field, cosmic rays would arrive at the same rate everywhere on the earth.

46.17.   in the interiors of stars

46.18.   If quasars were members of our galaxy, at least some of them would have components of motion toward the earth and would accordingly have Doppler shifts in their spectral lines toward the blue end of the spectrum.

Appendix A

PHYSICAL CONSTANTS AND QUANTITIES

Quantity	Symbol	Value
Absolute zero	0 K	$-273\ ^{\circ}\text{C}$
Acceleration of gravity at earth's surface	g	$9.81\ \text{m/s}^2 = 32.2\ \text{ft/s}^2$
Avogadro's number	N	6.023×10^{23} formula units/mole (or atoms/gram-atom or molecules/mole)
Boltzmann's constant	k	$1.38 \times 10^{-23}\ \text{J/K}$
Coulomb constant	k	$8.99 \times 10^{9}\ \text{N-m}^2/\text{C}^2$
Earth: mass	m_{earth}	$5.98 \times 10^{24}\ \text{kg}$
radius	r_{earth}	$6.38 \times 10^{6}\ \text{m}$
orbit radius (average)	r_{orbit}	$1.49 \times 10^{11}\ \text{m}$
Electron: charge	e	$1.60 \times 10^{-19}\ \text{C}$
mass	m_e	$9.11 \times 10^{-31}\ \text{kg} = 0.00055\ \text{u}$
Faraday constant	F	$9.65 \times 10^{4}\ \text{C/mole}$
Gravitational constant	G	$6.67 \times 10^{-11}\ \text{N-m}^2/\text{kg}^2 = 3.44 \times 10^{-8}\ \text{lb-ft}^2/\text{slug}^2$
Molar volume at STP	V_0	22.4 liters/mole
Neutron rest mass	m_n	$1.675 \times 10^{-27}\ \text{kg} = 1.008665\ \text{u}$
Planck's constant	h	$6.63 \times 10^{-34}\ \text{J-s}$
Proton rest mass	m_p	$1.673 \times 10^{-27}\ \text{kg} = 1.007277\ \text{u}$
Universal gas constant	R	$8.31 \times 10^{3}\ \text{J/mole-K} = 0.0821\ \text{atm-liter/mole-K}$
Velocity of light in free space	c	$3.00 \times 10^{8}\ \text{m/s} = 1.86 \times 10^{5}\ \text{mi/s}$

Appendix B

CONVERSION FACTORS

Time

1 day $= 1.44 \times 10^3$ min $= 8.64 \times 10^4$ s

1 year $= 8.76 \times 10^3$ hr $= 5.26 \times 10^6$ min $= 3.15 \times 10^7$ s

Length

1 meter (m) $= 100$ cm $= 39.4$ in. $= 3.28$ ft

1 centimeter (cm) $= 10$ millimeters (mm) $= 0.394$ in.

1 kilometer (km) $= 10^3$ m $= 0.621$ mi

1 foot (ft) $= 12$ in. $= 0.305$ m $= 30.5$ cm

1 inch (in.) $= 0.0833$ ft $= 2.54$ cm $= 0.0254$ m

1 mile (mi) $= 5280$ ft $= 1.61$ km

1 light year (ly) $= 9.46 \times 10^{15}$ m $= 5.88 \times 10^{12}$ mi $= 0.307$ pc

1 parsec (pc) $= 3.26$ ly $= 3.086 \times 10^{16}$ m

Area

1 m^2 $= 10^4$ cm^2 $= 1.55 \times 10^3$ in^2 $= 10.76$ ft^2

1 cm^2 $= 10^{-4}$ m^2 $= 0.155$ in^2

1 ft^2 $= 144$ in^2 $= 9.29 \times 10^{-2}$ m^2 $= 929$ cm^2

Volume

1 m^3 $= 10^3$ liters $= 10^6$ cm^3 $= 35.3$ ft^3 $= 6.10 \times 10^4$ in^3

1 ft^3 $= 1728$ in^2 $= 2.83 \times 10^{-2}$ m^3 $= 28.3$ liters

Velocity

1 m/s $= 3.28$ ft/s $= 2.24$ mi/hr $= 3.60$ km/hr

1 ft/s $= 0.305$ m/s $= 0.682$ mi/hr $= 1.10$ km/hr
(*Note*: It is often convenient to remember that 88 ft/s = 60 mi/hr.)

1 km/hr $= 0.278$ m/s $= 0.913$ ft/s $= 0.621$ mi/hr

1 mi/hr $= 1.47$ ft/s $= 0.447$ m/s $= 1.61$ km/hr

Mass

1 kilogram (kg) $= 10^3$ grams (g) $= 0.0685$ slug
(*Note*: 1 kg corresponds to 2.21 lb in the sense that the *weight*
of 1 kg at the earth's surface is 2.21 lb.)

1 slug $= 14.6$ kg
(*Note*: 1 slug corresponds to 32.2 lb in the sense that the *weight*
of 1 slug at the earth's surface is 32.2 lb.)

1 atomic mass unit (u) $= 1.66 \times 10^{-27}$ kg $= 1.49 \times 10^{-10}$ J $= 931$ MeV

Force

1 newton (N) $= 0.225$ lb $= 3.60$ oz

1 pound (lb) $= 16$ ounces (oz) $= 4.45$ N
(*Note*: 1 lb corresponds to 0.454 kg = 454 g in the sense that the *mass*
of something that weighs 1 lb at the earth's surface is 0.454 kg.)

Pressure

1 N/m^2 $= 2.09 \times 10^{-2}$ lb/ft^2 $= 1.45 \times 10^{-4}$ lb/in^2

1 lb/in^2 $= 144$ lb/ft^2 $= 6.90 \times 10^3$ N/m^2

1 atm $= 1.013 \times 10^5$ N/m^2 $= 14.7$ lb/in^2

Energy

1 joule (J) $= 0.738$ ft-lb $= 2.39 \times 10^{-4}$ kcal $= 6.24 \times 10^{18}$ eV

1 foot-pound (ft-lb) $= 1.36$ J $= 1.29 \times 10^{-3}$ Btu $= 3.25 \times 10^{-4}$ kcal

1 kilocalorie (kcal) $= 4185$ J $= 3.97$ Btu $= 3077$ ft-lb

1 Btu $= 0.252$ kcal $= 778$ ft-lb

1 electron volt (eV) $= 10^{-6}$ MeV $= 10^{-9}$ GeV $= 1.60 \times 10^{-19}$ J

Power

1 watt (W) $= 1$ J/s $= 0.738$ ft-lb/s

1 kilowatt (kW) $= 10^3$ W $= 1.34$ hp

1 horsepower (hp) $= 550$ ft-lb/s $= 746$ W

Temperature

$$T_C = \frac{5}{9}(T_F - 32°)$$

$$T_F = \frac{9}{5}T_C + 32°$$

$$T_K = T_C + 273°$$

Appendix C

PERIODIC TABLE OF THE ELEMENTS

The number above the symbol of each element is its atomic mass in u, and the number below the symbol is its atomic number. The elements whose atomic masses are given in parentheses do not occur in nature, but have been prepared artificially in nuclear reactions. The atomic mass in such a case is the mass number of the most long-lived radioactive isotope of the element.

Group / Period	I	II	III	IV	V	VI	VII	VIII
1	1.008 H 1							4.00 He 2
2	6.94 Li 3	9.01 Be 4	10.82 B 5	12.01 C 6	14.01 N 7	16.00 O 8	19.00 F 9	20.18 Ne 10
3	22.99 Na 11	24.31 Mg 12	26.98 Al 13	28.09 Si 14	30.98 P 15	32.06 S 16	35.46 Cl 17	39.95 Ar 18
4	39.10 K 19	40.08 Ca 20	69.72 Ga 31	72.59 Ge 32	74.92 As 33	78.96 Se 34	79.91 Br 35	83.8 Kr 36
5	85.47 Rb 37	87.62 Sr 38	114.82 In 49	118.69 Sn 50	121.76 Sb 51	127.61 Te 52	126.90 I 53	131.30 Xe 54
6	132.91 Cs 55	137.34 Ba 56	204.37 Tl 81	207.19 Pb 82	208.98 Bi 83	(209) Po 84	(210) At 85	(222) Rn 86
7	(223) Fr 87	226.05 Ra 88						

Transition elements (Period 4–6)

Period 4	Period 5	Period 6
44.96 Sc 21	88.91 Y 39	* 57-71
47.90 Ti 22	91.22 Zr 40	178.49 Hf 72
50.94 V 23	92.91 Nb 41	180.95 Ta 73
52.00 Cr 24	95.94 Mo 42	183.85 W 74
54.94 Mn 25	(99) Tc 43	186.2 Re 75
55.85 Fe 26	101.1 Ru 44	190.2 Os 76
58.93 Co 27	102.91 Rh 45	192.2 Ir 77
58.71 Ni 28	106.4 Pd 46	195.09 Pt 78
63.54 Cu 29	107.87 Ag 47	196.97 Au 79
65.37 Zn 30	112.40 Cd 48	200.59 Hg 80

Period 7: † 89-103

* Rare earths

138.91 La 57	140.12 Ce 58	140.91 Pr 59	144.24 Nd 60	(147) Pm 61	150.35 Sm 62	151.96 Eu 63	157.25 Gd 64	158.92 Tb 65	162.50 Dy 66	164.93 Ho 67	167.26 Er 68	168.93 Tm 69	173.04 Yb 70	174.97 Lu 71

† Actinides

(227) Ac 89	232.04 Th 90	(231) Pa 91	238.03 U 92	(244) Np 93	(244) Pu 94	(243) Pu 95	(247) Cm 96	(247) Bk 97	(251) Cf 98	(254) Es 99	(257) Fm 100	(256) Md 101	(254) No 102	(257) Lr 103

280

INDEX

Absolute magnitude of star, 264
Absolute pressure, 45
Absolute temperature, 56
Absorption spectrum, 118
Acceleration, 9
  centripetal, 23
  and force, 16
  of gravity, 10
Acid, 163
  organic, 194
Actinide elements, 124
Action and reaction forces, 17
Activated complex, 188
Activation energy, 188
Alcohol, 194
Aldehyde, 194
Aliphatic compound, 193
Alkali metals, 125
Alluvial fan, 225
Alluvium, 225
Alpha particle, 113
Alternating current, 89
  and power transmission, 91
Amide, 197
Ammonia, 163
Ampere, 74
Ampere-hour, 77
Angular momentum, 36
  of atom, 119
Annihilation of matter, 114
Anode, 176
Antarctic Circle, 252
Antibonding orbital, 131
Anticline, 232
Anticyclone, 211
Antimatter, 114
Antiparticle, 114
Apparent magnitude of star, 264
Archimedes' principle, 45
Arctic Circle, 252
Aromatic compound, 193
Asteroids, 255
Asthenosphere, 235
Astronomical unit, 251
Atmosphere (unit of pressure), 44
Atmosphere of earth, 204
  composition of, 204, 246
  general circulation of, 210
  moisture content of, 205
Atom, 58
  Bohr theory of, 118
  energy levels of, 118
  ground state of, 118
  ionization energy of, 118, 127, 130
  quantum theory of, 119
  spectrum of, 118
  structure of, 67, 124

Atomic mass unit, 58, 142
Atomic number, 107
Atomic orbital, 120
Atomic shells and subshells, 124
Aurora, 259, 262
Avogadro's number, 142

Baryon, 113
Baryon number, 116
Base, 163
Batholith, 230, 231
Battery, capacity of, 77
  storage, 177
Benzene molecule, 192
Bernoulli's principle, 45
Beta particle, 113
Big-bang model of universe, 272, 274
Binding energy, 107
"Black hole," 268
Blue color of sky, 97, 207
Bode's law, 258
Bohr model of hydrogen atom, 118
Boiling point, 52
  of solution, 158
Bonding orbital, 131
Bonding in solids, 131
Boyle's law, 56
Breccia, 230
British thermal unit (Btu), 51
Buoyant force, 45

Caldera, 231
Calorie, 51
Cancer, Tropic of, 252
Capricorn, Tropic of, 252
Carbon cycle, 259
Carbonic acid, 228
Carboxyl group, 194
Carnot engine, 63
Catalyst, 188
Cathode, 176
Celsius temperature scale, 50
Cenozoic era, 244
Centigrade temperature scale, 50
Centripetal acceleration, 23
Centripetal force, 23
Chain reaction, 108
Change of state, 51
Charge, electric, 67
  magnetic force on, 82
Charles's law, 56
Chemical bond, 130
Chemical energy, 183
Chemical equation, 137
  balanced, 138
Chemical formula, 136
  empirical, 149

Catalog

If you are interested in a list of SCHAUM'S
OUTLINE SERIES in Science, Mathematics,
Engineering and other subjects, send your name
and address, requesting your free catalog, to:

SCHAUM'S OUTLINE SERIES, Dept. C
McGRAW-HILL BOOK COMPANY
1221 Avenue of Americas
New York, N.Y. 10020